Cooling Flows in Clusters and Galaxies

NATO ASI Series

Advanced Science Institutes Series

A Series presenting the results of activities sponsored by the NATO Science Committee, which aims at the dissemination of advanced scientific and technological knowledge, with a view to strengthening links between scientific communities.

The Series is published by an international board of publishers in conjunction with the NATO Scientific Affairs Division

A Life Sciences   Plenum Publishing Corporation
B Physics   London and New York

C Mathematical   Kluwer Academic Publishers
   and Physical Sciences   Dordrecht, Boston and London
D Behavioural and Social Sciences
E Applied Sciences

F Computer and Systems Sciences   Springer-Verlag
G Ecological Sciences   Berlin, Heidelberg, New York, London,
H Cell Biology   Paris and Tokyo

Cooling Flows in Clusters and Galaxies

edited by

A. C. Fabian

Institute of Astronomy,
University of Cambridge, Cambridge, U.K.

Kluwer Academic Publishers

Dordrecht / Boston / London

Published in cooperation with NATO Scientific Affairs Division

Proceedings of the NATO Advanced Research Workshop on
Cooling Flows in Clusters and Galaxies
Cambridge, U.K.
22–26 June 1987

Library of Congress Cataloging in Publication Data

```
NATO Advanced Research Workshop on Cooling Flows in Clusters and
   Galaxies (1987   Cambridge, Cambridgeshire)
    Cooling flows in clusters and galaxies / edited by A.C. Fabian.
       p.    cm. -- (NATO ASI series. Series C, Mathematical and
physical sciences ; vol. 229)
    "Proceedings of the NATO Advanced Research Workshop on Cooling
Flows in Clusters and Galaxies, Cambridge, U.K., 22-26 June 1987"-
-Verso t.p.
    "Published in cooperation with NATO Scientific Affairs Division."
    Includes indexes.
    ISBN-13: 978-94-010-7828-3      e-ISBN-13: 978-94-009-2953-1
    DOI: 10.1007/ 978-94-009-2953-1
    1. Galaxies--Clusters--Congresses.  2. Galaxies--Congresses.
3. Cooling--Congresses.  4. Gas dynamics--Congresses.  5. Radio
astrophysics--Congresses.   I. Fabian, A. C., 1948-     II. North
Atlantic Treaty Organization.  Scientific Affairs Division.
III. Title.  IV. Series  NATO ASI series. Series C, Mathematical
and physical sciences . no. 229.
QB858.7.N38 1987
523.1'12--dc19                                          88-4467
                                                        CIP
    ISBN-13: 978-94-010-7828-3
```

Published by Kluwer Academic Publishers,
P.O. Box 17, 3300 AA Dordrecht, The Netherlands.

Kluwer Academic Publishers incorporates the publishing programmes of
D. Reidel, Martinus Nijhoff, Dr W. Junk, and MTP Press.

Sold and distributed in the U.S.A. and Canada
by Kluwer Academic Publishers,
101 Philip Drive, Norwell, MA 02061, U.S.A.

In all other countries, sold and distributed
by Kluwer Academic Publishers Group,
P.O. Box 322, 3300 AH Dordrecht, The Netherlands.

TABLE OF CONTENTS

PREFACE

X-ray astronomers discovered the diffuse gas in clusters of galaxies about 20 years ago. It was later realized that the central gas density in some clusters, and in elliptical galaxies, is so high that radiative cooling is a significant energy loss. The cooling time of the gas decreases rapidly towards the centre of the cluster or galaxy and is less than a Hubble time within the innermost few hundred kiloparsecs. This results in a cooling flow in which the gas density rises in order to maintain pressure to support the weight of the overlying gas. The rate at which mass is deposited by the flow is inferred to be several hundreds of solar masses per year in some clusters. The fraction of clusters in which cooling flows are found may exceed 50 per cent. Small flows probably occur in most normal elliptical galaxies that are not in rich clusters.

The implications of this simple phenomenon are profound, for we appear to be witnessing the ongoing formation of the central galaxy. In particular, since most of the gas is undetected once it cools below about 3 million K, it appears to form dark matter. There is no reason why it should be detectable with current techniques if each cooling proton only recombines once and the matter condenses into objects of low mass. Although the process offers insight into the origin of at least some dark matter and the envelopes of some giant elliptical galaxies, it has apparently not been welcomed by many non-X-ray astronomers. Perhaps non-X-ray astronomers can see no need of cooling flows in the interpretation of their own data and ideas and are reluctant to accept star formation rates 10 - 100 times that of our Galaxy taking place in a novel manner. In my view (also expressed by Bob O'Connell in this book) we may just be chauvinists with respect to star formation, only recognizing it when it resembles Galactic star formation - which principally means the formation of massive stars.

The subject of cooling flows was therefore ripe for a NATO Advanced Research Workshop, which took place at the Institute of Astronomy in the University of Cambridge, from 22 to 26 June 1987. The participants were drawn from most wavebands as well as theory. The talks and contributions are presented here in the order in which they were given. The X-ray evidence for cooling flows, including the very important X-ray line spectroscopic observations that show excellent agreement with the simplest cooling flow interpretation, is first. Heat sources and conduction are considered in several papers and, although some small reduction in the cooling rate is possible, the X-ray data appear not to allow any heat flow strong enough for an order-of-magnitude reduction.

There is evidence for extended optical and UV line emission from the central galaxies of many cooling flows and for some massive star formation. These were a major feature of many papers. The source of excitation of the emission lines is not yet securely identified. The optical line-emitting gas and its luminosity are a small fraction (less than a per cent) of the cooling luminosity of the flow, so whilst its diagnostic properties are important, it may play only a small part in the overall picture. Radio-emission, too, is common (Cygnus A is surrounded by a cooling flow) and this was discussed in several papers. Finally the implications of cooling flows for active galaxies, quasars and of galaxy formation were outlined. The views on the importance of cooling flows ranges from them being recent or much over-estimated to the strong conjecture of some of us that they were more common in the past, show us what galaxy formation was like, are responsible for baryonic dark matter, and are implicated in the evolution of quasars and powerful radio sources. Finding the correct view is going to require new data and in particular, new X-ray data.

Lively discussion periods were a significant part of the Workshop. All participants were encouraged to present as many single-viewgraph presentations as they wished. In order to retain the informality of these periods, no formal account was made. However, each participant was invited to provide one page of notes and these are collected at the end of the book as the Discussion Papers.

I am very grateful to the NATO Science Committee for funding this Workshop, to Professors Donald Lynden-Bell and Martin Rees for permitting it to be held at the Institute, to the Scientific Organizing Committee (Claude Canizares, John Danziger, Richard Mushotzky and Craig Sarazin), to the Local Organizing Committee (Carolin Crawford, Roderick Johnstone, Michael Loewenstein, Peter Thomas and Ray White) to Michael Ingham, Judith Moss and their colleagues for organizational and secretarial help, and finally to The Royal Society for my support.

A. C. Fabian

LIST OF PARTICIPANTS

K. A. Arnaud, Center for Astrophysics, 60 Garden Street, Cambridge, MA 02138, USA.

K. M. Ashman, Theoretical Astronomy Unit, School of Mathematical Sciences, Queen Mary College, Mile End Road, London, E1 4NS.

F. Bertola, Osservatorio Astronomico, 35100 Padova, Italy.

E. Bertschinger, Department of Physics, Massachusetts Institute of Technology, room 6-207, Cambridge, MA 02139, USA.

L. Binette, European Southern Observatory, Karl Schwarzschild Strasse 2, D-8046 Garching bei Munchen, West Germany.

J. J. Binney, Department of Theoretical Physics, University of Oxford, 1 Keble Road, Oxford, OX1 3NP, UK.

H. Bohringer, M.P.I. fur Extraterrestrische Physik, Karl Schwarzschild Strasse 1, D-8046 Garching bei Munchen, West Germany.

C. Boisson, Observatoire de Meudon, 92195 - Meudon Principal Cedex, France.

R. G. Bower, Physics Department, University of Durham, South Road, Durham, DH1 3LE, UK.

G. Branduardi-Raymont, Mullard Space Science Lab, Holmbury St Mary, Dorking, Surrey, RH5 6NT, UK.

J. N. Bregman, National Radio Astronomy Observatory, Edgemont Road, Charlottesville, VA 22903-0235, USA.

L. M. Buson, Osservatorio Astrofisico, Universita di Padova, 36012 Asiago (VI), Italy.

C. R. Canizares, Department of Physics, Massachusetts Institute of Technology 37-501, Cambridge, MA 02139, USA.

B. Carr, School of Mathematical Sciences, Queen Mary College, Mile End Road, London, E1 4NS, UK.

L. Cowie, Institute for Astronomy, University of Hawaii, 2680 Woodlawn Drive, Honolulu HI 96822, Hawaii, USA.

C. S. Crawford, Institute of Astronomy, Madingley Road, Cambridge, CB3 OHA, UK.

I. J. Danziger, European Southern Observatory, Karl Schwarzschild Strasse 2, D-8046 Garching bei Munchen, West Germany.

L. P. David, Leander McCormick Observatory, University of Virginia, P O Box 3818, University Station, Charlottesville, VA 22903, USA.

R. Ellis, Physics Department, University of Durham, South Road, Durham, DH1 3LE, UK.

A. C. Fabian, Institute of Astronomy, Madingley Road, Cambridge, CB3 OHA, UK.

W. Forman, Center for Astrophysics, 60 Garden Street, Cambridge, MA 02138, USA.

R. A. E. Fosbury, ST-ECF, European Southern Observatory, Karl Schwarzschild Strasse 2, D-8046 Garching bei Munchen, West Germany.

S. Garrington, Nuffield Radio Astronomy Labs, Jodrell Bank, Macclesfield, Cheshire, SK11 9DL, UK.

P. Hintzen, NASA-Goddard Space Flight Center, Code 681, Greenbelt, MD 20771, USA.

A. Hornstrup, Danish Space Research Institute, Lundtoftevej 7, DK-2800 Lyngby, Denmark.

E. Hu, Institute for Astronomy, University of Hawaii, 2680 Woodlawn Drive, Honolulu, HI 96822, Hawaii, USA.

W. Jaffe, Sterrewacht te Leiden, P B 9513, Leiden 2300 RA, The Netherlands.

C. Jenkins, Royal Greenwich Observatory, Herstmonceux Castle, Hailsham , East Sussex, BN27 1RP, UK.

R. M. Johnstone, Institute of Astronomy, Madingley Road, Cambridge, CB3 0HA, UK.

G. R. Knapp, Department of Astrophysical Sciences, Princeton University, Princeton, NJ 08544, USA.

T. R. Lauer, Peyton Hall, Princeton University Observatory, Princeton, NJ 08544, USA.

M. Loewenstein, Institute of Astronomy, Madingley Road, Cambridge, CB3 OHA, UK.

P. J. McCarthy, Department of Astronomy, University of California, Berkeley, CA 94720, USA.

B. McNamara, Department of Astronomy, University of Virgignia, Box 3818, Charlottesville. VA 22903, USA.

D. Machetto, Space Telescope Science Institute, 3700 San Martin Drive, Baltimore, MD 21218, USA.

W. G. Mathews, Lick Observatory, University of California, Santa Cruz, CA 95064, USA.

A. Meiksin, Department of Astronomy, University of California, Berkeley, CA 94720, USA.

L. Miller, Royal Observatory, Blackford Hill, Edinburgh, EH9 3HJ, UK.

J. Mittaz, Mullard Space Science Lab, Holmbury St. Mary, Dorking, Surrey, RH5 6NT, UK.

G. Morfill, M.P.I. fur Extraterrestrische Physik, Karl Schwarzschild Strasse 1, D-8046 Garching bei Munchen, West Germany.

M. Morini, Istituto di Fisica Cosmica ed Applicazion dell'Informatica, CNR, via M. Stabile 172, 90139 Palermo, Italy.

R. F. Mushotzky, NASA Goddard Space Flight Center, Code 661, Greenbelt, MD 20771, USA.

H. U. Norgaard-Nielsen, Danish Space Research Institute, Lundtoftevej 7, DK-2800 Lyngby, Denmark.

P. Nulsen, Mt Stromlo Observatory, Private Bag, Woden P O, ACT 2606, Australia.

R. W. O'Connell, Department of Astronomy, University of Virginia, Box 3818, Charlottesville, VA 22903, USA.

A. Pedlar, Nuffield Radio Astronomy Labs, Jodrell Bank, Macclesfield, Cheshire, SK11 9DL, UK.

M. J. Rees, Institute of Astronomy, Madingley Road, Cambridge, CB3 OHA, UK.

A. Robinson, European Southern Observatory, Karl Schwarzschild Strasse 2, D-8046 Garching bei Munchen, West Germany.

W. Romanishin, Physics Department, Arizona State University, Tempe AZ 85287, USA.

E. Sadler, Kitt Peak National Observatory, P O Box 26732, Tucson, AZ 85726, USA.

C. L. Sarazin, Department of Astronomy, University of Virginia, P O Box 3818, Charlottesville, VA 22903, USA.

N. Soker, Department of Astronomy, University of Virginia, P O Box 3818, Charlottesville, VA 22903, USA.

G. Stewart, X-ray Astronomy Group, University of Leicester, University Road, Leicester LE1 7RH, UK.

D. Sumi, Theoretical Astrophysics, California Institute of Technology, Pasadena, CA 91125, USA.

C. Tadhunter, University of St Andrews, University Observatory, Buchanan Gardens, St Andrews, Fife, KY16 9LZ, UK.

P. Thomas, Institute of Astronomy, Madingley Road, Cambridge, CB3 OHA, UK.

G. Trinchieri, Osservatorio di Arcetri, Largo Enrico Fermi 5, 50125 Firenze, Italy.

E. A. Valentijn, Kapteyn Laboratorium, P O Box 800, NL-9700 Groningen, The Netherlands.

W. van Breugel, Radio Astronomy Laboratory, 601 Campbell Hall, University of California, Berkeley, CA 94720, USA.

N. J. Westergaard, Danish Space Research Institute, Lundtoftevej 7, DK -2800, Lyngby, Denmark.

R. E. White, Institute of Astronomy, Madingley Road, Cambridge, CB3 OHA, UK.

NATO ARW
Cooling Flows in Clusters and Galaxies
Cambridge, England 22-26 June 1987

Row 1: L. Binette, M. Morini, R. White, N. Soker, M. Whittle, M. Loewenstein, W. Mathews, T. Lauer, R. Bower,
G. Morfill, H. Böhringer, B. Carr, R. Ellis, H. Norgaard-Nielsen, A. Hornstrup, E. Valentijn, W. Jaffe, M. Ingham
Row 2: C. Boisson, G. Trinchieri, R. O'Connell, D. Sumi, B. McNamara, L. David, J. Bregman, E. Bertschinger, L. Millar,
D. Macchetto, J. Binney, P. McCarty, C. Crawford, P. Hintzen, W. Romanishin, S. Garrington, K. Ashman, J. Mittaz, P. Nulsen
Row 3: A. Pedlar, J. Danziger, R. Fosbury, E. Sadler, A. Robinson, A. Fabian, C. Canizares, W. van Breugel, W. Forman,
P. Thomas, R. Johnstone, R. Mushotzky, K. Arnaud, W. Saslaw, F. Bertola, L. Buson, A. Meiksin, C. Sarazin, J. Moss, A. Julier

PROPERTIES OF CLUSTERS OF GALAXIES

Craig L. Sarazin
Department of Astronomy
University of Virginia
P. O. Box 3818 University Station
Charlottesville, VA 22901 U. S. A.

ABSTRACT. Recent results on clusters are reviewed and discussed in relation to cooling flows. Mergers of subclusters are apparently common in clusters. Such mergers will tend to disrupt cooling flows. I suggest that the recently observed luminous arcs in clusters may be shocks generated by a subcluster merger in a cluster with a cooling flow. The peaked galaxy distributions and radial orbits suggested by recent studies of clusters will lead to a larger number of rapidly moving galaxies passing through the cooling flow region, and this may affect the cooling flow. Recent studies of the mass distributions in clusters suggest that the dark matter is centrally condensed; this may indicate that it is dissipative and baryonic. I suggest that this dark matter and the ultimate reservoir of the cooling gas in cooling flows may both be low mass star or planetary objects. Finally, I discuss various ways in which the cooling flow environment, which appears to be ubiquitous in the early–type galactic systems which form the hosts for strong radio sources, may affect the origin and structure of the radio emitting regions.

1. INTRODUCTION

In this paper, I will review some recent developments in the study of clusters of galaxies. The topics were chosen to include an number of discoveries or ideas which may have some remote connection to or implication for cooling flows. In order to promote a lively discussion of the possible relationships between cooling flows and other cluster phenomena, I will propose several rather speculative connections between cooling flows and other cluster properties. Many of these phenomena have alternative (and generally more plausible) explanations. Specifically, I will discuss the dynamics of clusters and subcluster merging, the distribution of the galaxies and central "cusps", ther distribution of the hot gas in clusters, the distribution and origin of the dark matter in clusters, the radio emission of first rank cluster galaxies and other ellipticals, and the recently discovered luminous arcs in clusters. All comparisons to observations in this paper assume a Hubble constant of $H_o = 50$ km/s/Mpc.

1

A. C. Fabian (ed.), Cooling Flows in Clusters and Galaxies, 1–15.
© 1988 by Kluwer Academic Publishers.

(Fitchett and Webster 1987). It is possible that the two subclusters contained cooling flows centered on (and possibly helping to form) the two D galaxies in Coma, but that these focussed cooling flows were disrupted by the collision of the two subclusters (Stewart *et al.* 1984; McGlynn and Fabian 1984).

While the galaxy distribution given by equation (1) contains a central core with a uniform density, many clusters show peaked density distributions at their centers. In many cases, equation (1) is a very poor fit to the galaxy distribution unless such cores are removed (Sarazin 1980). The inferred central distribution in a cluster depends very strongly on the location of the cluster center. If the center is taken to be the maximum in the X-ray surface brightness or the position of a central cD or D galaxy, then in many clusters the surface density increases into the center roughly as r^{-1} (Beers and Tonry 1986; Oegerle *et al.* 1986,1987). This is the projected distribution of a singular isothermal sphere (core radius $a = 0$), and is similar to the central regions of a de Vaucouleurs profile.

Several methods have been used to determine the velocity distribution and the shapes of the galaxy orbits (radial, circular, isotropic, triaxial) in clusters. The standard technique (Kent and Gunn 1982) is to use the positions and radial velocities of the galaxies to determine the best fit model from among a set of self-consistent, equilibrium phase-space distributions. For spherical clusters, these can depend only on the galaxy orbital energy and on the magnitude of its angular momentum. Recent analyses (Merritt 1987; The and White 1986) show that it is not possible to uniquely determine the velocity distribution of the galaxies unless the gravitational potential (the total mass distribution) is known. Since much of the mass is due to dark matter, there is no compelling reason why the mass distribution should be the same as the distribution of galaxies.

As an alternative, Pryor and Geller (1984) used the distribution of gas–poor spirals and of truncated ellipticals to attempt to infer the orbits of galaxies in clusters, on the assumption that these distortions resulted from ram pressure stripping and tidal interactions, respectively. This technique is primarily sensitive to the orbits of outlying galaxies. In the Coma cluster, their results were consistent with isotropic orbits. Because these results depend on the poorly understood physics of these stripping processes, they are probably rather uncertain.

Recently, O'Dea, Sarazin, and Owen (1987) have used the orientations of the tails of Narrow–Angle–Tail (NAT) radio sources to determine the orbital distribution of galaxies in clusters. This method makes two major assumptions; the first is that the orientation of the tails reflects the velocity of the parent galaxy. There is a very large body of evidence supporting the theory that these tails are produced by ram pressure as the galaxy moves through the intracluster medium (O'Dea 1985). The second assumption is that NATs represent a fair sample of galaxy orbits in the cluster. While the distributions of NAT positions and velocities are indistinguishable from those of other cluster galaxies, this assumption is very uncertain. The attractive feature of this method is that the tail of an NAT gives a direct indication of its direction of motion. If the tails mainly point towards or away from the cluster center, the orbits are mainly radial. If the tails mainly point at 90° to the cluster center, the orbits are mainly circular.

When this technique is applied to the orientation of all NATs, one finds a slight preference for radial orbits. For NATs in the outer parts of clusters (at projected distances $d > 0.5$ Mpc), there is a slight preference for circular orbits. However, in the inner parts of the clusters ($d < 0.5$ Mpc) the orbits are strongly radial, with the best–fit ratio of radial to tangential velocity dispersions being $\sigma_r/\sigma_t = 3.5$. A

2. GALAXIES, GAS, AND DARK MATTER

The three main components of clusters of galaxies are the galaxies themselves, the intracluster medium, and the dark matter which makes up the majority of the mass. A typical cluster has several hundred galaxies, which are mainly ellipticals and S0s in regular clusters. The luminous portions of galaxies contribute only about ~5% of the mass in the cluster. The intracluster medium is mainly hot gas, with typical proton densities of $n \sim 10^{-3}$ cm^{-3} and temperature of $T \approx 8 \times 10^7$ K. As discussed in Sections 2.2 and 2.3 below, the total mass of this gas is very uncertain because the observed distribution gives a mass which diverges at large radii where the surface brightness of the X-ray emission from the gas is very uncertain. The total cluster masses are also very uncertain at large radii, with the result that the gas may contribute from 5% to 50% of the total cluster mass (Cowie et al. 1987). The intracluster medium also contains magnetic field and relativistic particles. The present estimates of the magnetic field strength and the energy density of relativistic particles suggests that these components are not very important dynamically in the cluster, but their influence should grow dramatically as the gas cools in cooling flows (Soker and Sarazin 1988). Finally, the dark matter contributes (by definition) the remainder of the mass, roughly 50–90% of the total cluster mass.

2.1 The Distribution of the Galaxies

Regular clusters of galaxies have centrally condensed galaxy distributions. Following Zwicky (1957), the galaxy distributions in clusters have most often been fit with an isothermal sphere distribution. Analytic formulae which approximate an isothermal sphere are often used, such as

$$\rho_{gal} \propto \left[1 + \left(\frac{r}{a} \right)^2 \right]^{-x} \tag{1}$$

where $x = 1$ fits the asymptotic form of an isothermal sphere at large radii, while $x = 3/2$ provides a good fit to the inner regions (King 1962). The core radius a is typically $a \approx 0.25$ Mpc (Bahcall 1977), but there is a considerable variation in the published values.

In recent years, it has become clear that many clusters are not really centrally condensed, but instead form several subclusters (Forman et al. 1981; Geller and Beers 1982; Baier and Oleak 1983; Fabricant et al. 1986). The subclusters may be separated in space and/or in radial velocity. If present in the cluster, supergiant cD or D galaxies are found at the density maxima of the subclumps. The subclumps often appear to be bound together, and have short enough orbital times to merge in a fraction of a Hubble time.

This leads to a more complex picture for the evolution of clusters and of cooling flows. Cooling flows may exist in the subclusters. When subclusters collide and merge to form larger clusters, pre-existing subcluster cooling flows may be disrupted (McGlynn and Fabian 1984). It is possible that cooling flows were even more prevalent in the past. This also means that the fact that a cluster lacks a well-established cooling flow now doesn't mean that it never had one. The Coma cluster, often regarded as the archetypical regular cluster, may be an example of just such a subcluster merger. Recent analyses of the galaxy positions and velocities in Coma suggest that it consists of two subclusters, associated with the two D galaxies

4

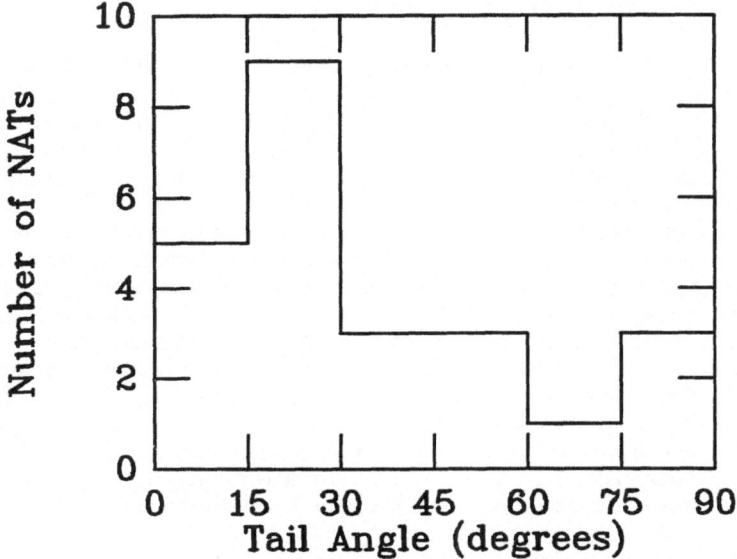

Figure 1. The distribution of the tail angles of NAT radio sources near the centers of clusters (inner 0.5 Mpc), showing the tendency for the orbits of these galaxies to be predominantly radial.

histogram of the observed NAT tail angles (measured relative to the direction to the cluster center) is shown for the inner regions of cluster in Figure (1).

This indicates that the orbits of cluster galaxies may be fairly radial near the centers of clusters. This would be consistent with the observations of galaxy density "cusps" at the centers of many clusters, as discussed above (Beers and Tonry 1986). From the point–of–view of cooling flows, the presence of cusps and of radial orbits suggests that many rapidly moving galaxies may pass through the innermost regions of clusters where the cooling flows are located. These galaxies might have an influence on the dynamics and energetics of of the cooling flow gas.

2.2 The Distribution of the Gas

Observations with the *Einstein* X-ray observatory have provided detailed images of the X-ray emission of clusters. In general, the X-ray morphology of clusters parallels their optical morphology. Forman and Jones (1982) proposed a two-dimensional classification scheme for the X-ray morphology of galaxies, which they relate to the evolutionary state of the cluster as determined by its optical properties. This scheme is presented in Table 1, which is taken in large part from Forman and Jones (1982). First of all, Forman and Jones classify the clusters as being irregular ('early') or regular ('evolved'), based on their overall X-ray distribution. The early clusters have irregular X-ray surface brightnesses. Their X-ray luminosities L_x and X-ray spectral temperatures T are low. On the other hand, the evolved clusters have regular, centrally condensed X-ray structures. The X-ray distribution is smooth. The evolved clusters have high X-ray luminosities and gas temperatures. The second determinant of the X-ray morphology of clusters is the presence or absence of a

Table 1
Morphological Classification of X-ray Clusters

	No X-ray Dominant Galaxy (nXD)	X-ray Dominant Galaxy (XD)
Irregular	Low $L_x \lesssim 10^{44}$ erg/s Cool gas $T = 1\text{-}4$ keV X-rays from many galaxies Irregular X-ray distribution High spiral fraction > 40% Low central galaxy density Prototype: A1367	Low $L_x \lesssim 10^{44}$ erg/s Cool gas $T = 1\text{-}4$ keV Central galaxy X-ray halo Irregular X-ray distribution High spiral fraction > 40% Low central galaxy density Prototype: Virgo/M87
Regular	High $L_x \gtrsim 10^{44}$ erg/s Hot gas $T \gtrsim 6$ keV No cooling flow Smooth X-ray distribution Low spiral fraction $\lesssim$ 20% High central galaxy density Prototype: Coma (A1656)	High $L_x \gtrsim 3 \times 10^{44}$ erg/s Hot gas $T \gtrsim 6$ keV Cooling flow onto central galaxy Compact, smooth X-ray emission Low spiral fraction $\lesssim$ 20% High central galaxy density Prototype: Perseus (A426)

central dominant galaxy in the cluster. The X-ray emission from a cluster tends to peak at the position of such a galaxy. Clusters containing such central dominant galaxies are classified by Forman and Jones as X-ray Dominant (XD); those without such a galaxy are classified as non-X-ray Dominant (nXD). The nXD clusters have larger X-ray core radii $r_x \approx 500$ kpc. There is no strong X-ray emission associated with any individual galaxy in these systems. The XD clusters have small X-ray core radii $r_x \approx 250$ kpc. The X-ray emission is strongly peaked on the central dominant galaxy. In many cases, X-ray spectral observations show that a cooling flow is present in these clusters. The data of Jones and Forman (1984) suggest that *all* cooling flows are centered on the dominant galaxies in XD clusters.

Double peaked X-ray emission has been seen in a number of clusters, of which the best studied example is A98 (Henry *et al.*, 1981; Forman *et al.*, 1981). In most of these systems, the galaxy distribution is also double peaked (Henry *et al.*, 1981; Beers *et al.*, 1983). These double systems may represent the merger of subclusters, as discussed above.

In the regular X-ray clusters, the gas distribution is smooth and centrally condensed. The observed X-ray surface brightnesses (averaged azimuthally) are well-fit by the radial gas density profile (Jones and Forman 1984)

$$\rho_{gas}(r) = \rho_o \left[1 + \left(\frac{r}{r_c} \right)^2 \right]^{-3\beta/2}, \tag{2}$$

where ρ_o is the central gas density and r_c is the core radius. Jones and Forman fit ρ_o, r_c, and β from the X-ray data. Equation (2) is a good fit to the majority of clusters, but fails in the central regions of some clusters, because these clusters contain cooling flows. The average value of β determined by fits to the X-ray surface brightness of a large number of clusters was found to be (Jones and Forman 1984)

$$\langle \beta_{fit} \rangle \approx \frac{2}{3}. \tag{3}$$

Thus the implied gas density varies at large radii as $\rho_{gas} \sim r^{-2}$.

The gas distribution in equation (1) follows from a very simple model for the hydrostatic equilibrium of the gas; this model is often called the "beta" model or the self-consistent isothermal model (Cavaliere and Fusco-Femiano 1976; Sarazin and Bahcall 1977). If one assumes that the gas is isothermal, and that the galaxy velocity distribution is isothermal and isotropic (the velocity dispersion is the same in all directions and at all positions), then the gas and galaxy distributions are related by

$$\rho_{gas} \propto \rho_{gal}^{\beta}, \tag{4}$$

where ρ_{gal} is the galaxy density,

$$\beta \equiv \frac{\mu m_p \sigma^2}{kT}, \tag{5}$$

μ is the mean atomic mass, T is the gas temperature, and σ is the one-dimensional velocity dispersion of the cluster. For convenience, the analytic King approximation to the isothermal sphere is often assumed (King 1962); this is equation (1) with $x = 3/2$. Equations (1) and (4) lead to equation (2).

If the arguments leading to this simple "beta" model are essentially correct, the observed value of β (equation 3) implies that the gas density falls off less rapidly with radius than the galaxy density (in possible agreement with other observations, such as Abramopoulos and Ku 1983), and that the energy per unit mass is higher in the gas than in the galaxies (Jones and Forman 1984).

Unfortunately, equations (3) and (5) do not agree with the determinations of the X-ray spectral temperatures and the galaxy velocity dispersions of clusters (Mushotzky 1984). From a sample of clusters with well-determined spectra, Mushotzky finds

$$\langle \beta_{spect} \rangle \approx 1.2, \tag{6}$$

which he notes is about twice the value determined from observations of the X-ray surface brightness.

A number of suggestions have been made as to the origin of this discrepancy. First, the gas may very well not be isothermal. However, Mushotzky (1984) has argued that the same problem occurs for other thermal distributions in the gas. Second, it may be that the line-of-sight velocity dispersion does not represent accurately the energy per unit mass of the galaxies. The distribution could be anisotropic, either if the cluster is highly flattened or if the cluster is spherical but galaxy orbits are largely radial. Third, it may be that many of the galaxy velocity dispersions measured for clusters are contaminated by foreground or background groups (Geller and Beers 1982). The velocity dispersions may also be affected by subclustering or nonvirialization of the cluster. All of these effects will cause the data to overestimate the actual velocity dispersion (the cluster potential), and thus to overestimate β_{spect}. Perhaps one indication that such systematic errors in the velocity dispersion might be occurring is that Coma, the best studied regular cluster, does not show a β discrepancy.

Finally, let me point out a trivial (but possibly relevant) explanation of the β discrepancy. The particular form of the gas distribution in the self-consistent isothermal model as given above depends on assuming that the galaxy distribution is fit by the King approximation to the isothermal sphere. This approximation breaks down at large radii, where the King model density varies as $\rho_{gal} \propto r^{-3}$ while a real isothermal sphere density varies as $\rho_{gal} \propto r^{-2}$. Of course, it is the gas distribution at large radii which has the greatest leverage in affecting the fit to equation (2). If the galaxy density in clusters is *really* isothermal, then equation (2) will not fit the observed gas distribution for the correct value of β. As an alternative, let us assume that the galaxy distribution can be fit by the simple analytic form (equation 1) with $x = 1$. This is similar to the King analytic fit, but has the correct asymptotic form for an isothermal sphere. Then all the formulae for the self-consistent isothermal model remain unchanged if we substitute $\beta \rightarrow (2/3)\beta$. Retaining the present definition of β, this is equivalent to $\beta_{fit} = (2/3)\beta_{spect}$, which is essentially consistent with the observations. Another way of expressing all this is to say that the observed gas distribution (equation 2) is essentially that of an isothermal sphere. If this argument is correct, then there is no need for any ongoing heating process for the intracluster medium, since any reasonable method of introducing the gas into the cluster will give $\beta \approx 1$ (Sarazin 1986).

2.3 The Distribution of Mass and Dark Matter

Zwicky (1933) first applied the virial theorem to the positions and velocities of galaxies in the Coma cluster and showed that the mass of the cluster was considerably higher than could be attributed to the sum of all the galaxy masses. More recent mass determinations have been based on fitting the measured galaxy positions and radial velocities in clusters to a self-consistent phase space distribution function for a spherical cluster (Kent and Gunn 1982). In the Coma cluster, Kent and Gunn assumed that the mass and the galaxies were distributed in the same manner (a constant mass-to-light ratio), and found that the best fitting models had more or less isotropic galaxy orbits. Their derived mass and mass-to-light ratio for Coma was

$$M/L_B \approx 180 M_\odot/L_\odot, \qquad (7)$$
$$M(\leq 2\,\text{Mpc}) \approx 1 \times 10^{15} M_\odot.$$

Unfortunately, it has now become clear that there is an ambiguity between the mass distribution and the galaxy orbits (The and White 1986; Merritt 1987; Bailey 1982). If one knows whether the orbits are isotropic, circular, or radial, one can determine the mass distribution. If one knows the mass distributions, one can determine the shapes of the orbits. Table 2 demonstrates this ambiguity. The models with circular galaxy orbits imply a dark matter distribution which is strongly centrally condensed, with the mass-to-light ratio reaching a maximum at the cluster center, and these models have the lowest total mass at large radii. The models with radial galaxy orbits are consistent with a nearly uniform dark matter distribution (the mass-to-light ratio increases rapidly outwards), and have the largest total mass.

Problems with the shapes of galaxy orbits, the small number of galaxies in a cluster, and the contaminating effects of background and foreground galaxies can be avoided by using the gas distribution in clusters to determine their mass distributions (Bahcall and Sarazin 1977; Fabricant and Gorenstein 1983). This method

Table 2
Mass Distributions and Galaxy Orbits in the Coma Cluster

Dark Matter Distribution	Orbits		Total Mass ($\lesssim 5$ Mpc)
Centrally Condensed	Circular	Minimum	$\sim 2 \times 10^{15}\,M_\odot$
Constant M/L	Isotropic		$\sim 4 \times 10^{15}\,M_\odot$
Uniform	Radial	Maximum	$\sim 12 \times 10^{15}\,M_\odot$

assumes that the gas in clusters is hydrostatic. This is a reasonable approximation as long as the cluster is stationary (the gravitational potential does not change on a sound crossing time), forces other than gas pressure and gravity (magnetic fields, for example) are not important, and any motions in the gas are subsonic. The cluster mass $M(r)$ can be determined if the density and temperature of the intracluster gas are known, through the hydrostatic equilibrium equation

$$M(r) = -\frac{kT(r)r}{\mu m_p G}\left(\frac{d\ln\rho_{gas}}{d\ln r} + \frac{d\ln T}{d\ln r}\right). \tag{8}$$

Note that the mass depends only weakly on the gas density ρ_{gas} (only its logarithmic derivative enters), but depends strongly on the gas temperature.

Unfortunately, the limited spectral response of the *Einstein* X-ray Observatory has prevented the direct determination of temperature profiles for the intracluster gas. Accurate profiles of the gas density are known. In order to apply the hydrostatic method to clusters, some simple assumption must be made about the temperature distribution $T(r)$. Unfortunately, because the mass is strongly affected by T, this means that the resulting mass profiles will be uncertain. Accurate global X-ray spectra exist for a number of clusters from the HEAO-1 A-2 detectors. These spectra generally cannot be fit by emission at a single temperature (Cowie *et al.* 1987). The spectra can be used to determine the amount of gas (or, more precisely, the amount of $EI \equiv \rho_{gas}^2 V$, where V is the volume) as a function of temperature, but cannot tell us where the gas is located because of their poor spatial resolution. The *Einstein* imaging observations give $\rho_{gas}(r)$ (which can be integrated to give $EI = \rho_{gas}^2 V$), but give no information about the temperature structure. However, the comparison of these two results [(EI vs. T) and (EI vs. r)] allow the determination of (T vs. r), if one assumes that T is a monotonic function of the radius r. This method has been applied to determine the mass profile in the Coma cluster by Cowie *et al.* (1987), The and White (1987), and Hughes *et al.* (1987). All of these authors assumed a polytropic equation of state for the gas $T \propto \rho_{gas}^{\gamma-1}$.

Since the observed gas densities vary at large distances like an isothermal sphere $\rho_{gas} \propto r^{-2}$, while these mass determinations require that the gas temperature decrease with radius, the total mass density will always decrease with radius more rapidly than the gas density. Since the total density is the sum of the gas density, the galaxy density, and the missing mass density, these determinations suggest that the missing mass is concentrated towards the cluster center. For the Coma cluster, this method gives a missing mass density which is more centrally concentrated than that of the galaxies or the intracluster gas. Thus, the mass–to-light ratio increases towards the cluster center. These analysis also suggest lower values for the total cluster mass and higher values for the total gas mass than have previously been assumed. In fact, the gas might contribute as much as about 30% of the mass, although these results are very uncertain since they require that the *Einstein* X-ray surface brightness profiles be extended to larger radii than those at which they are actually observed.

If these monotonic temperature mass determinations are correct, one is lead to a different picture of the missing mass than has usually been assumed. If the missing mass is more concentrated than the visible matter in the cluster, it suggests that the missing mass has undergone dissipation. Combined with the smaller ratio of missing mass to visible mass, this suggests that the missing mass is baryonic matter, and not some weakly interacting species. Since a large amount of material is currently cooling out at the centers of many clusters, such baryonic matter might be related to cooling flows (although the observed cooling rates are too small to produce much missing mass in a Hubble time). A major uncertainity in this is the assumption of a monotonically decreasing temperature gradient. While this may be plausible, there is no compelling physical reason why this must be so, and recent observations of the Perseus cluster suggest that it has an *increasing* temperature even outside of the cooling flow (Ulmer *et al.* 1987).

3. RADIO SOURCES AND COOLING FLOWS

It is well–known that strong, extended extragalactic radio sources are generally associated with elliptical galaxies. The energy which is observed in the large scale radio emission is believed to ultimately arise from the nucleus of the galaxies, and is delivered to the large scale radio lobes by jets. In addition to the radio sources associated with normal large elliptical galaxies, distorted radio sources, called wide–angle–tail (WAT) sources, are found to be associated with central dominant cluster galaxies. The X–ray and optical observations which form the main topic of this meeting indicate that large amounts of gas are cooling at and flowing into the centers of clusters of galaxies, and that normal elliptical galaxies also contain large quantities of hot gas, which probably forms cooling flows.

Thus, the interstellar medium of the host galaxies of strong radio sources generally consists of hot gas in the form of a cooling inflow. These cooling flows provide a unique physical environment for radio galaxies. It is possible that the cooling flow environment affects the formation and propagation of radio jets. There are several ways that this might occur. First, cooling inflows could provide the fuel to power the central engines of radio sources (Fabbiano *et al.* 1987; Valentijn and Bijleveld 1983; Jones and Forman 1984). Second, as suggested by Begelman (1986), accretion from a cooling flow may lead to the formation of radio jets in a galactic nucleus; this might explain why radio sources are associated with elliptical galaxies. Third, the cooling flow environment may affect the propagation of radio jets. For example, the rather high interstellar gas pressures in cooling flows ($P/k \sim 10^6$ cm^{-3} K) may help to confine radio sources.

Of particular interest in this regard are the radio sources associated with central dominant cluster galaxies. Many of these are wide–angle–tail (WAT) sources, which have strongly bent jets. WATs are typically somewhat smaller that the double radio sources associated with normal elliptical galaxies. Recently, O'Dea and Baum (1987) observed a number of radio sources associated with central dominant cluster galaxies in clusters with cooling flows, and compiled their measurements with previous results. Their result show that large radio sources are only associ-

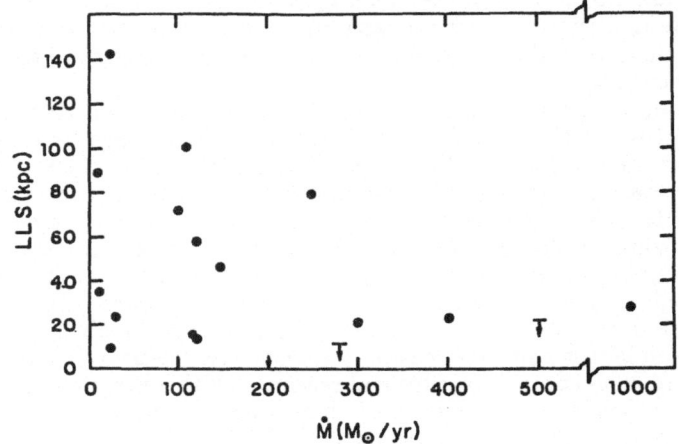

Figure 2. The Largest Linear Size (LLS) of radio sources associated with cluster cooling flows are shown as a function of the cooling rate. The data are from O'Dea and Baum (1987) and the figure is taken from Soker and Sarazin (1987).

ated with clusters with small inflow rates, suggesting some connection between the origin of WAT radio galaxies and cooling flows. Figure (2) shows the Largest Linear Size (LLS) of the radio sources associated with cooling flow cluster central galaxies as a function of the cluster cooling rate (Soker and Sarazin 1987).

Is there any way that cooling flows could bend and disrupt the jets in WATs? Sumi and Smarr (1984) suggested that the gradually decreasing pressure in the outer parts of a cooling flow may destabilize jets. Another possibility occurs if most of the hot gas in cooling flows is *not* removed by thermal instabilities at large radii. Then, the cooling flow will pass through a sonic transition and the inflow will become supersonic. In this case, one can generally show that the temperature and gas pressure will drop to nearly zero just within the sonic radius (Sarazin and White 1987; Soker and Sarazin 1987). Figure (3) shows the variation of the gas pressure in two simple numerical models for cooling flows; the model for an individual elliptical galaxy is taken from Sarazin and White (1987), while the cluster cooling flow models uses the numerical code developed by White and Sarazin (1987). In general, the pressure drops to essentially zero at a radius $r_c \approx 0.7r_s$, where r_s is the sonic radius of the cooling flow. There are two reasons for this drop in the pressure. First, within the sonic radius the inflow is supersonic, and conditions downstream can have no influence on the properties of the gas at the sonic radius. Thus, the drop in pressure doesn't lower the gas density and pressure or alter the position at the sonic radius. Second, at the sonic radius in a cooling flow, the cooling time and inflow time are comparable. At the temperatures of interest ($T \sim 10^6 - 10^7$ K), as the temperature decreases cooling accelerates rapidly, and the cooling time gets much shorter. On the other hand, within the sonic radius, the gas velocity is roughly constant, so the flow time does not decrease rapidly. As a result, the gas cools almost completely before it has flowed in very far from the sonic radius. [We note, however, that the *total* pressure may not show such an extreme drop because the magnetic pressure doesn't decrease with decreasing temperature (Soker and Sarazin

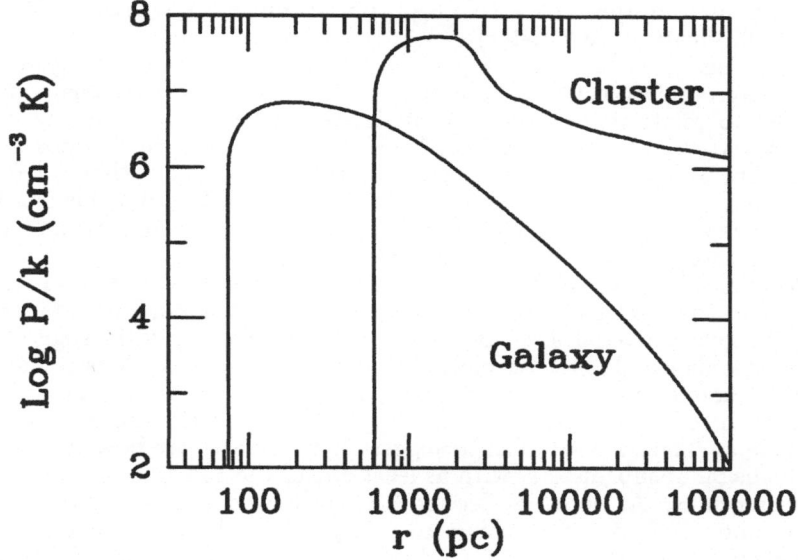

Figure 3. The variation of the gas pressure in the inner portions of a homogeneous model for a cooling flow in a cluster (labeled Cluster and taken from White and Sarazin 1987) and in an individual elliptical galaxies (labeled Galaxy and taken from Sarazin and White 1987). The figure shows the rapidly increasing gas pressure just within the sonic radius of the cooling flow.

1988).] Soker and Sarazin (1987) have shown that an outwardly moving jet will tend to be disrupted when it hits this "pressure wall" in the cooling flow. In general, the sonic radii of cooling flows are probably too small to explain the observed sizes of WAT radio sources, although Sumi *et al.* (1987) show that this mechanism may apply to the radio source associated with the cD in A2029.

Another possibility is that radio jets might be disrupted at the transition region between the intragalactic cooling flow gas and the surrounding intracluster medium. Since the central dominant galaxy will generally be moving at $\sim 10^2$ km/s relative to the center–of–mass of the cluster, there will probably be a considerable velocity shear in this layer. Finally, the gas in cooling flows is thermally unstable, and jets may be deflected or disrupted when they collide with thermally unstable clumps (Burns *et al.* 1986).

4. LUMINOUS ARCS IN CLUSTERS

Recently, Lynds and Petrosian (1986) and Soucail *et al.* (1987) have found luminous, narrow, blue arcs projected within the central regions of the two moderately high redshift clusters A370 and C2242–02. These arcs are approximately 150 kpc in length but only about 20 kpc wide. They are roughly 80° segments of nearly circular arcs of radius ~100 kpc, centered on bright cluster galaxies. A370 has a redshift of $z = 0.373$ and is a very luminous X-ray cluster ($L_X \approx 9.7 \times 10^{44}$ erg/s); C2242–02 has a similar redshift, but doesn't have a published X-ray flux. The arcs have visual luminosities $L_V \sim 10^{11} L_\odot$, and are quite blue (about 1/2 mag bluer than the ellipticals in the cluster).

A number of authors have suggested that the arcs are either a gravitationally lensed background object or a light echo within the cluster. In this paper, I have attempted to relate many other recent developments in the study of clusters to cooling flows. Along the same line, I will now speculate on the possibility that the luminous arcs are shocks in cluster cooling flows. I note that A370 is a very luminous X-ray cluster, and may have a cooling flow, although this is not known. As I was preparing my talk I received a preprint from Begelman and Blandford (1987), who propose basically the same idea.

What would cause such a shock to form? It might be due to an eruption of some sort in the nucleus of the central galaxy, but it is hard to understand why this would give a partial arc. Begelman and Blandford (1987) suggest that the arcs are bow shocks due to the motion of the central galaxy. A problem with this suggestion, noted by the authors, is that it is difficult to make a galactic bow shock as large as the observed arcs; one would typically expect the shock to be only ~ 20 kpc in length. Thus, I would like to suggest that the arcs are bow shocks produced by the collision of subclusters, with at least one of the subclusters having a cooling flow at its center. As noted in Section 2.1, such subcluster mergers appear to be fairly common. Of course, in this case a larger bow shock is expected than for a single galaxy.

The cluster must have a cooling flow in order that the shock produce a narrow feature and optical emission. If the gas were not cooling prior to the passage of the shock, the density increase in the shock (a factor of four in a strong shock) and offsetting temperature increase would not result in rapid cooling after the shock. Then, the shocked region would be broad, and most of the emission from the shock would be in X-rays. Even if the shock is radiative, it is difficult to see how it could produce so much optical emission. The energy flux through the shock is

$$\dot{E}(shock) \lesssim 3 \times 10^{11} L_\odot \left(\frac{n}{0.01\,\mathrm{cm}^{-3}}\right) \left(\frac{v}{1000\,\mathrm{km/s}^{-1}}\right)^3, \tag{9}$$

where n is the proton density ahead of the shock and v is the shock velocity. This is comparable to the observed visual luminosity. However, most of the emission produced by gas cooling from X-ray emitting temperatures does not occur in the optical, so it is difficult to understand how such a shock could produce the observed luminosity directly. The emission could be synchrotron emission by relativistic particles accelerated in shock shock, although this requires a very high efficiency for particle acceleration, and it is unlikely that most of the synchrotron emission would be in the optical part of the spectrum.

The most plausible emission mechanism may be emission following star formation in the cooling material behind the shock (Begelman and Blandford 1987). If a significant part of the cooling material were converted into massive stars, then the observed optical emission could be due to these stars and their H II regions. The narrow width of the arcs requires that the timescale for this star formation be less than about $2 \times 10^7 \mathrm{yr}(v/1000\,\mathrm{km/s})$, which may be just marginally possible for O stars.

This theory obviously has a number of problems. In any case, it should be possible to measure a spectrum for the arcs. If their redshift is much greater than that of the cluster, they must be the scattered or imaged light of a background object. If the arcs are due to shocks in cluster cooling flows, their redshifts must be the same as those of the clusters to within $\sim 10^3$ km/s. A second test of this theory concerns the surface brightness within the arcs. If they are shocks, they are actually limb–brightened spherical caps, and they should have some surface brightness in their interior. The interior surface brightness would be roughly the arc surface brightness reduced by the ratio of the arc thickness to its radius of curvature. If the arcs are lenses or light echos, they are really one dimensional and should have no interior surface brightness.

5. CONCLUSIONS

In recent years, a more complex picture of the dynamics of clusters has emerged. In many cases, irregular clusters consist of several bound subclusters, which are in the process of merging. As a result, in many cases it is a great oversimplification to treat the evolution of the gaseous component of clusters and of cooling flows as occurring in a fixed and spherical potential. The collisions of subclusters will tend to disrupt any focused cooling flows that might have existed in the subclusters. In a cluster with a fixed, spherical potential, a cooling flow once established will tend to continue until most of the intracluster gas is exhausted. If this were true of clusters, then any cluster which lacks a focused cooling flow today never had one in the past. With the merging of subclusters, the lack of a cooling flow today doesn't establish that the cluster has never had such a flow in the past. It is possible that many of the effects of cooling flows, such as their possible contribution to the formation of D and cD galaxies, were more significant in cooling flows in subclusters which preceded the presently observed clusters.

I have suggested that the luminous arcs recently observed in two moderate redshift clusters (Lynds and Petrosian 1986; Soucail *et al.* 1987) may be shocks associated with the merger of subclusters with cooling flows.

Recent studies of the distribution and orbits of galaxies in clusters suggest that the galaxies may have a centrally peaked density distribution and radial orbits. This would result in a relatively large number of high velocity galaxies passing through the cooling flow region. These galaxies might influence the dynamics of the cooling gas. Drag from the galaxies could provide a source of heat in this region, and the galaxies might help to generate the density perturbations necessary for the thermal instability of the gas and its rapid cooling at large radii.

Recent studies of the mass distribution in clusters using the distribution of both the galaxies and the hot gas suggest that the dark matter is more centrally condensed than the galaxies or gas. This argues that at least this component of the dark matter is dissipative (baryonic) matter. Since large amounts of baryonic matter are currently disappearing from the centers of clusters in cooling flows, perhaps a similar process could have formed the dark matter. (However, I note that the current rates of cooling in clusters are too small to produce any significant portion of dark matter in clusters in a Hubble time.) The leading suggestion for the ultimate reservoir of the gas cooling in cooling flows is that it forms low mass stars or planetary bodies (Fabian *et al.* 1982; Sarazin and O'Connell 1983). Perhaps similar objects form the centrally condensed dark matter in clusters.

Observations of normal elliptical galaxies and of central dominant cluster ellipticals indicate that the gaseous environments of these galaxies consist of high pressure hot gas, often in the form of cooling flows. These galaxies are the host

14

systems of powerful radio sources. It is important to consider the effect that this unique gaseous environment has on the origin and structure of the radio sources. Accretion of a portion of the cooling gas by the galactic nucleus may power the radio emission, and the form of the accreting gas may influence whether the nuclear activity in the galaxy produces radiative luminosity or the kinetic luminosity of jets. The confinement, propagation, and stability of the jets will affected by the high pressure cooling flow gas. Of particular interest in this regard are the smaller, distorted Wide–Angle–Tail (WAT) radio sources associated with central dominant cluster galaxies. Many of these galaxies occur at the centers of cluster cooling flows. I discussed the possibility that radio jets would be disrupted by the "pressure wall" at the sonic radius of the cooling flow (if it exists), by the shear in the flow when it moves from the intracluster medium to being bound to the central galaxy, or by collisions with thermally unstable blobs of gas in the cooling flow.

This work was supported in part by NASA Astrophysical Theory Center Grant NAGW-764.

REFERENCES

Abramopoulos, F., and Ku, W. 1983, *Ap. J.*, **271**, 446.
Bahcall, J. N., and Sarazin, C. L. 1977, *Ap. J. Lett.*, **213**, L99.
Bahcall, N. A. 1977, *Ann. Rev. Astr. Ap.*, **15**, 505.
Baier, F. W., and Oleak, H. 1983, *Astr. Nach.*, **304**, 277.
Bailey, M. E. 1982, *M. N. R. A. S.*, **201**, 271.
Beers, T. C., Huchra, J. P., and Geller, M. J. 1983, *Ap. J.*, **264**, 356.
Beers, T. C., and Tonry, J. L. 1986, *Ap. J.*, **300**, 557.
Begelman, M. C. 1986, *Nature*, **322**, 614.
Begelman, M. C.,and Blandford, R. 1987, preprint.
Burns, J. O., O'Dea, C. P., Gregory, S. A., and Balonek, T. J. 1986, *Ap. J.*, **307**, 73.
Cavaliere, A., and Fusco-Femiano, R. 1976, *Astr. Ap.*, **49**, 137.
Cowie, L. L., Henriksen, M. J., and Mushotzky, R. 1987, *Ap. J.*, **317**, 593.
Fabian, A. C., Nulsen, P. E., and Canizares, C. R. 1982, *M. N. R. A. S.*, **201**, 933.
Fabricant, D., Beers, T. C., Geller, M. J., Gorenstein, P., Huchra, J. P., and Kurtz, M. J. 1986, *Ap. J. Lett.*, **243**, L133.
Fabricant, D., and Gorenstein, P. 1983, *Ap. J.*, **267**, 535.
Fitchett, M., and Webster, R. 1987, *Ap. J.*, **317**, 653.
Forman, W., Bechtold, J., Blair, W., Giacconi, R., Van Speybroeck, L., and Jones, C. 1981, *Ap. J. Lett.*, **243**, L133.
Forman, W., and Jones, C. 1982, *Ann. Rev. Astr. Ap.*, **20**, 547.
Geller, M. J., and Beers, T. C. 1982, *Publ. Astr. Soc. Pac.*, **94**, 421.
Henry, J. P., Henriksen, M. J., Charles, P. A., and Thorstensen, J. R. 1981, *Ap. J. Lett.*, **243**, L137.
Hughes, J. P., Yamashita, K., Okumura, Y., Tsunemi, H., and Matsuoka, M. 1987, preprint.
Jones, C., and Forman, W. 1984, *Ap. J.*, **276**, 38.
Kent, S. M., and Gunn, J. E. 1982, *Astr. J.*, **87**, 945.
King, I. R. 1962, *Astr. J.*, **67**, 471.
Lynds, R., and Petrosian, V. 1986, *Bull. Amer. Astr. Soc.*, **18**, 1014.
McGlynn, T. A., and Fabian, A. C. 1984, *M. N. R. A. S.*, **208**, 709.

Merritt, D. 1987, *Ap. J.*, **313**, 121.
Mushotzky, R. F. 1984, *Phys. Scripta*, **T7**, 157.
O'Dea, C. P. 1985, *Ap. J.*, **295**, 80.
O'Dea, C. P. and Baum, S. A. 1987, in *Continuum Radio Processes in Clusters of Galaxies: NRAO-Green Bank Workshop # 16*, ed. C. P. O'Dea and J. M. Uson (Greenbank: NRAO), p. 141.
O'Dea, C. P., Sarazin, C. L., and Owen, F. N. 1987, *Ap. J.*, **316**, 113.
Oegerle, W. R., Hoessel, J. G., and Ernst, R. M. 1986, *Astr. J.*, **91**, 697.
Oegerle, W. R., Hoessel, J. G., and Jewison, M. S. 1987, *Astr. J.*, **93**, 519.
Pryor, C. and Geller, M. 1984, *Ap. J.*, **278**, 457.
Sarazin, C. L. 1980, *Ap. J.*, **236**, 75.
Sarazin, C. L. 1986, *Rev. Mod. Phys.*, **58**, 1.
Sarazin, C. L., and Bahcall, J. N. 1977, *Ap. J. Suppl.*, **34**, 451.
Sarazin, C. L., and O'Connell, R. W. 1983, *Ap. J.*, **268**, 552.
Sarazin, C. L., and White, R. E. 1987, *Ap. J.*, **320**, 37.
Soker, N., and Sarazin, C. L. 1987, *Ap. J.*, in press.
Soker, N., and Sarazin, C. L. 1988, paper in this proceedings.
Soucail, G., Fort, B., Mellier, Y., and Picat, J. P. 1987, *Astr. Ap.*, **172**, L14.
Stewart, G. C., Fabian, A. C., Jones, C., and Forman, W. 1984, *Ap. J.*, **285**, 1.
Sumi, D. M., Norman, M. L., Smarr, L. L., and Owen, F. N. 1987, preprint.
Sumi, D. M. and Smarr, L. L. 1984, in *Physics of Energy Transport in Extragalactic Radio Sources: NRAO Workshop No. 9*, ed. A. H. Bridle and J. A. Eilek (Greenbank: NRAO), p. 168.
The, L. S., and White, S. D. M. 1986, *Astr. J.*, **92**, 1248.
The, L. S., and White, S. D. M. 1987, preprint.
Ulmer, M. P., Cruddace, R. G., Fenimore, E. E., Fritz, G. G., and Snyder, W. A. 1987, preprint.
Valentijn, E. A., and Bijleveld, W. 1983, *Astr. Ap.*, **125**, 223.
White, R. E. III, and Sarazin, C. L. 1987, *Ap. J.*, **318**, 629.
Zwicky, F. 1933, *Helv. Phys. Acta*, **6**, 110.
Zwicky, F. 1957, *Morphological Astronomy*, (Berlin: Springer).

CLUSTERS OF GALAXIES AND COOLING HOT GAS

W. Forman
Harvard-Smithsonian Center for Astrophysics
60 Garden Street
Cambridge, Massachusetts 02138
U.S.A.

ABSTRACT. This review describes the information derived from Einstein x-ray imaging observations which is relevant to cooling flows in clusters of galaxies. It describes the properties of those clusters which have cooling flows and their relative proportions compared to other non-cooling flow clusters. A method for deriving quantitative parameters (gas densities and cooling times) is described and the results are compared with other techniques for estimating the importance of cooling. Finally, several effects of cooling flows around galaxies in clusters are discussed.

1. INTRODUCTION

This review emphasizes three aspects of cooling gas and clusters. First, we place clusters with cooling gas in a larger astronomical context by reviewing cooling in individual galaxies and in clusters at all stages of dynamical evolution. Second, we discuss the derivation of basic gas parameters obtained by modelling the x-ray surface brightness distribution and third, several of the effects of the observed cooling gas are mentioned.

2. RADIATIVE COOLING OF HOT GAS AROUND GALAXIES

We discuss hot gas around individual early-type galaxies not at the centers of clusters and the properties of clusters where cooling is observed.

2.1. Early Type Galaxies and Hot X-ray Coronae

Hot gaseous coronae around individual galaxies (not at the centers of groups or clusters) were first detected around several galaxies in the Virgo cluster (Forman et

A. C. Fabian (ed.), Cooling Flows in Clusters and Galaxies, 17–29.
© 1988 by Kluwer Academic Publishers.

al. 1979). Nulsen et al. (1984) first noted that some early-type galaxies in the field also contained significant amounts of hot gas. More recently, it has become clear that hot coronae are a general property of bright, early-type galaxies (see paper by Thomas in this workshop for a detailed review) outside rich clusters. The x-ray properties of these coronae include luminosities up to a few 10^{42} ergs sec^{-1}, gas temperatures around 10^7K (within a factor of two) for about 20 galaxies that have been measured, gas masses in the range 5×10^8 to $5 \times 10^{10} M_\odot$, and surface brightness profiles which are similar in form ($S_x \propto r^{-2}$ for about 20 measured galaxies).

From the observed parameters it is found that the cooling time of the gas is relatively short at the centers of these galaxies ($\tau_{cool} \leq 10^9$ years) and that even throughout much of the corona the cooling time is less than the Hubble time. This suggests (e.g., Nulsen et al. 1984) that cooling of the coronae may be an important and interesting phenomenon.

In addition, to the presence of a considerable hot interstellar medium in early-type galaxies, IRAS observations have shown evidence for the existence of a cool component (Jura 1986, Jura et al. 1986) with a mass of about $10^8 M_\odot$. Optically, spectroscopic surveys (e.g., Phillips et al. 1986) typically find $10^3 - 10^4 M_\odot$ of warm material and sensitive photometry (Sadler and Gerhard 1985) has shown that $10^4 M_\odot$ of dust is often found in elliptical galaxies.

Thus bright early-type galaxies *not* at the centers of clusters, have complex interstellar media consisting of cool gas and dust (T~30K, IR-optical), warm gas (T~ 10^4K; optical) and hot gas (T~ 10^7K; x-ray). The study of this complex medium has only just begun and will undoubtedly provide new insights on galaxy evolution and star formation which will be relevant to the processes occurring in clusters of galaxies.

2.2. Properties of Nearby Clusters with Cooling

2.2.1. Dynamical Timescales, Cooling, and Central, Massive Galaxies. Individual galaxies represent one extreme of the range where radiatively cooling gas is important. For individual galaxies the cooling rates are estimated to be up to ~ 1 $M_\odot$ per year. Clusters of galaxies represent the other extreme with estimated rates of cooling ranging up to many hundreds of solar masses per year. What can we say about the properties of those clusters where cooling is observed to be an important phenomenon?

First, present epoch clusters display a wide variety of dynamical states. This was emphasized by Gunn and Gott (1972) who noted that while the dynamical timescale of the Coma cluster (a rich, relaxed system) was less than a Hubble time, most other less dense clusters would have dynamical timescales greater than a Hubble time and, hence, could not be relaxed. Thus, one of the necessary parameters common to most cluster classification systems (see Bahcall 1977 and Forman and Jones 1982) is a parameter or property relating to the dynamical timescale (e.g., galaxy distribution, galaxy population).

However, one property that does not neatly fit into this scenario is the presence of a massive, centrally located galaxy. It had been suggested (e.g., Hausman and Ostriker 1978) that the importance of a central galaxy was directly related to the dynamical stage of a cluster as given by the Bautz-Morgan classification system, for example. However, the x-ray observations show that there is a class of clusters whose properties are those of dynamically young systems – high spiral fractions, low x-ray temperatures, low velocity dispersions, irregular galaxy distributions – and which nevertheless display the presence of a massive, centrally located galaxy. The two best examples of these systems are M87 in Virgo and NGC4696 in Centaurus (Fabricant and Gorenstein 1983 and Matilsky et al. 1985). Other examples are NGC703 in A262 and NGC3311 in A1060. Thus, the suggestion has arisen that a second parameter – the importance of a central galaxy – be added to the dynamical timescale to generate a two dimensional cluster classification system (Forman and Jones 1982).

The independence of the importance of the central galaxy and the cluster dynamical timescale agrees with more recent theoretical studies of cluster dynamics. These show that the importance of the central galaxy is determined at early stages of the cluster evolution (Carnevali et al. 1981 and Merritt 1985). Following cluster collapse, little further growth of the central galaxy occurs by mergers.

The importance of the two dimensional cluster classification is that clusters with cooling flows appear to be exclusively associated with those having central dominant galaxies, regardless of the dynamical stage of evolution. Thus, dynamically young and old clusters like A1367 or A2256 – with no central galaxy – show no evidence for cooling gas in their cores in spite of their very different dynamical timescale indicators while systems like A262 (and others mentioned above) and A85, again with very different dynamical indicators, both show strong evidence of radiative cooling (around the central galaxies).

Figure 1 graphically illustrates the cluster classification system by comparing the optical image and the x-ray contours. The clusters on the top of the figure (A1367—left and A262—right) have low velocity dispersions, low x-ray temperatures, and high spiral fractions which are indicators of dynamically young systems. By contrast, the systems at the bottom of the figure (A2256—left and A85—right) have high velocity dispersions, high gas temperatures, and low spiral fractions which are indicators of dynamically more evolved systems (those with shorter dynamical timescales).

The clusters on the left have no central dominant galaxy while those at the right have a bright galaxy around which the x-ray emission is clearly centered and concentrated. The central concentration in the clusters at the right of the figure is, as described in section 3, evidence for the existence of cooling flows.

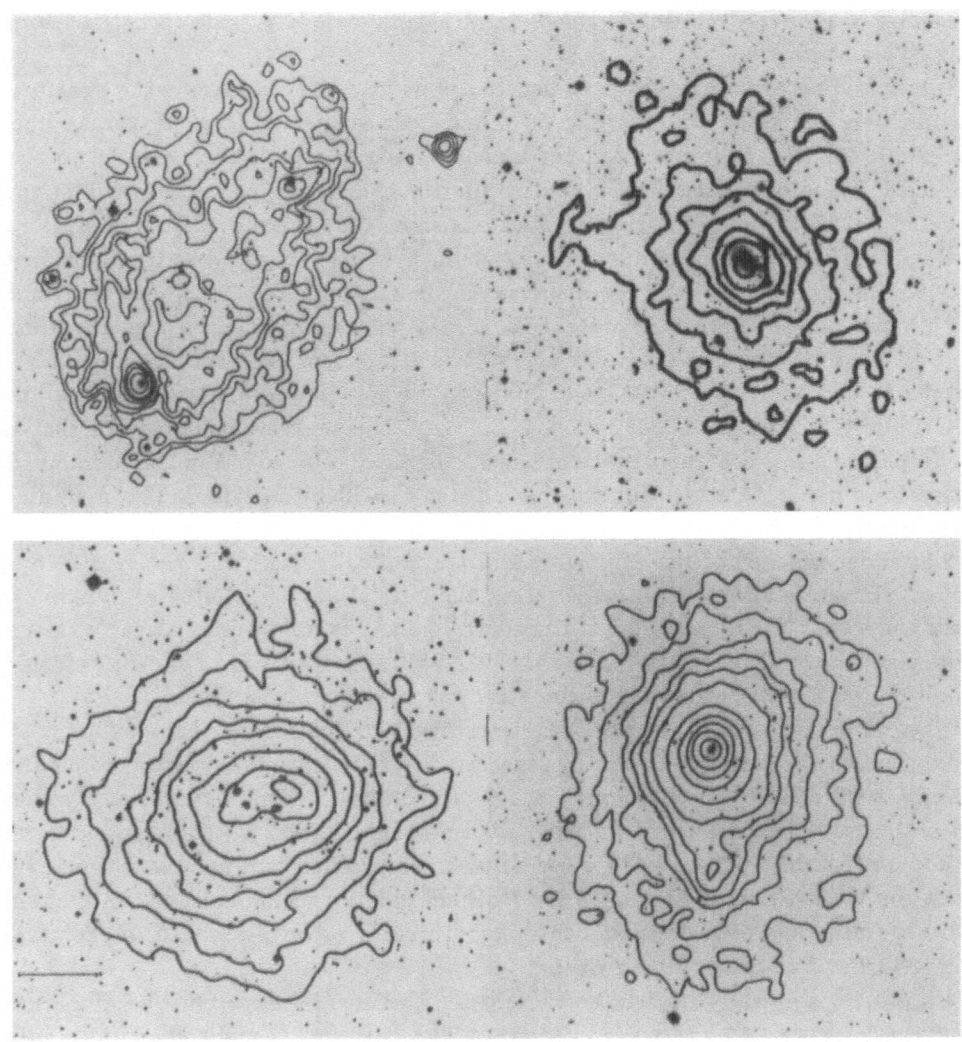

Figure 1: The 0.5–4.5 keV isointensity contours of A1367 (upper left), A2256 (lower left), A262 (upper right), and A85 (lower right) are shown superposed on optical prints. A1367 and A2256 represent the dynamically unevolved and evolved states for the class of clusters with no central dominant galaxies (nXD systems) while A262 and A85 represent these same dynamical states for the class of clusters with central dominant galaxies (XD systems).

2.2.2. Frequencies in A Sample of 250 Clusters. Jones (1987) has studied the x-ray properties of a sample of $\sim$ 250 clusters ($z < 0.15$) observed with the Einstein imaging detectors. Of these 250 clusters, 185 are detected and 149 are bright enough to attempt to "classify" through the morphology of the x-ray gas distribution.

Of the 149 clusters, 70% appear to have single intensity maxima. Of the remaining 30%, 2/3 are double (two maxima) while 1/3 are more complex. Of the 70% of the clusters with single intensity maxima, nearly half (40%) have bright galaxies at their centers and have small core radii (see Section 3 for a discussion of the quantitative analysis). By contrast, in the Abell sample surveyed by Einstein only 20% of the clusters are Bautz-Morgan type I or I-II. Thus, the x-ray observations indicate that twice as many clusters have a central, dominant galaxy as would be expected from optical classification. These clusters, with central dominant galaxies, are the candidates for cooling flows and their large representation in the cluster population indicates the potential importance of this phenomenon.

3. QUANTITATIVE ANALYSIS OF CLUSTER SURFACE BRIGHTNESS PROFILES

3.1. Method

We have derived quantitative parameters of cluster emission by modelling the x-ray surface brightness distribution as observed with the Einstein IPC. This requires that the distribution be (approximately) spherically symmetric. We omit serendipitous (foreground/background) sources, subtract an appropriate (energy, detector pulse height, and position dependent) background, and properly model the point response function of the detector (pulse height and energy dependent). For details of the analysis see Jones and Forman (1984).

We use a simple model for the radial surface brightness distribution

$$S(r) = S(o)[1 + (r/a)^2]^{-3\beta + 1/2}$$

$S(o)$ is the central surface brightness, a is a characteristic scale (core radius), β characterizes the radial fall off of the distribution, and r is the radial distance from the cluster center. This formulation can conveniently be inverted to give the underlying gas density distribution (assuming rough isothermality):

$$n(r) = n(o)[1 + (r/a)^2]^{-3\beta/2}$$

These distributions can be associated with the physical models of a hydrostatic-isothermal gas suggested by Cavalieri and Fusco-Femiano (1976) and Bahcall and Sarazin (1977) where β is the ratio of the specific energy of the galaxies to that of the gas and the galaxies are assumed to follow a King model. Within this model, the characteristic length scale a is identified with the core radius of the galaxy distribution. For the purposes of discussing cooling flows and deriving density distributions and cooling times, the surface brightness model can be considered as a convenient empirical form to derive the gas density distribution and other physical quantities of the gas.

Figure 2 shows the results of several fits to clusters both with and without bright, central galaxies. As the figure shows, acceptable fits are obtained for clusters with no central, dominant galaxy with all data points included (A2255 and A2256–top). For clusters with central bright galaxies (A262 and A2199–bottom) acceptable fits are not obtained (χ^2 values are too large) unless data points in the core are omitted. For these galaxies, we define the best fits by omitting successive data points from the center until acceptable values of χ^2 are obtained.

Given the fitted parameters, one can derive the density distribution of the gas which is relatively insensitive to the gas temperature (since the count rate in the IPC band is roughly constant for temperatures above 2.5 keV; see Fabricant et al. 1980). Next we derive the cooling time of the gas as a function of radius which does depend on our assumption about the temperature distribution (but only as $T^{1/2}$).

3.2. Small Core Radii and Central Galaxies

By applying the model described above to a sample of clusters observed by Einstein, Jones and Forman (1984) derived a set of core radii which could be used to investigate the correlation of core radius with the presence of a central dominant galaxy (defined as the brightest galaxy within the larger of 2 core radii or 0.5 Mpc of the peak of the x-ray distribution). They found that the sample of clusters for which the central, dominant galaxy was coincident with the x-ray peak had smaller core radii (as a class) than the sample for which the brightest galaxy in the core was not at the cluster center.

This provides quantitative evidence for the existence of the two-dimensional classification scheme suggested in Section 2. It also relates cluster structure to the existence of cooling flows as is discussed below.

3.3. Central Surface Brightness Excesses

As noted in Figure 2 some, *but not all*, spherically symmetric clusters are not adequately described by the simplest model for their surface brightness but show an excess in flux at the center. Using the radius, r_{max}, of the last data point which was omitted to yield an acceptable value of χ^2, one can determine the density at that distance from the cluster center, $n(r_{max})$, and also the cooling time, $\tau(r_{max})$. For about 20 clusters (with good statistics), we find that

$$\tau(r_{max}) \sim \tau_{Hubble}.$$

This provides evidence that these central excesses are produced by the

23

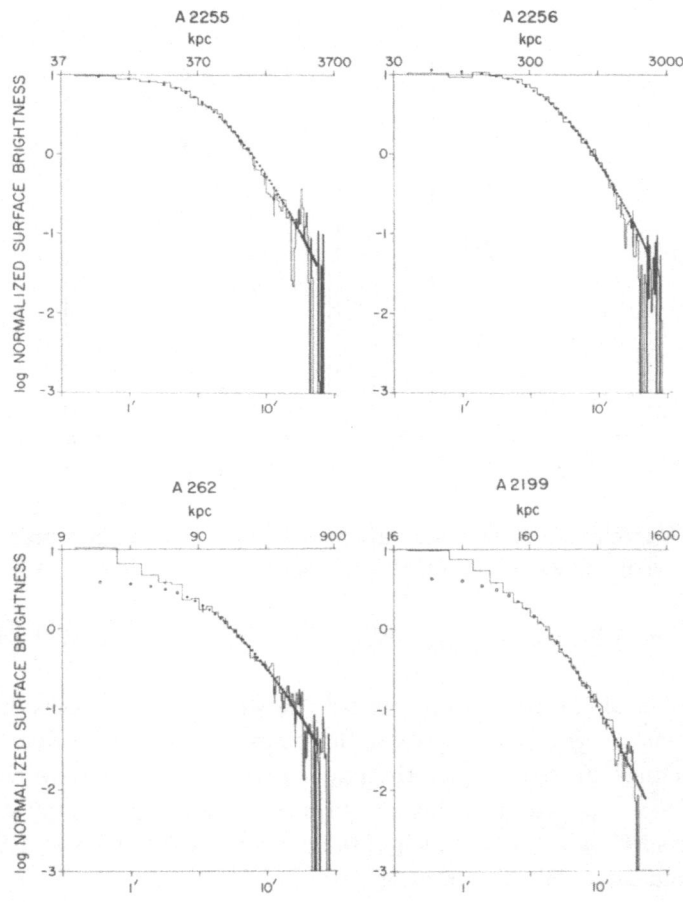

Figure 2: The radial surface brightness profiles of A2255 and A2256 are shown for the 0.5–4.5 keV energy band. Error bars (1σ) are shown for every fourth data point. The best fitting isothermal-hydrostatic model is shown as symbols. The radial surface brightness profiles of A262 (bottom left) and A2199 (bottom right) are shown with open circles indicating those points not included in the model fit in order to obtain acceptable values of χ^2.

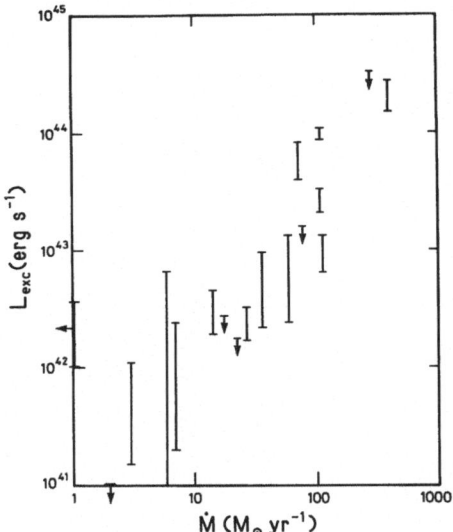

Figure 3: Correlation of derived values of the mass accretion rate from Stewart et al. (1984) with the excess, central luminosity.

effects of radiative cooling/cooling flows as described by Fabian and Nulsen (1977).

Stewart et al. (1984) also compared the values of the excess luminosity (above the fitted curve) with the values of the mass accretion rate determined from the "onion peeling" procedure (see the paper by Arnaud at this workshop). Figure 3 shows a clear correlation of the two parameters and again supports the idea that the excesses indicate the presence of radiatively cooling flows and that the excesses are good indicators of the accretion rate.

In summary, we have shown that the core radius, derived from the surface brightness distributions, indicates that, as a class, clusters with bright central galaxies are more centrally concentrated than those clusters having no central galaxy. Also, these clusters with central galaxies often show evidence for central luminosity excesses (deviations from the model) and that these excesses correlate well with mass accretion rates derived from a different technique with different assumptions.

4. EFFECTS OF COOLING FLOWS

The variety of effects of cooling flows around galaxies in and out of clusters have been summarized by Fabian at this workshop. Notably, cooling flows may add significantly to the mass of the central galaxy if they last for significant fractions of the age of the cluster. The nature of the star formation process for converting large amounts of hot gas into low mass stars with often little indication (at wavelengths

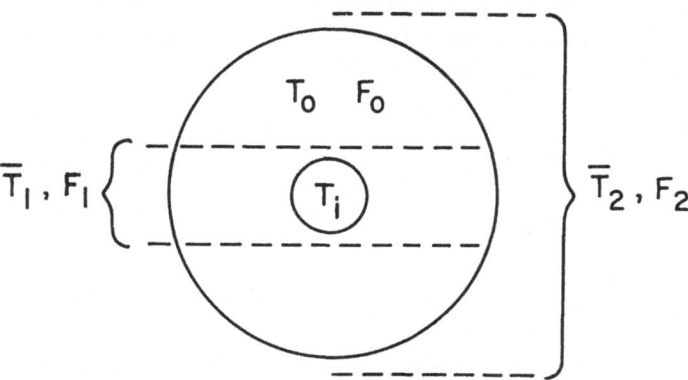

Figure 4: Schematic diagram of the fields of view of two detectors observing a cluster.

other than x-ray) for cooling and the rather unusual mass function required are not well understood.

These are two effects of the radiatively cooling gas which present observations can address. One relates to the determination of the cluster gas temperature and the applicability of the hydrostatic isothermal model and the second to the relationship between cooling flows and radio activity.

4.1. Temperature Determinations in Cooling Clusters

The cooling flow produces a region of enhanced emission with a temperature lower than that of the gas which is outside the cooling core. This can be important when the cooling flow (excess) is sufficiently large. A particularly interesting example is NGC1275 in the Perseus cluster. Flanagan et al. (1987) have used several different, large field of view detectors to estimate temperatures inside and outside the cooling core. Figure 4 shows an idealized cluster consisting of an outer (isothermal) region at temperature, T_0 with a flux, F_0, and an inner region of mean temperature T_i. Consider two different detectors with fields of view such that:

1. The smaller FOV detector sees all of the cooling region and detects a flux F_1 with a mean temperature $\bar{T}_1$.

2. The larger FOV detector sees a region of mean temperature $\bar{T}_2$ with flux F_2. If we define T_0 as the temperature of the outer isothermal cluster volume whose flux is F_0 then we have:

$$\bar{T}_2 = (\bar{T}_1 F_1 + T_0 F_0)/(F_0 + F_1)$$

Solving for T_0 and using $F_2 = F_1 + F_0$ we have

$$T_0 = (\bar{T}_2 F_2 - \bar{T}_1 F_1)/(F_2 - F_1)$$

where all the quantities on the right are observables and T_0 is the temperature of the isothermal outer region.

Applying this technique to Perseus one must first correct for the nuclear power law component which Flanagan et al. assumed to be given by the Spartan observations described by Ulmer et al. (1987). By combining OSO-8 and Einstein MPC spectra, an outer temperature for the Perseus cluster of $\sim 9.5 \pm 1$ keV was derived which represents a 40% increase in the "cluster" temperature compared to that measured over the entire cluster.

This large increase in temperature is particularly interesting in the context of discussions of the applicability of the hydrostatic-isothermal model to Perseus (see Section 3.1). The parameter β can be determined both by fitting the x-ray surface brightness profile and by calculation from the velocity dispersion and x-ray gas temperature ($\beta \propto v^2/T$). For Perseus these two independent estimates (0.57 ± 0.03 from the surface brightness fitting and 1.55 ± 0.27 from the computation) were in poor agreement.

The combination of the correction for the temperature for the central cooling flow and a correction for a system with radial rather than isotropic orbits brings the two values into acceptable agreement (with the two values of β being 0.57 ± 0.03 and 0.69 ± 0.17).

4.2. Radio Emission from Cluster Galaxies with Cooling Flows

One of the interesting properties of central galaxies in clusters is their radio activity. Since one of the models for producing this activity is by accretion onto a compact object, it is natural to look to the cooling flows as a potential source of accreting matter.

The sample of radio sources used to test for correlations of accretion rate (or central excess) with radio luminosity must be carefully constrained. One needs to use only galaxies which are at rest (or nearly so) with respect to the cluster gas, since only they could actually accrete any material. A galaxy passing through the cluster core may be randomly superposed on the cluster center and hence those galaxies used should not exhibit head-tail radio structures (as indicators at high velocity with respect to the ICM).

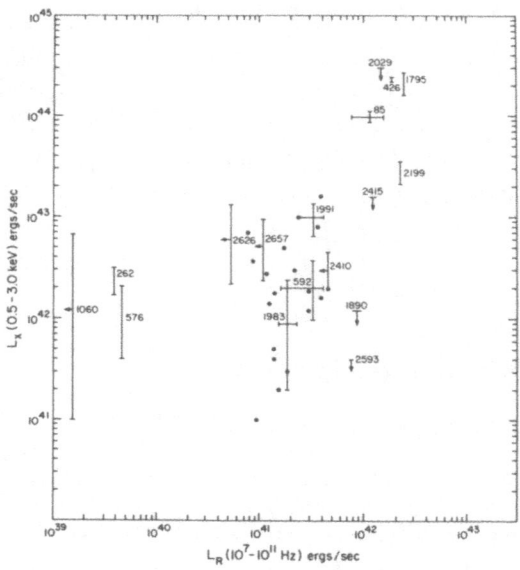

Figure 5: The luminosity of the excess x-ray emission above the hydrostatic isothermal model is shown plotted against radio luminosity. All clusters with either detected radio emission within a 0.2 Abell radius or central x-ray excesses are named. Clusters with upper limits on both an x-ray excess and radio luminosity are shown with solid dots.

With a relatively small sample of galaxies (restricted as above) Figure 5 (taken from Jones and Forman 1984) shows a reasonable correlation of excess luminosity (above the simple fit) with the radio luminosity. Thus, there appears to be a relation between these two parameters. However, since only a very small fraction of the calculated accretion rate needs to reach the galaxy center (much less than 1%) the actual connection is not obvious and could be related to the external pressures seen by the ejected radio emitting plasma. Larger samples with improved observations could provide some insight into the onset of activity in central cluster galaxies by detailed comparisons of cooling flow rates, radio luminosities, and radio morphologies.

5. CONCLUSIONS

We have summarized the observations that show that clusters exhibit two basic forms — with and without central dominant galaxies. Within both types there is a broad range of evolutionary states. The class of clusters with central dominant galaxies have smaller core radii and often exhibit excess emission in their cores (compared to a hydrostatic isothermal model). These central excesses appear to be good indicators of cooling gas since they correlate well with cooling rates derived in other ways and the excesses become significant (as a function of radius) compared to the model, where the cooling time of the gas falls below the Hubble time. Cooling flows appear to be present in a large fraction of clusters and they may be important for the growth of the stellar matter in the central galaxy and the nature of its nuclear activity.

REFERENCES

Bahcall, J. and Sarazin, C. 1977, *Ap.J. (Lett.)*, **213**, L99.

Bahcall, N. 1977, *Ann.Rev. of Astr. and Astr.*, **15**, 505.

Carnevali, P., Cavalieri, A., and Santangelo, P. 1981, *Ap.J.*, **249**, 449.

Cavaliere, A. and Fusco-Femiano, R. 1976, *A&A*, **49**, 137.

Fabian, A.C. and Nulsen, P.E.J. 1977, *MNRAS*, **180**, 479.

Fabricant, D., Lecar, M., and Gorenstein, P. 1980, *Ap.J.*, **241**, 552.

Fabricant, D. and Gorenstein, P. 1983, *Ap.J.*, **267**, 535.

Flanagan, J. and Jones, C. 1987, in preparation. **20**, 547.

Forman, W. and Jones, C. 1982, *Ann. Rev. of Astr. and Astr.*, **20**, 547.

Forman, W., Schwarz, J., Jones, C., Liller, W., and Fabian, A.C. 1979, *Ap.J. (Lett.)*, **234**, L27.

Gunn, J.E. and Gott, J.R. 1972, *Ap.J.*, **176**, 1.

Hausman, M. and Ostriker, J. 1978, *Ap.J.*, **224**, 320.

Jones, C. 1987, in preparation.

Jones, C. and Forman, W. 1984, *Ap.J.*, **276**, 38.

Jura, M. 1986, *Ap.J.*, **306**, 483.

Jura, M., Kim, D.W., Knapp, G.R., and Guhathakurta, P. 1986, preprint.

Matilsky, T., Jones, C., and Forman, W. 1985, *Ap.J.*, **291**, 621.

Merritt, D. 1985, *Ap.J.*, **289**, 18.

Nulsen, P.E., Stewart, G.C., and Fabian, A.C. 1984, *MNRAS*, **208**, 185.

Phillips, M.M., Jenkins, C.R., Dopita, M.A., Sadler, E.M., and Binetta, L. 1986, *A.J.*, **81**, 1062.

Sadler, E.M. and Gerhard, O.E. 1985, *MNRAS*, **214**, 177.

Stewart, G.C., Fabian, A.C., Jones, C., and Forman, W. 1984, *Ap.J.*, **285**, 1.

Ulmer, M., Cruddace, R., Fenimore, E., Fritz, G., and Snyder, W. 1987, *Ap.J.*, **319**, 118.

A CATALOG(UE) OF COOLING FLOWS

K.A. Arnaud
Harvard-Smithsonian Center for Astrophysics,
Cambridge, MA, U.S.A.

ABSTRACT. I describe the analysis of a sample of 106 clusters of galaxies observed by the Einstein Observatory with particular emphasis on the effects of statistical and systematic error on the derived physical parameters. Of these 106 clusters, 43 have central cooling times $< 2 \times 10^{10}$ years and hence are designated 'cooling flows'. Mass accretion rates range from $\sim 1 - 500$ $M_\odot$ yr^{-1}.

1. INTRODUCTION

The *Einstein Observatory* observed over 250 optically-catalogued clusters with the Imaging Proportional Counter (IPC) and High Resolution Imager (HRI) detectors (Forman, this workshop) and has also detected over 60 serendipitous clusters (Gioia *et al.* 1987). I have considered a subsample of all targeted clusters defined as those which have $\gtrsim 300$ counts, which are circularly symmetric and are without significant substructure.

I follow Fabian *et al.* (1981) and Stewart *et al.* (1984) in assuming a gravitational potential and then using the surface brightness data and the assumption of hydrostatic equilibrium to calculate the density and temperature profiles in a non-parametric form. The combination of an emission mechanism that depends on the square of the density and on the square root of the temperature (thermal bremsstrahlung) with a telescope mirror that has little effective area above 4.5 keV means that this method determines the density much more accurately than the temperature. In fact, as I will show, the surface brightness provides an almost direct measurement of the density profile once a geometry is defined. This direct deprojection method has the advantage of working for all clusters. However, this is done at the expense of increasing the error bars on individual points and an inability to correct for the smoothing due to the instrumental point response function.

Previous use of this method has suffered from an inability to assign statistical errors to the derived radial profiles as may be done with the parametric methods. I introduce a method of assigning statistical errors to the calculated density and temperatures by Monte Carlo regenerations of the original data. I also consider the magnitude of systematic errors due to our incomplete knowledge of the gravitational potential and mean gas temperature.

Given the temperature and density profiles it is possible to calculate a cooling time profile and if the cooling time is $< 2 \times 10^{10}$ yr, a mass accretion rate. I describe the method and some of the results of this further analysis.

A. C. Fabian (ed.), Cooling Flows in Clusters and Galaxies, 31–40.

2. METHOD

2.1 The deprojection technique

The surface brightness profiles used were extracted using the procedure of Jones & Forman (1984). This assumes that the X-ray emission is circularly symmetric. In general this is not the case; there is often ellipticity or clumping. I have ignored the effects of ellipticity, which will lead to a smearing out of the profile. For clusters with a severe departure from circular symmetry at some point in the image I have only used data from the circularly symmetric region of the cluster.

I now assume that the emission from the cluster is spherically symmetric. There is then a simple linear relation between the emissivity in any radial shell and the surface brightness in any circular ring. If S_i is the surface brightness in the ith bin (i = 1 is outer bin) in cts s^{-1} arcmin^{-2} and C_i is the emissivity in the ith shell in counts s^{-1} cm^{-3} then (e.g. Kriss, Cioffi & Canizares 1983) :

$$S_i = \sum_{j=1}^{i} A_{ij} C_j$$

where A_{ij} is a matrix of geometrical factors with amplitude determined by the source distance (I assume $H_0 = 50$ km s^{-1} throughout). Since only shells with $j \leq i$ contribute to the surface brightness of bin i the matrix A_{ij} is triangular and it is possible to iteratively solve for the C_i by starting at the outside bin and working in. This is equivalent to inverting the matrix A_{ij}. Since the relation between S_i and C_i is linear, errors on the count emissivity can be calculated from those on the surface brightness. However, these errors are not independent. If the surface brightness is depressed at some radius then the count emissivity at that radius will be decreased but that at interior radii will be increased thus exaggerating the difference.

The method detailed in this section implicitly assumes that the observations extend to the edge of the cluster. In general this will not be true (in particular for HRI images). The principal effect is to assign too large an emissivity to the outer radial shell rather as if the background was underestimated.

2.2 Calculation of density and temperature profiles

Using the Raymond & Smith (1977) code it is possible to derive the spectrum incident on the detector from a gas of given density and temperature. I shall assume that the gas has a metal abundance of approximately 0.4 times Solar as is indicated by the spectral observations (Mushotzky 1984) and that any absorption on the line-of-sight is given by the HI measurements. This incident spectrum is folded through the detector response to give an emissivity which can be compared with the deprojected data.

Assuming that the cluster gas is ideal and in hydrostatic equilibrium the density, temperature, pressure and gravitational potential are related by :

$$\frac{dP}{dr} = -\rho \frac{d\phi}{dr}$$

$$P = \frac{\rho k T}{\mu m_H}$$

where ρ, T, P and μ are respectively the density, temperature, pressure and molecular weight of the gas. ϕ is the gravitational potential, k is Boltzmann's constant and m_H is the atomic mass of hydrogen. The radial profile of any two (of ρ, T, P, ϕ) thus defines the profile of the others. Since the observed emissivity profile provides another constraint it is possible to derive three of the four quantities by assuming a profile for one of them. I have chosen to assume a gravitational potential.

While this may not be well known there is certainly more data available on it than there is on the gas temperature profile (which is the other profile that I could use - assuming a pressure profile would be equivalent to assuming a temperature profile because the density is so well determined by the surface brightness).

The detailed method for deriving gas radial profiles is as follows. The first requirement is to determine the emissivity profile from the azimuthally averaged surface brightness profile as described above. A gravitational potential must then be selected. I use a King (1966) approximation to an isothermal sphere to describe the cluster potential. The gravitating matter is distributed with a density profile :

$$\rho_g = \left(9\sigma^2/4\pi Gr_c^2\right)\left(1 + r/r_c\right)^{-3/2}$$

where r_c is the cluster core radius and σ is the line-of-sight velocity dispersion. I add to this another King potential to give a central galaxy contribution (this latter potential has very little effect - see §3.2.4. Thus the potential is described in terms of two core radii and two velocity dispersions. The galaxy portion of the potential is kept fixed with a core radius of 250 pc and a line-of-sight velocity dispersion of 300 km s^{-1}. The analysis procedure starts at the outer edge of the data and requires an outer pressure to be set (this is essentially the integration constant for the equation of hydrostatic equilibrium – I could also have used the central temperature). This pressure is assumed to be due to gas which is not detected because its surface brightness is too low. The observed emissivity from this shell now determines the temperature and, through the ideal gas law, the density. The pressure is stepped inwards using the equation of hydrostatic equilibrium and the procedure is repeated.

The main technical problem is to set the outer pressure to the appropriate value. If there is a published temperature for the cluster (usually from the HEAO-1 A-2 experiment) then for each run with a different outer pressure an emission-weighted temperature is generated and compared with that observed. Unfortunately the required data is only available for a few clusters so, in general, possible outer pressures can span a wide range.

Outer pressures for HRI observations of clusters for which IPC data exist are taken from the appropriate radius in the IPC pressure profile.

3. THE EFFECTS OF ERRORS

3.1 Statistical errors

Because the original surface brightness profile has an associated statistical error, the derived physical quantities must also have statistical errors. The analysis procedure is sufficiently complex that there is no analytic expression for the errors on the derived quantities in terms of the errors on the observations. Consequently I have used a Monte Carlo technique. The original radial surface brightness profile and errors give an estimate of the mean and standard deviation of the observed counts at each radius. I assume that the counts have a Normal distribution and randomly regenerate the profile. I have performed 100 regenerations for each cluster and analyzed the profiles using the same assumptions that were used for the original data (including the same outer pressure). A check on the method is provided by the fact that it gives the same errors on the emissivity as the analytic result.

3.2 Systematic errors

The errors calculated using the Monte Carlo method described in the previous section are only statistical. As such they are almost certainly underestimates of the true error. In this section I shall consider the effects of the various assumptions that have been made and show how they introduce systematic errors. These should be added in quadrature to the statistical errors. In order to demonstrate the various systematics I use the observations of Abell 2199. The X-ray data on this cluster are sufficiently good that the differing effects of the statistical and systematic errors can be clearly seen. The gravitational potential used has a core radius of 0.22 Mpc and a line-of-sight velocity dispersion of 784 km s^{-1}. I have chosen the outer pressure to give an emission-weighted temperature of 3.6 keV In figure 1 I show density, temperature, pressure and cooling time profiles for this choice of input parameters. These solutions will be referred to below as the nominal ones.

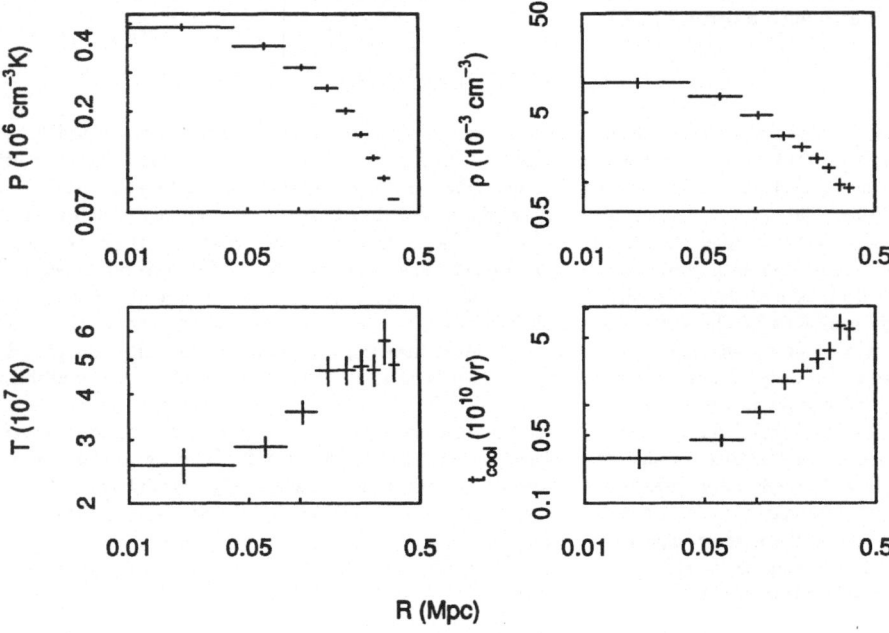

Figure 1. Pressure, electron density, temperature and cooling time profiles from the IPC observation of Abell 2199. In all these plots the vertical error bar gives the 1 sigma statistical error and the horizontal error bar gives the radial range of the bin.

3.2.1 <u>Outer pressure uncertainty</u>. Figure 2 shows the results of taking the outer pressure as half nominal and twice nominal. There is very little change in the density profile but the other profiles are changed by more than the statistical errors. These changes correspond to a variation in mean temperature of 2.8 to 4.9 keV. The integrated spectral data give a 90 % confidence interval on the temperature of 3.0 - 4.2 keV (Mushotzky 1984). Note that, in general, the imaging data does not extend to the total flux observed by the large field-of-view detectors. I assume that all the flux above that measured by the imaging detectors comes from gas at the temperature determined from the integrated spectra. This is a conservative assumption as may be seen from considering figure 2. If, for instance, the low value of the outer pressure is chosen then (unless the temperature rises again) the gas exterior to that imaged has a lower temperature than the calculated emission-weighted temperature. Therefore, if I had included this exterior gas at its correct temperature then the calculated emission-weighted temperature would have been lower than the mean temperature observed and I would have used a higher outer pressure. So, for a given range in mean temperature the range in outer pressures is smaller than I have assumed.

3.2.2 <u>Velocity dispersion uncertainty.</u> Figure 3 shows the effect of changing the line-of-sight velocity dispersion from 400 to 800 to 1000 km s^{-1}. In each case the outer pressure has been adjusted so that the integrated spectrum has the correct temperature (although it is interesting to note that the integrated spectra are not isothermal and could, in principle, be distinguished). Again the density profile is unchanged with the result that the change in the depth of the potential drives a change in the pressure profile which drives a change in the temperature profile. In particular, using a model with the wrong velocity dispersion can produce a wildly inaccurate temperature profile at large radii.

3.2.3 <u>Core radius uncertainty.</u> Figure 4 illustrates the results for three different choices for the core radius (0.12, 0.22 and 0.32 Mpc). As expected this mainly affects the pressure, temperature and cooling time profiles for the inner shells.

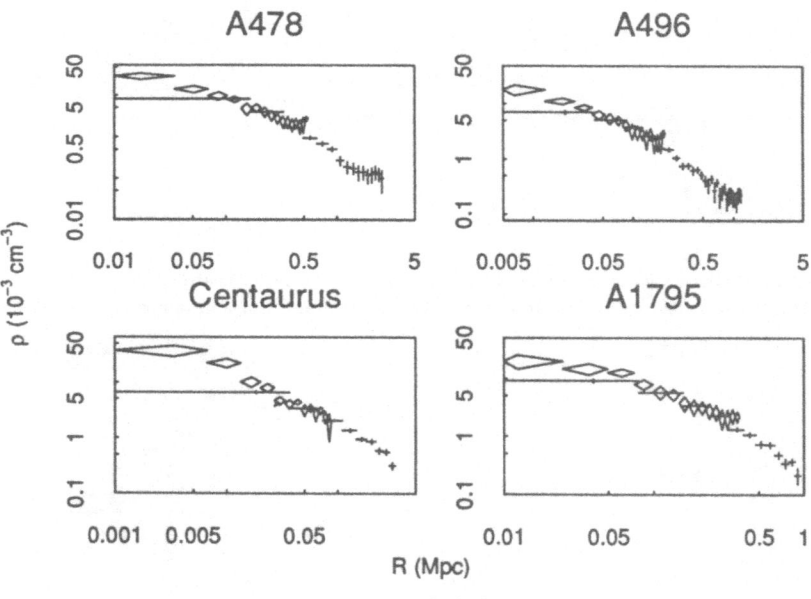

Figure 6. Electron density profiles for four clusters with both IPC (crosses) and HRI (diamonds) data.

3.2.4 <u>Potential shape uncertainty.</u> As a test of the systematics introduced by our assumption of two King profiles I have also used: a De Vaucouleurs' potential with an effective radius of 3 Mpc and the nominal velocity dispersion (giving a total mass of 2×10^{15} M$_\odot$); and a single King profile (i.e. no central galaxy potential). The results for these potentials and for the nominal one are shown in figure 5. The principal difference between the nominal and De Vaucouleurs cases is that the pressure does not flatten off in the center (because the potential doesn't) and consequently the temperature is higher there. The change in the density profile is less than the statistical errors. The differences between the nominal and single King profile potentials are negligible.

To sum up the effects of systematics, the density profile is very robust and systematic errors are

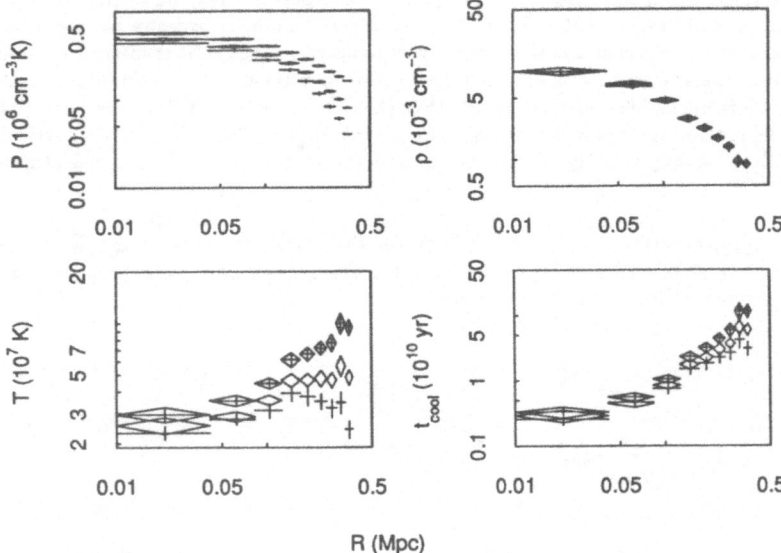

Figure 2. An illustration of the effect of changing the outer pressure used. The crosses are half nominal pressure; the diamonds are nominal; and the crosses and diamonds are twice nominal.

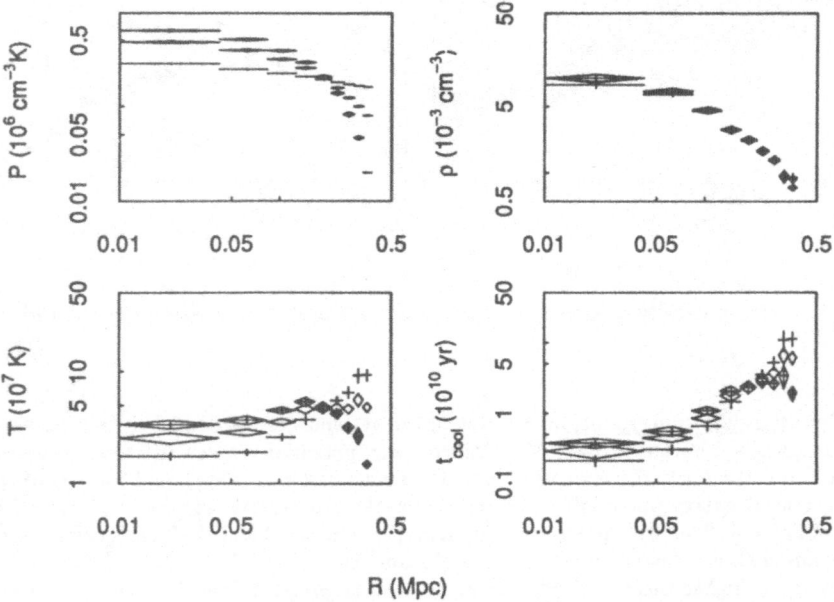

Figure 3. As figure 2 but for velocity dispersions. The crosses are 400 km s^{-1}; the diamonds are 800 km s^{-1}; and the crosses and diamonds are 1000 km s^{-1}.

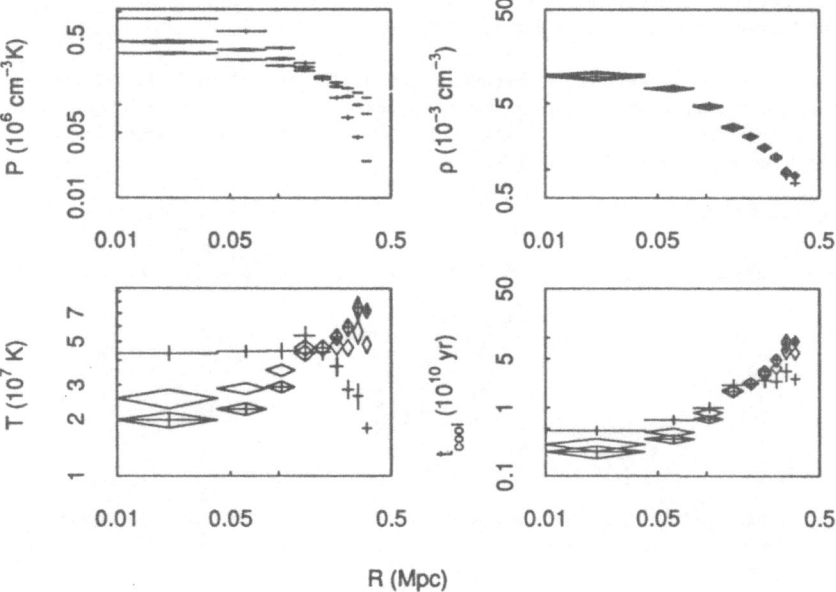

Figure 4. As figure 2 but for optical core radius. The crosses are 0.12 Mpc; the diamonds are 0.22 Mpc; and the crosses and diamonds are 0.32 Mpc.

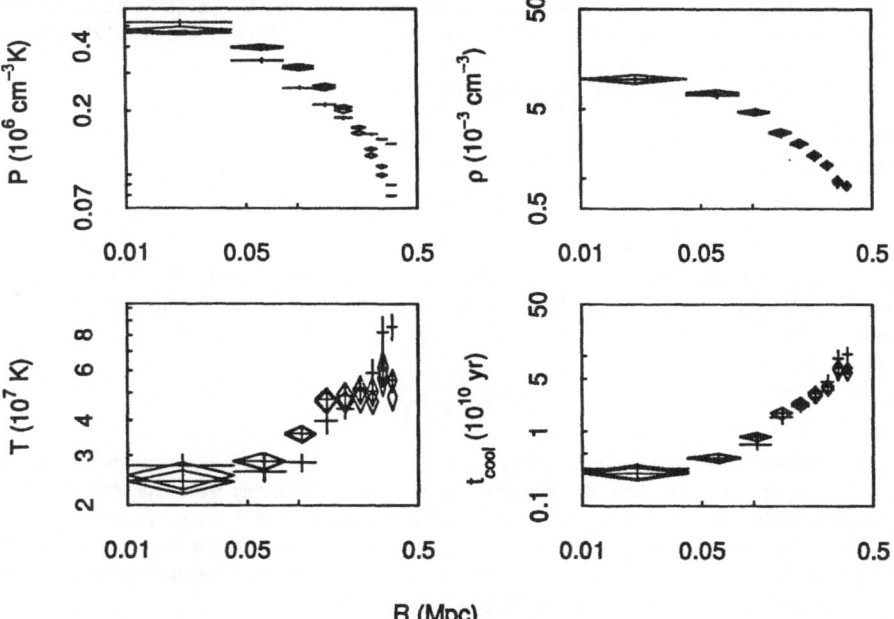

Figure 5. As figure 2 but for different gravitational potentials. The crosses are DeVaucouleurs; the diamonds are nominal and the crosses and diamonds are nominal without the central galaxy.

generally of the same order or less than the statistical errors. The same is not true of the other profiles, particularly with respect to the gravitational potential.

Finally in this section I demonstrate the effects due to the spatial resolution of the detectors. Figure 6 shows density profiles for four clusters which were observed by both the IPC and the HRI. The central density measured by the HRI (with its higher spatial resolution) can be five times as high as that measured by the IPC.

4. THE MASS FLOW RATE.

The general equations for a static, spherically-symmetric subsonic flow in a perfect gas are :

$$\frac{1}{r^2}\frac{d}{dr}(r^2\rho v) = \alpha(r)$$

$$\frac{dP}{dr} = -\rho\frac{d\phi}{dr}$$

$$\frac{1}{r^2}\frac{d}{dr}(r^2\rho v(H+\phi) + r^2\kappa\frac{dT}{dr}) = \alpha(r)\phi(r) + h(r)$$

where v is the bulk velocity, ϕ is the gravitational potential, $\alpha(r)$ is the net mass density added to the gas, $h(r)$ is the net heating function, κ is the thermal conductivity and H is the enthalpy :

$$H = \frac{5}{2}\frac{kT}{\mu m_H}$$

Now assume that there is no mass added to the gas, that conductivity is negligible and that there is no significant heating mechanism. Then the general equations can be reduced to a single equation giving the mass accretion rate in terms of the luminosity, temperature and gravitational potential profiles :

$$\Delta L_j = \Delta \dot{M}_j H_j + \dot{M}_{j-1}(\Delta H_j + \Delta\phi_j) + f_j\Delta\dot{M}_j\Delta\phi_j$$

where $\Delta \dot{M}_j$ is the mass dropping out of the flow in the jth shell, $\dot{M}_j$ is the mass flow rate in the jth shell :

$$\dot{M}_j = \sum_{i=1}^{j} \Delta\dot{M}_i$$

ΔL_j, H_j, ΔH_j and $\Delta\phi_j$ are the bolometric luminosity from the jth shell, the enthalpy in the jth shell and the enthalpy and potential differences across it. f_j is a factor of order unity to take into account the fact that the mass drops out in a volume-averaged manner across any one shell.

$$f_j = \frac{6j^2 - 8j + 3}{4(3j^2 - 3j + 1)}$$

Physically the three terms that equate to the luminosity are the thermal energy release from the mass which cools out of the flow, the thermal and gravitational energy from the mass which flows through the shell but does not drop out and finally the gravitational energy from the mass that cools out of the flow (under the assumption that the mass drops out uniformly over the shell). This equation takes into account the fact that if matter leaves the flow before reaching the center then it will contribute a smaller amount of gravitational energy.

These equations only hold for a steady-state flow. Such a flow can only be set up in the region where the cooling time is less than the Hubble time so I only calculate mass flow rates out to this point.

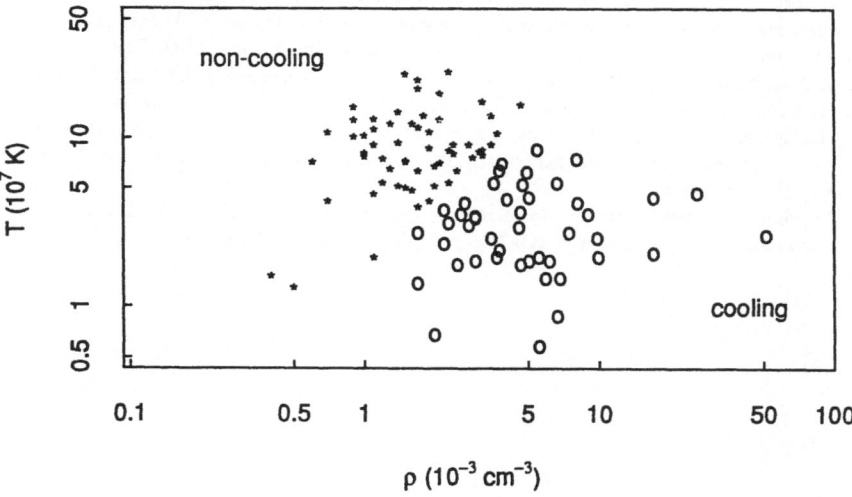

Figure 7. Central densities and temperatures for the total sample. Circles indicate cooling flows.

5. COOLING FLOW RESULTS

Figure 7 shows the central density and temperature for all clusters in the sample (from the IPC observations only). The stars and circles divide the clusters into those with central cooling times longer and shorter than $t_h = 2 \times 10^{10}$ yr respectively. There are 43 cooling flows. However, because the central temperature is generally poorly constrained this is a statistical statement only. There are a number of clusters with cooling times close to t_h which cannot be reliably placed in either class. It should also be born in mind that these cooling times are overerestimates because the central densities are underestimated by the IPC observations.

Table 1
Cooling flow systematics

Cluster	Central cooling time (10^{10} yr)		Cooling radius (kpc)		Accretion rate ($M_\odot$ yr^{-1})	
Nominal	0.3	± 0.1	174	± 27	134	± 30
Low pressure	0.3	± 0.1	198	± 28	174	± 32
High pressure	0.3	± 0.1	140	± 20	94	± 13
Low sigma	0.2	± 0.1	181	± 21	141	± 21
High sigma	0.4	± 0.1	146	± 29	100	± 24
Low core radius	0.5	± 0.1	151	± 33	91	± 26
High core radius	0.2	± 0.1	174	± 27	135	± 27
de Vauc potl.	0.3	± 0.1	176	± 24	121	± 18
No galaxy potl.	0.3	± 0.1	160	± 23	125	± 21

I have also investigated the effects of the systematic errors considered in §3.2 on the total mass accretion rate. Table 1 lists the central cooling time, cooling radius and total mass flow rate for Abell 2199 under the these tests. None of the changes can deprive Abell 2199 of its status as a cooling flow. Steepening the central potential, whether by increasing the velocity dispersion, decreasing the core radius or using a De Vaucouleurs' law, decreases the total mass accretion rate with the velocity dispersion being the most important parameter. Note however that these changes are not independent, e.g. one cannot necessarily increase the velocity dispersion and decrease the core radius at the same time.

Finally, figure 8 shows a histogram of mass accretion rates for the 43 cooling flows. While most flows have an accretion rate of < 100 M$_\odot$ yr^{-1} there is a tail to very high accretion rates.

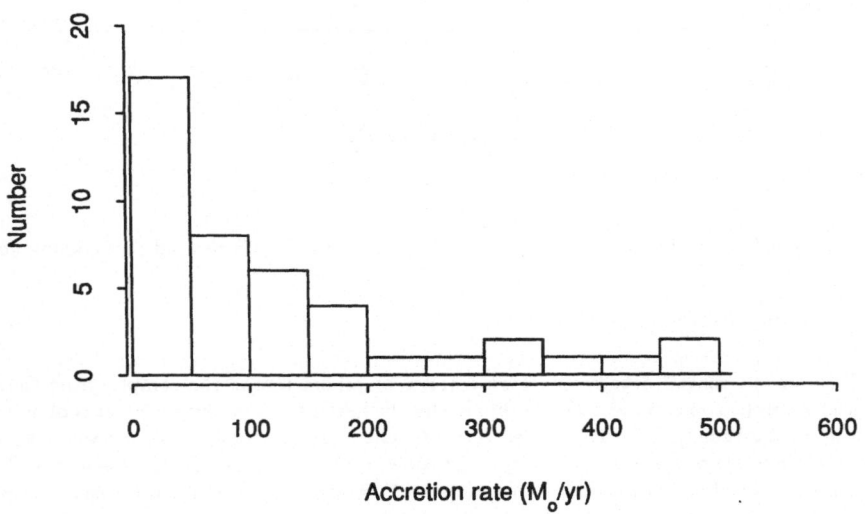

Figure 8. Histogram of accretion rates for the 43 cooling flows.

REFERENCES

Fabian, A. C., Hu, E. M., Cowie, L. L. & Grindlay, J., 1981. *Ap.J.* , **248**, 47.

Gioia, I.M., Maccacaro, T., Morris, S.L., Schild, R.E., Stocke, J.T. & Wolter, A., 1987. *Proc. Paris conference on "High redshift and primeval galaxies"*.

Jones, C. & Forman, W., 1984. *Ap.J.* , **276**, 38.

King, I. R., 1966. *A.J.* , **71**, 64.

Kriss, G. A., Cioffi, D. F. & Canizares, C. R., 1983. *Ap.J.* , **272**, 439.

Mushotzky, R. F., 1984. *Physica Scripta*, **T7**, 157.

Raymond, J. C. & Smith, B. W., 1977. *Ap.J.Suppl.* , **35**, 419.

Stewart, G. C., Fabian, A. C., Jones, C. & Forman, W., 1984. *Ap.J.* , **285**, 1.

ALTERNATIVE COOLING FLOW MODELS

Joel N. Bregman
NRAO
Edgemont Road
Charlottesville, VA   22903
USA

L.P. David
Astronomy Dept.
University of Virginia
Charlottesville, VA 22903
USA

ABSTRACT. We have examined models where the cooling or cooled cluster
gas is reheated by conduction, by galaxy motions, and by supernovae.
In each case, it is difficult or impossible to reproduce the observed
temperature gradient while also reducing the net rate at which cooled
gas is produced. None of the models examined lead to a viable
alternative cooling flow scenario.

1. INTRODUCTION

We have been examining alternatives to the "standard" cooling flow
scenario. The desire for an alternative model stems from an
uneasiness with the requirement in the standard model that the cooled
gas form stars with an initial mass function radically different from
the one we are familiar with. In particular, almost entirely low mass
stars must form with high efficiency. A successful alternative model
must be able to reproduce the X-ray observations yet produce a cooling
rate for the gas that is considerably lower than in the standard
scenario.

The cooling or accretion rate $\dot{M}$ is approximately the ratio of the
net cooling rate to the gas temperature. If this rate is to be
substantially reduced, a heat source must in large part balance
cooling. However, Field (1965) first showed that thermal
instabilities can grow in a fluid where heating balances cooling; this
leads to gas cooling in a clumpy fashion. Consequently, thermal
instabilities must be suppressed in addition to having a heat source
that can offset radiative cooling.

Several promising models have been suggested, such as reheating by
conduction (Takahara and Takahara 1981; Tucker and Rosner 1983;
Bertschinger and Meiksin 1986), drag heating by galaxies (Miller
1986), and supernovae (Silk et al. 1986). We have examined these
models through the use of numerical hydrodynamic simulations and find,
to our disappointment, that none provide a reasonable alternative
model.

A. C. Fabian (ed.), Cooling Flows in Clusters and Galaxies, 41–46.

2. THERMAL CONDUCTION

In a cluster of galaxies, only the gas within the central 100–300 kpc has a cooling time less than the Hubble time. The amount of mass and thermal energy constrained within the cooling radius is small compared to the total in the cluster. Consequently, only a modest amount of heat need be transferred between the inner and outer regions to balance radiative cooling. Conduction of heat by thermal electrons holds the promise of effecting this energy transfer (Takahara and Takahara 1981; Tucker and Rosner 1983; Bertschinger and Meiksin 1986).

Stewart et al. (1984) and Sarazin (1986) have argued that conduction would balance radiative losses throughout the flow only for a small range of cluster densities. However, there is no need for this balance to exist everywhere. Both the accretion rate and the ratio of the conduction to the radiative loss coefficient increase with radius. So conduction may be ineffective in the very central regions yet suppress the cooling rate at larger radii, which is where most of the gas cools. Along these lines, Meiksin and Bertschinger (1988) discovered a steady state solution in which M was significant reduced while still reproducing the observations. We have investigated whether gas in clusters of galaxies would naturally evolve to this solution.

Two models were considered, both of which have as initial conditions gas at a constant temperature (1×10^8 K) in hydrostatic equilibrium. The equations that are solved are similar to those described by White and Sarazin (1986) except that the time dependent terms were retained, a conductive heat flux term was added, and mass removal from the system occurs only when thermal instabilities can grow (for details, see Bregman and David 1988a). The normal Spitzer conductivity term is multiplied by an efficiency factor μ, which accounts for the reduction in the mean free path by tangled field lines. Thermal instabilities are suppressed by conduction for length scales smaller than λ_{crit}, which was first described by Field (1965). A one-dimensional implicit hydrodynamic code (Ruppel and Cloutman 1975) was used to follow the evolution of the cluster gas for 15×10^9 yr. In one class of models, μ is constant in time and space. In the second class of models, we assume that scattering centers in the fluid are stretched (compressed) according to the velocity derivative of the flow. Then μ, which is proportional to the mean free path of the electrons, is a function of time and space; as initial conditions, μ is constant everywhere.

In both models, we find that there was only a narrow range in the initial choice of μ in which $\dot{M}$ was reduced by a factor of three while also achieving the observed temperature gradient, $T(10 \text{ kpc})/T(r_{cool}) > 3$ (Fig. 1). The narrow range of values for μ are not naturally suggested by nature and they would need to be different in each cluster, which seems unlikely. We conclude that cooling flows with conduction only rarely evolve to solutions in which $\dot{M}$ is reduced while also achieving a significant temperature gradient. Perhaps two types of clusters exist, one in which conduction is efficient, leading to small accretion rates and small temperature gradients, and one in which conduction is inefficient and the standard cooling flow model exists.

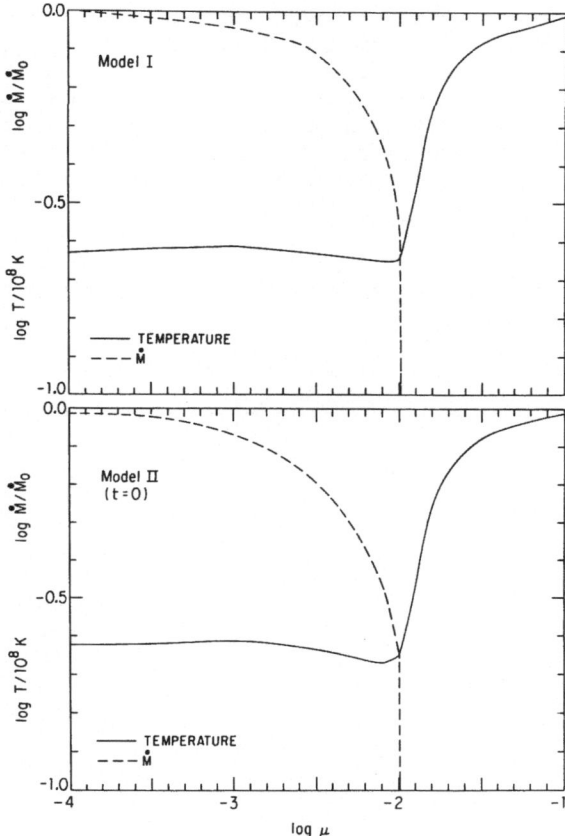

Figure 1. The accretion rate relative to the model with no reheating processes and the temperature contrast in the cluster are plotted as a function of conductive efficiency μ. In the top figure, μ is independent of time while in the lower frame, μ changes according to the flow properties. There is only a narrow range in μ for which both a reduction in the accretion rate and a significant temperature gradient results.

3. REHEATING BY GALAXY MOTIONS

The natural motion of galaxies through the cluster medium can create shocks that provide a significant source of heat (Miller 1985). We have expanded upon Miller's suggestion by making a distinction between heating by galaxies moving supersonically and subsonically. In particular, galaxies moving subsonically lead to no heating while supersonic galaxies lead to shocks and the temperature increase is given by the Rankine-Hugoniot jump conditions:

$$\frac{\Delta E}{E} = \frac{(5M^2 + 3)\,(M^2 - 1)}{16M^2}$$

If all galaxies are assumed to move a the velocity dispersion, the local heating rate is given by $H = \Delta E n \alpha_h$, where $\alpha_h = n_{gal}\,A\,v$; n is the gas density, n_{gal} is the density of galaxies, A is the cross section area of a galaxy that leads to a shock, and v is the velocity of the galaxy (Miller 1986). Because the heating rate is a sensitive function of temperature when the Mach number is near unity, the condition to suppress thermal instabilities may be satisfied:

$$2 - \frac{\partial \ln \Lambda}{\partial \ln T} < \frac{\partial \ln H}{\partial \ln n} - \frac{\partial \ln H}{\partial \ln T}$$

where Λ is the radiative cooling function. The heating mechanism holds the promise of balancing radiative cooling and suppressing thermal instabilities.

Reheating By Galaxy Motions

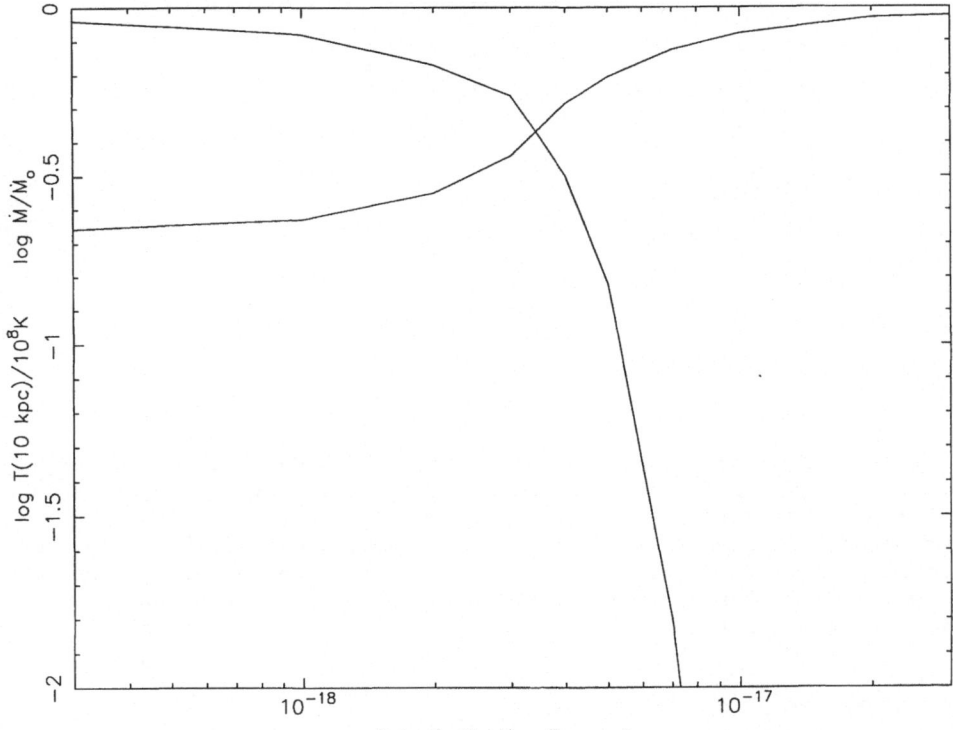

Galactic Heating Parameter

Fig. 2. Like figure 1, the accretion rate and temperature gradient are plotted as a function of α_h, the galaxy reheating parameter.

The equations that were solved are similar to the above model except that the conductive heating term is replaced with the galaxy heating term (for details, see Bregman and David 1988b). The free parameter in this case is α_h, which is a measure of merit by which galaxy motions can reheat cluster gas.

Once again, we find that $\dot{M}$ can be reduced and a significant temperature gradient will be achieved, but only for a narrow range of values of α_h (Fig. 2). In order for the cluster to adjust to the necessary value of α_h in a reasonable way, cooling flows would have to lead to the formation of new galaxies that increase N_{gal} and consequently α_h. Because this seems unlikely and because it is difficult to imagine how all clusters managed to choose the proper value of α_h, we suggest that this model is untenable. A further difficulty is that galactic reheating is unlikely to suppress thermal instabilities on all wavelengths.

4. REHEATING BY SUPERNOVAE

In this scenario, the hot gas cools normally and begins to form stars. The first few supernovae that occur in the star forming region reheat the cooled gas to the temperature of the ambient medium (a somewhat different model involving only type Ia supernovae was discussed by Silk et al 1986). We assume that the timescale for all of the gas to be converted into stars (probably 10^7–10^8 yr) is longer than the time required for the first supernovae to occur (a 5 $M_\odot$ star has a lifetime of 10^8 yr and a 20 $M_\odot$ star has a lifetime of 10^7 yr). Then, the first supernovae occur in the presence of cold gas. If the remaining gas is to be reheated to the ambient cluster temperature, the fraction f of cold gas first going into star formation must be

$$\frac{1}{f} = 0.82 \ \frac{nu}{1 \ SN/100 \ M_\odot} \ \frac{E_{sn}}{10^{51}erg} \ \frac{3\times10^7 \ K}{T_{cf}} \ - \ 1$$

where nu is the supernova yield per unit mass of star formation, E_{sn} is the energy released per supernova, and T_{cf} is the temperature of the ambient cluster medium.

If the accretion rate is to be suppressed by an order of magnitude (f = 0.1), one supernova must be produced for every 7.5 $M_\odot$ of gas converted into stars. For a normal initial mass function (Scalo 1986), 100–300 $M_\odot$ of star formation is required to produce one supernova, depending upon the assumed mass of the progenitor. So, to significantly reduce $\dot{M}$, only high mass stars may be formed. This would lead to a supernova rate of 3–5 SN/yr within the cooling radius of Perseus, which is not observed. Unless all supernovae are highly subluminous so as to avoid detection, this model is also untenable.

The authors wish to acknowledge support from the NASA Astrophysical Theory Center Grant NAGW-764 and from the NRAO. The NRAO is operated by Associated Universities, Inc., under contract with the National Science Foundation.

5. REFERENCES

Bertschinger, E., and Meiksin, A. 1986, Ap.J. (Letters), 306, L1.

Bregman, J.N., and David, L.P. 1988a, Ap.J., in press.
Bregman, J.N., and David, L.P. 1988b, in preparation.
Field, G.B. 1965, Ap.J., 142, 531.
Meiksin, A., and Berschinger, E. 1988, in "Cooling Flows in Galaxies and Clusters", NATO Workshop in Cambridge, UK, June 22-26, 1987.
Miller, L. 1986, M.N.R.A.S., 220, 713.
Ruppel, H.M., and Cloutman, L.D. 1975, Los Alamos National Laboratory Report LA-6149-MS.
Sarazin, C.L. 1986, Rev. Mod. Phys., 58, 1.
Scalo, J.M. 1986, Fundamentals of Cosmic Physics, 11, 1.
Stewart, G.C., Canizares, C.R., Fabian, A.C., and Nulsen, P.E.J. 1984, Ap.J., 278, 536.
Takahara, M., and Takahara, F. 1981, Progr. Theor. Phys. (Letters), 65, 369.
Tucker, W.H., and Rosner, R. 1983, Ap.J., 267, 547.
White, R.E., and Sarazin, C.L. 1987, preprint.

THE ROLE OF HEAT CONDUCTION IN COOLING FLOWS

Avery Meiksin
Departments of Physics and Astronomy
601 Campbell Hall
University of California
Berkeley, CA 94720

ABSTRACT. The role of heat conduction in the cooling flows of clusters of galaxies has been investigated by running a series of numerical hydrodynamic computations. It is found that flows with no source or sink develop a cold and dense condensing core inconsistent with X-ray observations, while including a sink for the cooling gas produces a gradual drop in temperature toward the cluster center in agreement with X-ray measurements. Heat conduction can suppress the accretion rate by a factor of a few to several, while larger reductions result in a nearly isothermal Intra-Cluster Medium.

1. INTRODUCTION

Early in the theoretical literature on cooling flows, it was realized that, in principle, heat conduction in the Intra-Cluster Medium (ICM) of a cluster of galaxies could suppress the large-scale flows which otherwise should arise from central cooling of the ICM (Takahara and Takahara 1981). It was argued by Binney and Cowie (1981), however, that a model which included thermal conductivity at a rate any higher than 10^{-3} of the Spitzer rate for a plasma (Spitzer 1962) would fail to reproduce the M87 data of Fabricant, Lecar, and Gorenstein (1980).

The need for a high heat conduction suppression factor arose from the high outer temperature of 10^8 K used by Binney and Cowie for M87, based on the *Ariel V* observations of Davison (1978) and Lawrence (1978). Since then, lower outer temperatures have been reported. Based on *Einstein* observations, Stewart *et al.* (1984) (SCFN) report a temperature at 50' of 2.2×10^7 K. At 75', *EXOSAT* finds a temperature of 3×10^7 K (Smith and Stewart 1985). Bertschinger and Meiksin (1986) (BM) were now able to construct steady-state cooling flow models including heat conduction at the Spitzer rate which were consistent with the data of SCFN. The mass flow rate they found was reduced by an order of magnitude to 1 $M_\odot$ yr^{-1}. They were similarly able to reduce the mass flow rate of NGC1275 by an order of magnitude to 30 $M_\odot$ yr^{-1} by including heat conduction.

A question not addressed by BM was whether the solutions are globally stable. Bulk motion may be preferred over a nearly static atmosphere that has a temperature gradient such that heat conduction nearly balances radiative losses. The results of a series of time-dependent hydrodynamic computations performed to investigate this question are reported here. The question was also addressed by Bregman and David (1987; this workshop) with fairly similar results.

A. C. Fabian (ed.), Cooling Flows in Clusters and Galaxies, 47–51.

2. ASSUMPTIONS AND METHOD

The ICM of a galaxy cluster is modeled as spherically symmetric with a time-independent gravitational potential. The self-gravity of the gas is neglected. The ICM is taken to radiate through optically thin line radiation and thermal bremsstrahlung at the rate given by Raymond, Cox, and Smith (1976). The Spitzer (1962) rate of heat conduction is used, with an allowance for saturation of the heat flux (Cowie and McKee 1977).

The fluid equations solved are the equations of mass, momentum, and energy conservation

$$\frac{\partial \rho}{\partial t} + \nabla \cdot (\rho \mathbf{v}) = -\dot{\rho}_*, \tag{1}$$

$$\frac{D\mathbf{v}}{Dt} + \frac{1}{\rho}\nabla p = -\mathbf{g}, \tag{2}$$

$$\rho \frac{D\theta}{Dt} - (\gamma - 1)\theta \frac{D\rho}{Dt} = (\gamma - 1)[-\mathcal{L} + f\nabla \cdot (\kappa \nabla \theta)] + \alpha_* \theta \dot{\rho}_*, \tag{3}$$

and the equation of state $p = \rho\theta$. Here, p is the pressure, ρ is the density, and θ is $k_B T/\mu m_H$, where T is the temperature and $\mu = 0.6$ is the mean molecular weight of a fully ionized gas of cosmic abundances. The gravitational acceleration $\mathbf{g}$ is taken to arise externally. The net cooling rate per unit volume is $\mathcal{L}$, κ is the coefficient of thermal conductivity, and f is a factor by which the coefficient is suppressed (e.g., by magnetic fields). The cooling rate $\mathcal{L}$ is set to zero below 10^6 K to avoid an excessive restriction on the time step imposed by extremely rapid cooling.

For the sake of generality, a star-formation term $\dot{\rho}_*$ is included. The star-formation rate is assumed to have the form $\dot{\rho}_* = q_* \rho/t_{cool}$, where $t_{cool} = \frac{p/(\gamma-1)}{\mathcal{L}}$. The factor α_* parametrizes the energetics of star-formation. In the results reported, the gas is taken to leave the flow isobarically, so that $\alpha_* = 0$.

The numerical hydrodynamics code used is a spherically symmetric explicit finite difference scheme second order accurate in time. A description can be found in Cioffi, McKee, and Bertschinger (1987). The code is run in its fully Eulerian mode. The boundary conditions used are that all vectors vanish at the center and that the pressure and density be held fixed at an outer radius of 2.5 Mpc.

The initial conditions used in the runs presented are that the gas be isothermal and in hydrostatic equilibrium in a two component gravitational potential. The temperature is 5.0×10^7 K and the central hydrogen density is 5.0×10^{-2} cm^{-3}. The temperature is high enough that heat conduction is significant compared to cooling for a fair range of heat conduction suppression factors f. This permits a study of the effect of heat conduction by varying f.

The gravitational potential consists of a cluster component of the form $\phi_{\text{cluster}}(r) = \sigma_c^2 \log(1 + x_c^2)$, where $x_c = r/a_c$, and a galaxy component given by $\phi_{\text{galaxy}}(r) = \sigma_g^2 \log\left[1 - \frac{1}{x_g}\log(x_g + \sqrt{1 + x_g^2})\right]$, where $x_g = r/a_g$. The values used for the parameters are $\sigma_c = 1000$ km s^{-1}, $a_c = 25$ kpc, $\sigma_g = 300$ km s^{-1}, and $a_g = 1$ kpc. The potential, gas density, and temperature are a fair approximation to M87, although the temperature is somewhat higher.

The runs were terminated when the total energy and momentum errors exceeded 1%. A typical run lasted 7 billion years, took 10^5 time steps, and 40 cpu hours on a *MicroVAX II*.

3. RESULTS AND DISCUSSION

Two sets of runs were executed to investigate the evolution of a cooling flow with heat conduction. One set allowed no sources or sinks for the gas (except for gas flow across the outer boundary), while the second set allowed a sink for the gas. The rate of heat conduction was varied for both through the parameter f. It was unfortunately necessary to include a small amount of heat conduction in the set of runs presented here to suppress an instability which otherwise forms at the outer boundary after a few billion years.

3.1 *No Source or Sink*

The evolution of a cluster without a source or sink ($\dot{\rho}_* = 0$) results in a dramatic drop in temperature and increase in density at the center of the cluster. The cluster gas evolves into two spatially distinct phases. After 10^9 yrs, the run with $f = 0.1$ forms a cold central core of very high density. The surrounding gas rushes and cools across the boundary into the central core, which thereby grows indefinitely. A similar result can be seen in the thermally unsteady wind solutions for elliptical galaxies of Mathews and Baker (1971). The physical reason for the behavior is clear. As the gas in the center of the cluster cools, its density grows and the gas cools all the faster until the cooling time is much less than the dynamical time. The central thermal pressure of the cluster gas then no longer supports the gas above it. The result is a cooling catastrophe which leads to the formation of the two phases. A similar result was found for a slightly different run with $f = 0$. Although a process similar to this may represent the early stages of galaxy formation, the structure is in gross conflict with the observations of present day cooling flows. Such a high central density would show up as a spectacular increase in X-ray surface brightness toward the cluster center.

Increasing the rate of heat conduction does not substantially alter the behavior, it merely delays the catastrophe. The results of a run with $f = 1$ are shown in Figure 1 at $t = 10^9$ yrs and at $t = 5 \times 10^9$ yrs. Heat conduction allows only a small drop in temperature by 10^9 yrs, in contrast to the $f = 0.1$ run; but by 5×10^9 yrs, the two phase structure develops with $\dot{M} \sim 100\ M_\odot\,\mathrm{yr}^{-1}$.

3.2 *Sink*

Since cooling and heat conduction alone cannot produce a cooling flow which matches the X-ray observations, it is necessary to incorporate additional processes into the model. Heating the central regions might prevent the cooling catastrophe. Tucker and Rosner (1983) have argued for relativistic electrons and Silk *et al.* (1986) for supernovae as heating mechanisms. Alternatively, a sink for the gas could be introduced. If the rate at which gas left the flow, *e.g.* by forming stars, increased with gas density, one might expect to have a self-regulating mechanism which would prevent the run-away growth of the central density. A second set of computations was performed including a local sink at the rate $\dot{\rho}_*$ described above with the intent of determining whether removing mass from the system can produce a cooling flow which agrees with X-ray observations.

Including even a small amount of gas loss prevents the formation of the cold core and results in temperature profiles similar to those inferred for cooling

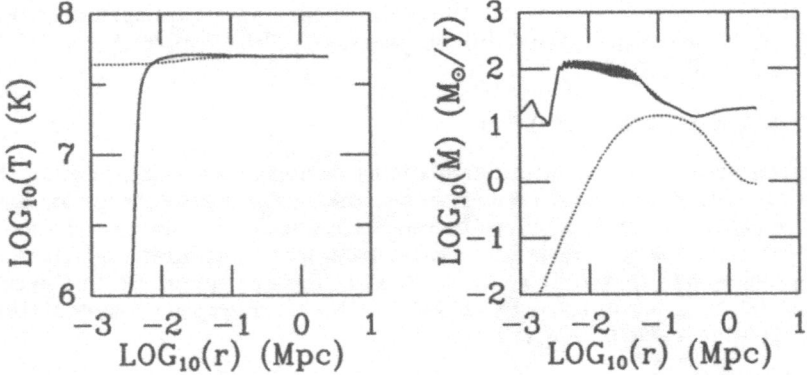

Figure 1.   Run for f=1.0, q.=0.0.   Shown at
t=1×10⁹ yrs (dotted) and at t=5×10⁹ yrs (solid).

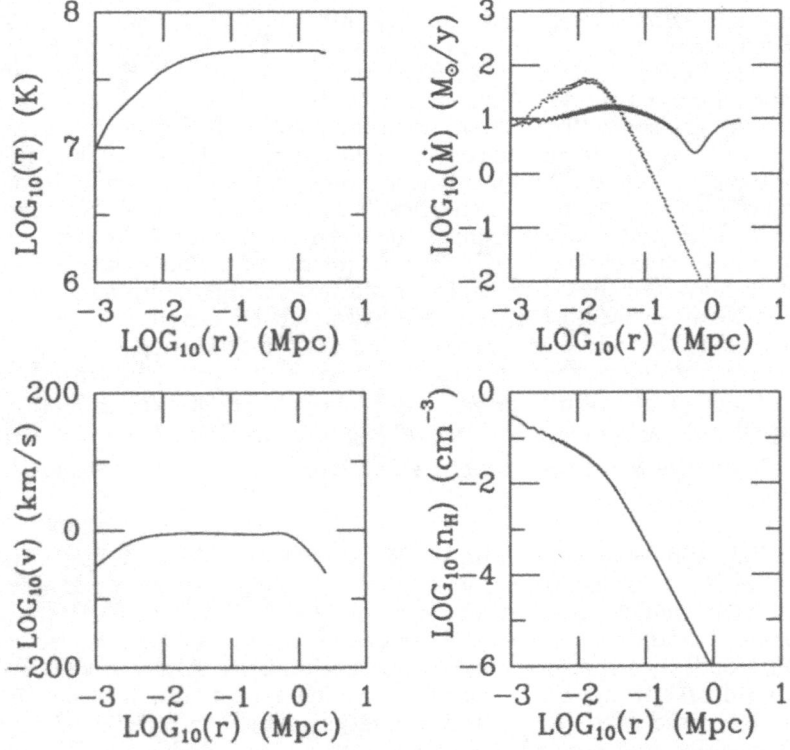

Figure 2.   Run for f=0.3, q.=0.1.   Accretion rates
plotted are true rate (solid) and inferred rate
assuming a steady–state system without heat
conduction (dotted).

flows. In Figure 2, the results of a run with $f = 0.3$ and $q_* = 0.1$ are presented at 7×10^9 yrs. The heavy $\dot{M}$ curve is the true mass flow rate while the dotted curve is the rate one would infer from the X-ray luminosity under the assumption of isobaric mass loss in a steady-state system without heat conduction. At 40 kpc, the cooling time matches the age. One sees that the peak inferred accretion rate exceeds the true rate of $\sim 12\ M_\odot\ \mathrm{yr}^{-1}$ by a factor of 4. The excess arises from the combined effect of neglecting heat conduction (50%) and the assumption of a steady-state (50%). For $f = 0.1$, $q_* = 0.1$, the discrepancy is only a factor of 2. For $f = 1.0$, $q_* = 0.1$, the discrepancy is a factor of 20. In this case, however, the central temperature decreases by only 15% from its asymptotic outer value. For $f = 1.0$ and $q_* = 0.05$, the accretion rate is reduced by a factor of 4 with a gradual temperature drop similar to that in Figure 2.

4. CONCLUSIONS

The energetics of cooling flows must involve more than a single gas component with heat conduction and cooling, otherwise a dropping central gas temperature runs away and a cold and dense central core develops. Systems which allow gas loss at the rate $\dot{\rho}_*$ discussed above do evolve to cooling flows in qualitative agreement with the structures inferred from X-ray data. Heat conduction can reduce the mass flow rate by factors of a few to several from the peak values which would be inferred from the X-ray data under the assumptions of a steady-state and no heat conduction. Reductions by as much as a factor of $10 - 15$, as in the steady-state solutions of BM, were not found. A rate of heat conduction which produces such a high reduction generally renders the gas nearly isothermal. Although steady-state solutions with a large decrease in accretion rate exist, they appear not to be stable except perhaps for fine-tuned matching of the heat conduction and star-formation rates to the cooling rate.

ACKNOWLEDGMENTS The author is grateful to Ed Bertschinger and Marc Davis for many helpful conversations. The research was supported in part by NSF grant AST-8614552. The AAS and NATO provided funds for attending the Workshop.

REFERENCES

Bertschinger, E., and Meiksin, A. 1986, *Ap.J.*, **306**, L1 (BM).
Binney, J., and Cowie, L. L. 1981, *Ap.J.*, **247**, 464.
Bregman, J. N., and David, L. P. 1987, *Ap.J.*, *preprint*.
Cioffi, D., McKee, C. F., and Bertschinger, E. 1986, *in preparation*.
Cowie, L. L., and McKee, C. F. 1977, *Ap.J.*, **211**, 135.
Davison, P. J. N. 1978, *M.N.R.A.S.*, **183**, 39P.
Fabricant, D., Lecar, M., and Gorenstein, P. 1980, *Ap.J.*, **241**, 552.
Lawrence, A. 1978, *M.N.R.A.S.*, **185**, 423.
Mathews, W. G., and Baker, J. 1971, *Ap.J.*, **170**, 241.
Raymond, J. C., Cox, D. P., and Smith, B. W. 1976, *Ap.J.*, **204**, 290.
Silk, J., Djorgovski, S., Wyse, R. F. G., and Bruzual, G. 1986, *Ap.J.*, **307**, 415.
Smith, A., and Stewart, G. 1985, *Sp.Sci.Rev.*, **40**, 661.
Spitzer, L. 1962, *Physics of Fully Ionized Gases* (New York: Wiley-Interscience).
Stewart, G. C., Canizares, C. R., Fabian, A. C., and Nulsen, P. E. J. 1984, *Ap.J.*, **278**, 536 (SCFN).
Takahara, M., and Takahara, F. 1981, *Prog. Theo. Phys. (Letters)*, **65**, 369.
Tucker, W. H., and Rosner, R. 1983, *Ap.J.*, **267**, 547.

Einstein Observatory Solid State Detector Observations of Cooling Flows in Clusters of Galaxies

R. F. Mushotzky and A. E. Szymkowiak
Laboratory for High Energy Astrophysics
Goddard Space Flight Center
Greenbelt Maryland 20771

ABSTRACT. We present SSS observations of cooling flows in 9 clusters of galaxies chosen to have cooling flows based on *Einstein Observatory* imaging results. We analyze the data with three models, 1) the addition of a cool to a hot component, 2) a power law distribution of emission measure with temperature and 3) a cooling flow model. We find strong evidence for the existence of cool components in the centers of these clusters. The derived distribution of emission measure versus. temperature for M87 agrees very well with that derived from the *Einstein Observatory* Focal Plane Crystal Spectrometer. Values of the cooling rate, $\dot{M}$, derived from the SSS data (in this paper, we will always quote values of $\dot{M}$ in solar masses per year and use a Hubble constant of 50 km/sec/Mpc) agree very well with those derived from analysis of the *Einstein Observatory* IPC and HRI data (K. Arnaud, this workshop). This agreement is not fortuitous since the methods of analysis and the free parameters involved are totally different. We feel that the agreement between the three *Einstein Observatory* experiments is good and gives weight to the the numerical values of $\dot{M}$ for the larger sample of objects observed only with the *Einstein Observatory* IPC and HRI.

I Introduction

The *Einstein Observatory* (hereafter *EO*) solid state spectrometer (hereafter SSS) was sensitive in the 0.6-4.5 keV range with a constant energy resolution of 160 eV and a 6' diameter field of view (Holt et al. 1979). It had a high quantum efficiency (essentially 100%) across this band. Many of the strongest lines expected from a cooling flow (the K lines of Mg, Si, S, and Ca and the L lines of Fe) appear in this energy range. However the lower energy limit does not allow observation of the K lines of C or O. This energy bandpass essentially limits the temperature sensitivity of the SSS to plasmas of $6 \times 10^6 < T < 5 \times 10^7$ degrees. While the spectral resolution of the SSS is quite high compared to the *EO* IPC it is poorer than that of the FPCS. The spectral resolution allows separation and measurement of the H and He like lines of S, Si, Ca and Ar but was not sufficient to separate the numerous Fe L lines or to separate the Fe L complex from Ne lines. There is also some confusion between certain Fe L lines and the K lines of Mg. For weak sources ($<\sim 2 \times 10^{-11}$ ergs/cm^2-sec in the 0.5-4.5 keV band), the SSS was background limited but relatively high signal-to-noise data were obtained in exposures of $\sim 10^4$ seconds for brighter sources.

There is an effect which limits the sensitivity of the SSS to the signature of distant cooling flows compared to the imaging detectors on the *EO*. Because (see figure 1) the SSS had a fixed angular beam size, as clusters get more distant the contribution of the cooling flow to the signal in the SSS field of view gets diluted by the total, non-cooling, cluster emission. Using the formula for r_{cool} from Fabian, Nulsen and Canizares (1984) $r_{cool} \sim 3'$ $\dot{M}_{100}/D_{100Mpc}$; where $\dot{M}_{100}$ is the cooling rate in 100 M_0/yr. and D_{100Mpc} is the distance in units of 100 Mpc.

If we can characterize the "size" of the cooling flow by the "cooling radius" r_{cool}

A. C. Fabian (ed.), Cooling Flows in Clusters and Galaxies, 53–62.
© 1988 by Kluwer Academic Publishers.

54

(which is typically on the order of ~200 kpc), then for clusters more distant than ~250 Mpc (z~0.04 for H_0 =50 km/sec/Mpc) the dilution becomes severe. Of course the SSS never detects only the signal from the cooling flow, there is always the "foreground/background" cluster emission to subtract. For the purpose of determining the parameters of cooling flow clusters the SSS database is quite limited due to the short duration of SSS operation (while it achieved its prelaunch design life of 10 months, this resulted in only ~50 days spent at the focus of the *EO*) and the "late" discovery of cooling flows. That is, the SSS did not live long enough for the "discovery" of cooling flows (Mushotzky *et al.* 1981, Fabian *et al.* 1981) to affect the observing program. Of course, cooling flows had been predicted (Fabian and Nulsen 1977, Cowie and Binney 1977) before the launch of the *EO*. The relatively small number of cooling flow clusters observed (~10)

Generic Cooling Flows with The Einstein Observatory Solid State Spectrometer

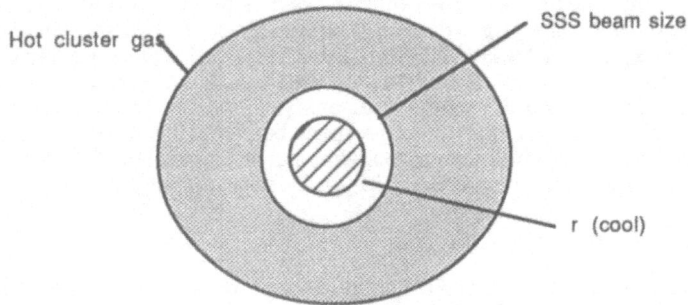

Figure 1. A schematic diagram of a typical cooling flow as "seen" by the SSS

was not limited by the sensitivity of the SSS. The relatively large error bars for the cooling parameters derived from the SSS data were limited by the rather short net exposures devoted to many of the clusters and modeling "errors". With 20/20 hindsight it is now clear that the SSS could have done an excellent job of determining the parameters in many cooling flow clusters if we had known of their existence while the SSS was operating.

II Two-Component Analysis

It was clear from the first SSS observations of the Perseus cluster (A426) and M87 that an isothermal model did not fit the data well (Mushotzky *et al.* 1979; Lea, Mushotzky and Holt 1982). The deviation from isothermality was due to 1) the presence of emission lines which could only arise from gas whose temperature was well below the effective temperature of the continuum (see figure 2, in Mushotzky *et al.* 1981) and 2) the shape of the continuum was not well fit by an isothermal bremsstrahlung spectrum. In addition the derived value of the temperature from an isothermal fit to the SSS data was in disagreement with the average temperature as derived from HEAO-1 data. In order to fit the "low temperature" lines, in particular the He like lines of Si and the Fe L lines which have the highest equivalent width in the temperature range to which the SSS was sensitive (see figure 2 below), an ad hoc model consisting of two isothermal plasmas of variable abundance was fit to the data. The fit was acceptable and physically reasonable in that the "high kT" component had an emission measure, kT and abundance consistent with that seen by the "big beam" HEAO-1 A-2 experiment and the "low kT" component had similarly reasonable values. The value of the temperature of the "low kT" component, log T~6.8, was driven, primarily, by the relative strength of different lines in the Fe L complex. Subsequently this model was fit to the the SSS data for 12 clusters (Mushotzky 1980). (see Table 1 for slightly revised values for 8 of these clusters). The SSS data is thus direct evidence for the existence of "cool" gas in the center of cooling flow clusters. To derive values of $\dot{M}$, we assigned all the cooling to the "low kT" component, calculated the cooling time from the derived temperature and emission measure assuming a density distribution. Since $t_{cool} \sim 5kT/n\Lambda$, where Λ is the cooling function, the derived $\dot{M}$ scales as the emission measure in the beam e.g. $\dot{M} = A \Lambda n^2 V/kT$ where A is a constant that depends, primarily on physical constants and weakly on the density distribution.

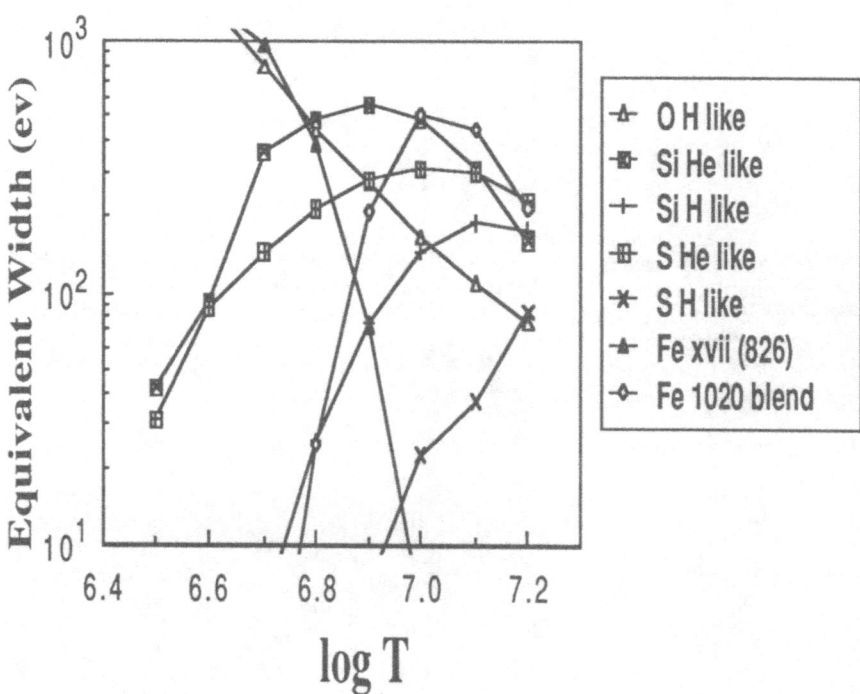

Figure 2. The equivalent width versus temperature of the H and He like lines from O, Si and S and selected Fe l lines; many of the Fe L lines have been omitted for clarity.

Since for temperatures less than 2 keV, $\Lambda \sim T^{-0.6}$, the dependence of $\dot{M}$ on T is relatively weak . If the true density distribution is smooth (e.g. a King model or a relatively flat power law similar to the fits of K. Arnaud (this workshop)) the dependence of A on the density distribution is also quite weak. There is, of course the assumption that the SSS beam sees the entire cooling flow (figure 1). If the size of the cooling region is larger than the SSS beam our calculated $\dot{M}$ are only lower limits.

TABLE 1

Two Component Model Estimates of $\dot{M}$ from SSS Data

NAME	$<n^2V>x10^{66}$ [†]	$\dot{M}(SSS)$	$\dot{M}(IPC)$ [††]
A85	≤7.5	≤450	100
A401	≤12	≤720	---
A426	4.8	290	250
A496	6.6	400	200
A1795	8	480	400
A2029	4.5	270	250
A2142	≤9	≤540	---
A2199	0.9	54	106

[†] The emission measure of the "low kT" component

[††] Values taken from Stewart et al. 1984

We note that there is good agreement between the values of $\dot{M}$ derived from the SSS data and other values of $\dot{M}$. The values of $\dot{M}$ derived from this simple method are stable and reasonable, however, this two component decomposition is not a sensible physical model for cooling in clusters. It is clear that any physically sensible cooling flow model must have a temperature gradient and thus the gas radiating in the SSS beam must have a range of temperatures. Unfortunately since the "two-temperature" models have good reduced χ^2 the SSS is not very sensitive to the presence of much of this gas.

III Cooling Flow Models

In this and the subsequent sections we report results from research in progress. We have not finished this work and thus these results should be taken as indicative and not final. In particular, we have not finished the error analysis for many of the objects in our sample. However, it is unlikely that our final conclusions will change substantially.

Because of the "non-physical" nature of the "two-temperature" models we have fit cooling flow models to the SSS data. As for the "two-temperature" models the signature of the "cooling flow" is the presence of "low kT" emission lines and the deviation of the continuum from isothermality. However, the observed equivalent width of many of these lines is quite low. This is due primarily to the dilution of the signature of the cooling component by the foreground hot gas emission. With a smaller beam size instrument the equivalent widths would have been very large. This is seen, easily, in the comparison of the SSS M87 spectrum with the spectrum of A426. The M87 spectrum is dominated by emission lines due to the fact that virtually all the gas seen by the SSS is cooling while in the case of A426 the observed emission measure is dominated by the hot gas component.

What is observed by the SSS is the sum over the emission versus T distribution in the range from 0.6-4.5 keV. That is the

$$\text{Observed flux in channel } E = \frac{1}{4\pi D^2}\int P(E,T)d(EM)$$

where D is the distance and P(E,T) is the photon spectrum of plasma at a temperature T in energy bin E. Assuming that the same mass flow rate pertains throughout the cooling flow, the emission measure for each temperature is determined by the time it takes for the matter to radiate away sufficient energy to cool down to the next temperature shell. The differential emission measure is thus proportional to the reciprocal of the bolometric luminosity at that temperature so that the

$$\text{Observed Spectrum} = \frac{C\dot{M}}{4\pi D^2}\int \frac{P(E,T)}{\varepsilon(T)_{Total}}dT$$

where $\varepsilon(T)_{Total}$ is the total (bolometric) emissivity from a shell at temperature T. In practice, we allowed the distribution of emission measures to vary from that determined for a constant pressure solution, such as would be the case for a situation where some material condenses out of the flow, and there is a smaller amount of matter flowing through the lower temperature shells. This calculation does not explicitly take into account the effect of the gravitational potential of the central galaxy (cf. Fabian, Nulsen and Canizares 1984 equations 7 and 8)

This cooling flow model has seven free parameters:

$\dot{M}$;   the mass accretion rate

s ;   the parameter describing the distribution of Em versus T; the emission measure for each shell was multiplied by $(T_{shell}/10^7 \text{ K})^s$ so that s = 0 corresponds to a constant pressure model. Values of s greater than zero correspond to more emission measure at higher T than in the constant pressure solution.

T_H;   the temperature the gas cools from
T_L;   the temperature the gas cools to
A;   the abundance of the metals (Fe, S, Si, Mg ...) relative to solar. We have fixed the ratios of the abundances of the different elements to be the solar value. The absolute abundances are a free parameter.

N_{High};   the amount of high temperature non-cooling gas
T_{High};   the temperature of the non-cooling gas.

While the column density to the source along the line of sight is also a free parameter its value, within the allowed fitted range, does not strongly affect the determination of $\dot{M}$.
    Note that with the exception of T_{High} none of these parameters are the same as those K. Arnaud (this workshop) had to assume to fit his cooling flow models. Thus the SSS and the EO imaging analysis are independent of each other. This is extremely important because the independence of the data sets and the modeling assumptions assures us that the good agreement of the results is not entirely due to modeling assumptions. In addition, it means that the error bounds in the SSS analysis are not due to the same modeling assumptions, as in the IPC/HRI case, but to statistical uncertainties and a different set of modeling uncertainties That is, the analysis of the SSS data to determine cooling rates does not depend on the assumptions of outer cluster pressure, the velocity dispersion of the galaxies, the form of the cluster potential or the value of the cluster core radius which are required in the IPC/HRI analysis (K. Arnaud this workshop). If for some reason we knew, a priori, some of these variables (such as T_{High} from the HEAO-1 A-2 data for a cluster, or N_{High} from the IPC image) we would be able to reduce our uncertainties in $\dot{M}$.

58

A major uncertainty in the SSS analysis is the value of **s**, the parameter describing the modification of the distribution of emission measures from constant mass flow rate. In figure 3 we show how changing **s** affects the emission. (Recall that values of **s** greater than zero corresponds to more emission measure at the higher temperatures). We note that as **s** becomes smaller the spectrum becomes "softer" and the strengths of various lines changes.

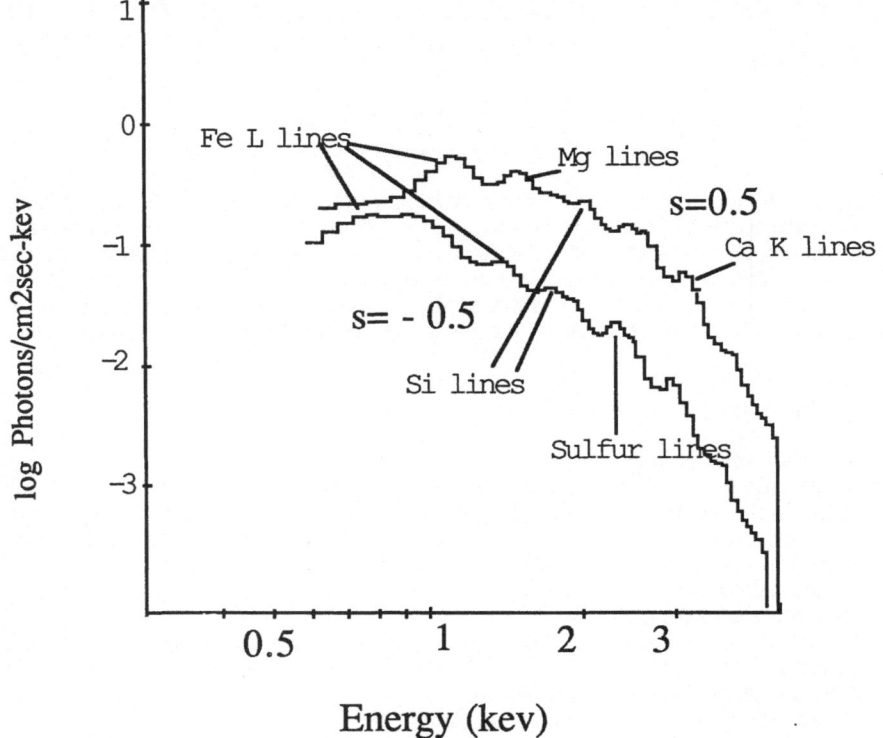

Figure 3. Representative cooling flow spectrum with 2 values of **s**. The strongest emission lines in the spectrum are indicated.

The solutions that we find have larger $\dot{M}$ associated with larger values of **s**. We have determined the range of acceptable values of **s** for a few clusters. For Perseus we find values of -0.3<s<0.7, while for M87 we find larger values, 0.4<s<0.9. For most of the clusters the best fit values for **s** are ~0.7. We also find that the abundance is strongly correlated with M, **s**, N_{High} and T_{High}. If the larger $\dot{M}$, larger **s** solutions are correct then the abundances tend to be lower than previously indicated (except for M87). T_L is not well determined by the SSS data as long as it is less than 700 eV.

The determination of T_H is a bit uncertain. If we assume that the outer atmospheres of clusters are isothermal then $T_H = T_{High}$. However, if the cluster atmosphere has a temperature gradient T_{High} represents the cluster temperature averaged over the SSS beam while T_H, the actual temperature the gas cools from in the SSS beam, could be higher (if the gas is polytropic or if there is a weak AGN in the center of the cluster) or could be lower if the cooling radius is larger than the SSS beam.

TABLE 2
Estimates of T_H from SSS Data

NAME	T_H(keV)	T_∞(keV)	Reference for T_∞
M87	3 (+0.5/-0.2)	3.2	Edge, Stewart and Smith 1987
A2199	<4 keV	3.6	Mushotzky 1984
A426	<5	4.15, 6.4	Ulmer et al. 1987, Mushotzky 1984
0335+096	>5	2.9, 2.0	Singh et al. 1987, Mushotzky 1987

T_∞(keV) is the average temperature seen by a "large" field of view instrument

In only a few of the clusters is T_H even marginally constrained by the SSS data (Table 2) Because of the correlation between $\dot{M}$ and these other free parameters (in particular s) the errors in $\dot{M}$ are not, primarily, statistical but modeling uncertainties. Alternatively one could decide that we know what s should be, from cooling flow theory and that the abundance in the cooling flow is fixed by the HEAO-1 A-2 and EXOSAT data (implicitly assuming no abundance gradient). We have chosen a very conservative course and have allowed all the free parameters to vary resulting in <u>maximal</u> error bars and have derived the following values of $\dot{M}$. (Table 3).

TABLE 3
Comparison of Values of the Mass Accretion Rate Between the SSS and
EO Imaging Experiment Determinations

NAME	$\dot{M}$ (SSS)	$\dot{M}$ (IPC)	Reference for IPC $\dot{M}$
A85	120	100	Stewart et al. 1984b
A426	120 (+65/-50)	(84,95,105)	K. Arnaud (this workshop)
A496	28 (+56/-17)	(70,101,121)	" " " "
A1795	425	400	Stewart et al. 1984b
A2029	260 (+260,-170)	250	Stewart et al. 1984b
0335+096	105 (+40,-50)	100-200	Singh et al. 1987
A2199	45(+24,-17)	(96,119,166)[†]	K. Arnaud (this workshop)
M87	4	4, 1.7**	Stewart et al. 1984a, Canizares et al. 1982

[†] K. Arnaud (table 1 of his paper this workshop) reports that for a "low core radius solution" the $\dot{M}$ for A2199 is 91±26 solar masses per year.

* This value comes from integrating the $\dot{M}$ versus r graph of Stewart et al. inside the SSS beam of 3' radius

** I have rescaled the value of $\dot{M}$ from Canizares et al. to a Virgocentric distance of 20 Mpc

In some sense the error bars quoted here correspond to the full range of systematic errors described by K. Arnaud in his table 1. If, for example, I restrict the value of N_{High} for A2199 to be consistent with the IPC values (K. Arnaud, private communication) then the value for $\dot{M}$ and its errors become $\dot{M}$=63(+16,-26) M_0/year.

With the possible exception of A2199 the agreement in the values of $\dot{M}$ derived from the SSS and the IPC/HRI is excellent. We believe that this result strongly confirms the general $\dot{M}$ values derived for over 30 other clusters from *EO* imaging data, the general method of analysis used by Fabian and colleagues to derive $\dot{M}$, and our own method of analysis. The fact that the values of $\dot{M}$ overlap so strongly also indicates that many of the assumptions used by Fabian and colleagues (and described in detail by K. Arnaud in this workshop) to analyse the *EO* data are, in general, correct.

In particular we hope that, in the future, the two data sets can be analyzed together and that this joint analysis will result in a general reduction in modeling uncertainties and in systematic errors. In addition, it may become possible to determine, with some fidelity, the form of the temperature profile for the cluster . This can be seen from K. Arnaud's figures 2,3 and 4 for A2199. If we assume that the SSS and IPC/HRI results must agree with each other then either the high outer pressure, low core radius or high velocity dispersion solutions of K. Arnaud are more appropriate. We see from the figures that these solutions require that T is a negative function of r at large radii. The required value of the maximum temperature reached in the central region of A2199 is ~5.5 keV from the simultaneous SSS and HEAO-1 analysis of Henriksen and Mushotzky (1988). This high a value is again only reached in the central 3' region sampled by the SSS for the high velocity dispersion, low core radius, low $\dot{M}$ solutions. Clearly more work is necessary before such conclusions can be confirmed.

IV Distribution of Emission Measure versus Temperature Models

In order to derive somewhat model independent results from the SSS data we have fit power models for the distribution of temperature versus emission measure. That is

$$EM=(EM)_0(T/T_0)^\omega$$

if we also assume that $n \propto r^{-\chi}$ and $T \propto r^\xi$ then $\omega=(3-2\chi-\xi)/\xi$

We have normalized the models at $T_0=10^7$ degrees; note the sensitivity of the values of ω to ξ =[dlogT/d log r]. With our ansatz above the value of χ can be determined directly from HRI/IPC imaging data since it is simply related to the slope of the x-ray surface brightness. The values of ξ can only be compared, directly, with the results of Canizares *et al.* 1982 for M87, derived from FPCS data. By using the value of χ from the IPC/HRI data and ω from the SSS data we can calculate an appropriate value of ξ, which we shall call $\xi_{predicted}$. Indirectly we can compare our results for, $\xi_{predicted}$ with the results of K. Arnaud's (this workshop and priv comm) models for cooling flows, that is values of EM(T) inferred from the deprojection analysis of IPC/HRI data.

TABLE 4
Comparison of SSS Values of the Slope of Emission versus Temperature
with Deprojection Analysis Results

NAME	ω (SSS)	$\chi^{\S}$(IPC/HRI)	$\xi_{predicted}$	$\xi^{\dagger}$"observed"
A426	0.65(+0.45,-0.65)	1.1±0.1	0.50±0.25	0.50
A496	0.9±.35[0.7±.35]^	~0.9	0.6,0.4	0.65
0335+096	0.9(+0.25,-0.7)	1.0±0.2	0.52	~0.4
A2199	0.9±0.45	~0.9	0.63	0.5
M87	1.5	~0.6	0.7	0.25?$^{\ne}$

$^{\ne}$ From Stewart et al. 1984a model C.

†All values unless indicated are from K. Arnaud (this workshop)

§ I have estimated these values from published HRI/IPC surface brightness versus radius plots.

$^\wedge$ The second values for A496 assumes that there is no hot gas component to the SSS spectrum, e.g. that all observed emission measure is due to the cooling flow

It is clear from table 4 that, within the errors, the SSS results and the deprojection analysis results for ξ agree quite well. Again this gives confidence not only to the values of $\dot{M}$ derived from the deprojection analysis analysis but also to the detailed distribution of temperature versus radius in the cooling flow. In addition to general agreement in slope we can also compare the normalizations, e.g. EM(0). I have parametrized K. Arnaud's analysis of A426 into power law distribution of density (n) and temperature versus r. Taking the best fit results gives a predicted function log EM= 66.4+0.70 log T_7. The best fit for the SSS fit to this model is log EM= 66.15+0.65 log T_7. Since the errors are "algebraic" (that is it is not clear that the best fit of EM versus T should be the algebraic manipulation of the best fit n versus r and T versus r) the agreement is excellent.

We show in figure 4 a direct comparison between the SSS EM(T) distribution with that derived from the FPCS results of Canizares et al. 1982

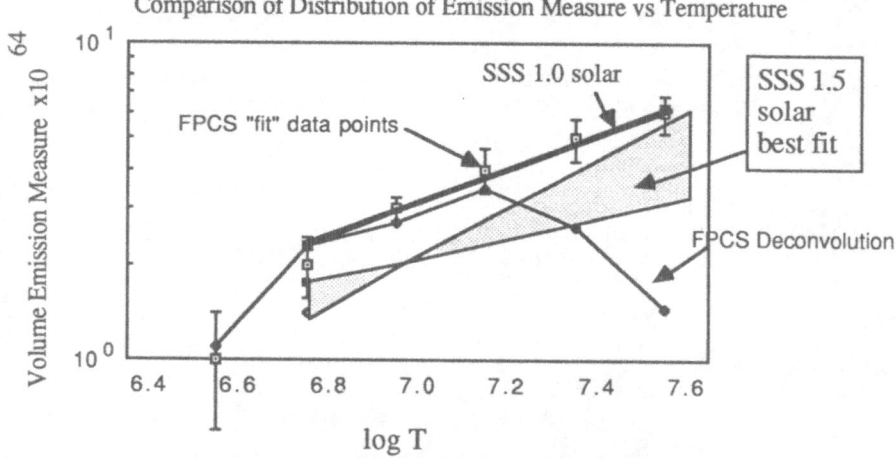

Figure 4. Comparison of the distribution of emission measure versus temperature in M87. The open squares and crosses are from Canizares et al. 1982 The stippled area is the best fit and error range for the SSS data with the best fit of 1.5 solar abundance. The bold line is the SSS best fit when we force the abundances to be solar. The SSS does not constrain the distribution below log T~6.8

We note that they are very consistent. In particular if we adjust the FPCS distribution for the 1.5 times solar abundances inferred from the SSS analysis of M87 the results are identical. It is entirely possible to perform an analysis similar to that which Canizares *et al.* (this workshop) presented for the Fe XXVII line with higher energy lines of Si and S with SSS quality data. However, due to the low equivalent width of these lines in the SSS data (due to the foreground cluster contamination) the fluxes in these lines are fairly uncertain. It is not clear if the results so derived are any less model dependent than the other results presented in this paper.

V Conclusions

The *Einstein Observatory* Solid State Spectrometer observations of cooling flows in clusters of galaxies provide good estimates of the mass accretion rates, $\dot{M}$, and the distribution of emission measure versus temperature. In general our estimates of $\dot{M}$ are in quite good agreement with those derived entirely independently from analysis of the *Einstein Observatory* imaging data. The method of analysis of the SSS data does not involve many of the assumptions that the imaging data require, in particular we do not have to assume the form or the depth of the cluster potential. On the other hand the SSS data require the measurement of the contribution to the SSS signal of the non-cooling gas which is measured directly by the imaging data. Thus the agreement of the SSS and imaging determinations of $\dot{M}$ is not fortuitous and gives weight to analysis of many clusters by Fabian and colleagues (K. Arnaud, this workshop). In addition for the 2 clusters for which there is relevant FPCS data the estimates of $\dot{M}$ are also in good agreement with the SSS results. In more detail the distribution of emission measure versus temperature derived from the SSS observations of M87 is in excellent agreement with that derived from FPCS observations.

The present analysis of the SSS data indicates that future spectroscopic observations of cooling flows, combined with the presently (and soon to be) available imaging will strongly constrain simple models of cooling flows. We have not attempted in this work to model the "multi-phase" cooling flows presented by Thomas and Nulsen (this workshop). Such work will probably require higher quality data of spectral resolution similar to that obtained by the SSS.

ACKNOWLEDGEMENTS: We would like to thank K. Arnaud for extensive discussion and communication of his results. We would like to thank A. Fabian for stimulating this work .

REFERENCES

Arnaud, K. 1988 this workshop
Canizares, C, Markert, T. and Donahue, M. 1988 this workshop
Canizares, C, Clark, G, Jernigan, G and Markert, T. 1982 *ApJ.* **262**, 33
Cowie, L and Binney, J. 1977 *ApJ.* **215**, 723
Fabian, A., and Nulsen, P 1977 *M.N.R.A.S.* **180**, 479
Fabian, A., Hu, E., Cowie, L., and Grindlay 1981, *ApJ.* **248**, 47
Fabian, A., Nulsen, P. and Canizares, C. 1984 *Nature* **310**, 733
Henriksen, M. and Mushotzky, R. 1988 *Ap. J.* submitted
Holt, S., White, N, Becker, R., Boldt, E., Mushotzky, R. Serlemitsos, P. and Smith, B. 1979 *Ap. J. (Letters*)**234**, L65
Lea, S. Mushotzky, R. and Holt, S. 1982 *ApJ.* **262**, 24
Mushotzky, R., Holt, S., Smith, B.W., Boldt, E., and Serlemitsos, P. 1981 *ApJ. (Letters)***244**, L47
Stewart, G., Canizares, C, Fabian, A. and Nulsen, P. 1984a *ApJ.* **278**, 536
Stewart, G., Fabian, A. Jones, C. and Forman, W. 1984b *ApJ.* **285**,1

X-RAY EMISSION LINES FROM COOLING FLOWS

Claude R. Canizares, Thomas H. Markert,
and Megan E. Donahue
Department of Physics and Center for Space Research
Massachusetts Institute of Technology
Cambridge, Massachusetts, U.S.A. 02139

ABSTRACT. Individual X-ray emission line strengths can be used to estimate the rate at which gas cools in the intracluster medium. We review the technique and show that departures from ionization equilibrium in the cooling plasma are not important at X-ray temperatures. We present data from 14 observations of 7 clusters performed with the Focal Plane Crystal Spectrometer on the *Einstein* Observatory, in addition to a brief review of earlier results on M87. About half the observations gave detections and half upper limits. Our estimates of the cooling rate $\dot{M}$ within the 3×30 arc min aperture are ≈4 $M_\odot$ yr^{-1} for M87, ≈20-30 $M_\odot$ yr^{-1} for Centaurus and ≈200 $M_\odot$ yr^{-1} for Perseus. For four other cooling flow clusters, 0335+096, A262, A496 and A1060 we obtain non-restrictive upper limits. For M87, Cen and Per, our values of $\dot{M}$ are similar to those found by completely independent methods based on imaging or other spectral data. Although our estimates are formally only upper limits to the actual values of $\dot{M}$ if the gas is being heated by some mechanism, this agreement is circumstantial evidence in favor of the existence of substantial cooling flows and against the dominance of conduction or other heat sources. For Perseus, an estimate of the volume emission measure required to produce the Fe XVII line leads to the conclusion that the intracluster gas must be a multi-phase plasma. The unusually strong oxygen lines from M87 and Perseus suggest that the O/Fe abundance ratio is larger in the intracluster gas than in the Sun.

1. INTRODUCTION

High resolution X-ray spectroscopy can be a very powerful tool for the study of cooling flows. Narrow emission lines account for more than 10% of the power radiated by plasma with cosmic abundances at temperatures of ≈ 10^8 K, and this fraction increases to >60% for T < 10^7 K. Furthermore, for T < 10^7 K as much as 10% of the luminosity may be radiated in a single line. Therefore, by measuring the strengths of a few lines, or even a single line, one can learn a lot about gas cooling through this temperature range.

The first spectral evidence of a cooling flow was just such a high resolution measurement of the OVIII Lyman α line in the central region of M87 made with the Focal Plane Crystal Spectrometer (FPCS) on the *Einstein* Observatory (Canizares *et al.* 1979). This line is produced primarily at T < 8×10^6 K, and it was much too strong to be due to the bulk of the plasma surrounding M87, which has a mean temperature of 3×10^7 K. It suggested the presence of a significant amount of cooler gas. This was demonstrated in a later analysis of eight narrow regions of the spectrum that isolated lines or blends due to various ionization stages of iron in addition to the OVIII line (Canizares *et al.* 1982). We detected lines from Fe XVII, Fe XX, and Fe XXI through Fe XXIV, which requires that M87 contain plasma at temperatures

63

A. C. Fabian (ed.), Cooling Flows in Clusters and Galaxies, 63–72.
© *1988 by Kluwer Academic Publishers.*

covering at least a decade, from 3×10^6 to 3×10^7 K. Because all the lines are from the same element, this conclusion is independent of assumptions about relative or absolute abundances. We used the measured line strengths to deconvolve an emission measure vs. temperature distribution that was consistent with gas cooling steadily through this temperature range at a rate of ≈ 4 $M_\odot$ yr^{-1}. These data were also used in the subsequent work of Stewart *et al.* (1984). Mushotzky (this meeting) has derived an emission measure vs. temperature distribution which is in excellent agreement with ours. It is important to emphasize that these two determinations are completely independent: Mushotzky's results are based on fits to the total line and continuum spectra over ≈ 0.6-4 keV measured with the Solid State Spectrometer (SSS) on *Einstein*, whereas our results come from an analysis of eight carefully selected slices of the spectrum, each only 20-40 eV wide.

The M87 spectra are discussed in detail in the literature cited above, so I will not address them further (except in the discussion of abundances below). Instead, I will concentrate on FPCS observations of other clusters. None of these is as extensive as the M87 observation, but they do give important information about cooling flows.

We performed fourteen observations of seven clusters with the FPCS in addition to M87. These are listed in Tables 1-4. Half of the observations yielded only upper limits, but as shown below, even these can be of some use. All observations were made through a 3×30 arc min aperture. A paper giving details of the observations and analysis is in preparation.

Before presenting the results I will address two general questions: whether cooling gas is in ionization equilibrium, which is central to any spectral analysis, and how to determine mass accretion rates from individual line strengths.

2.   IONIZATION EQUILIBRIUM IN COOLING FLOWS

In an optically thin plasma that is in statistical equilibrium at a given temperature, the relative population of ionization stages of each element is governed by the balance between ionization and recombination rates. In a plasma that is cooling or being heated, the temperature changes as a function of time $T(t)$. If the temperature changes sufficiently slowly, the ionization fractions at a given time t will be those of an equilibrium plasma at temperature $T = T(t)$, and the emitted spectrum can easily be synthesized by an appropriate superposition of equilibrium model spectra. On the other hand, if the temperature changes too rapidly, then the ionization fractions will depart from the equilibrium values and will depend in detail on the history of the plasma. Such is the case, for example, in supernova remnants, where the heating is sudden and the degree of ionization generally lags behind (e.g. Shull 1983) so many elements are under ionized for the given temperature. In cooling gas, elements could be over ionized if recombination lags behind cooling.

We have found that departures from instantaneous ionization equilibrium are not important for the radiatively cooling gas at X-ray temperatures (see also Fabian *et al.* 1984; significant departures from equilibrium can occur for $T < 10^6$ K; Shapiro and Moore, 1976; Edgar and Chevalier 1986). There are two relevant time scales: the cooling time Δt_{cool}, and the recombination time Δt_{rec}. We take Δt_{cool} to be the time over which radiative cooling will reduce the temperature of a plasma by an amount, $\Delta logT$, sufficient to change the *equilibrium* ionization fraction of a given element by a factor of ≈ 2. Typically this means $\Delta log(T) > 0.1$. For isobaric cooling,

$$\Delta t_{cool} = (5/2) \times kT \times 2.3\Delta logT/[n\Lambda(T)] \ , \tag{1}$$

$$\Delta t_{cool} \approx 4 \times 10^{14} \ [n/0.1 \ cm^{-3}]^{-1} \ [T/10^7 \ K] \ [\Delta logT/0.1] \ sec. \tag{2}$$

Here n is the density. We have assumed cosmic abundances and neglected the slow variation of the cooling function $\Lambda(T)$ near 10^7 K.

The recombination time scale is

$$\Delta t_{rec} = (n\alpha)^{-1}, \tag{3}$$

where α is the recombination rate coefficient. For recombination to multielectron ions α is generally dominated by the dielectronic recombination rate, which is typically $\approx 10^{11}$ cm^3 s^{-1}. Thus,

$$\Delta t_{rec} \approx 10^{12} [n/0.1 \text{ cm}^{-3}]^{-1} \text{ s (multi-electron ions)}. \tag{4}$$

The radiative recombination rate to hydrogenic ions is $\approx 2.5 \times 10^{-13}$ cm^3 s^{-1} so

$$\Delta t_{rec} \approx 4 \times 10^{13} [n/0.1 \text{ cm}^{-3}]^{-1} \text{ s (hydrogenic ions)}. \tag{5}$$

Comparison of eqs. 2, 4 and 5 shows that, independent of density, the cooling time typically exceeds the recombination time by ≈ 400 for multielectron ions and by ≈ 10 for hydrogenic ions. Therefore, we can expect the ionization fractions to follow their equilibrium values as the plasma cools. This is not surprising if one recalls that at the temperatures of interest the cooling is often dominated by line emission from these very ions: the ions must recombine and be collisionally excited several times in order to radiate the thermal energy of the plasma.

To verify these rough arguments a former student, Andrew Bernoff, and I performed a numerical calculation of the ionization fractions of iron in a cooling plasma (these are mentioned briefly in Canizares *et al.* 1982). We allowed the plasma to cool at its equilibrium rate but solved the ionization balance equations explicitly for iron. The initial temperature is 3×10^7 K and the initial ionization fractions are the equilibrium values for this temperature. By using $\tau = tn$ in place of t, the density is removed from the problem. In Figure 1a we plot the ionization fraction vs. T for the cooling gas. This can be compared with Figure 1b, which shows the same quantities for the case of statistical equilibrium. There are very slight differences that could easily be attributed to the numerical accuracy of the model. We conclude that departures from equilibrium are negligible for X-ray emission from cooling flows.

3. ESTIMATING $\dot{M}$ FROM A SINGLE EMISSION LINE

Cowie (1981) pointed out that the luminosity in a single emission line can be used to estimate the rate at which matter is cooling, $\dot{M}$ (this is usually referred to as the mass accretion rate, but for the present argument the important fact is that matter *cools*, whether or not it *accretes*). Consider a volume V in which plasma cools from T to T-Δ T by radiating luminosity ΔL. For isobaric cooling,

$$\Delta L = n^2 V \Lambda(T) = (5/2) \times (\dot{M}/\mu m) \times k\Delta T, \tag{6}$$

where μm is the mean mass per particle (typically 0.6). The luminosity of a single line (denoted by the subscript i) is

$$\Delta L_i = n^2 V \varepsilon_i(T), \tag{7}$$

where $\varepsilon_i(T)$ is the emissivity of the ith line. Substituting for $n^2 V$ from eq. 6 gives

$$\Delta L_i = \dot{M} \times (5k/2\mu m) \times [\varepsilon_i(T)/\Lambda(T)] \Delta T. \tag{8}$$

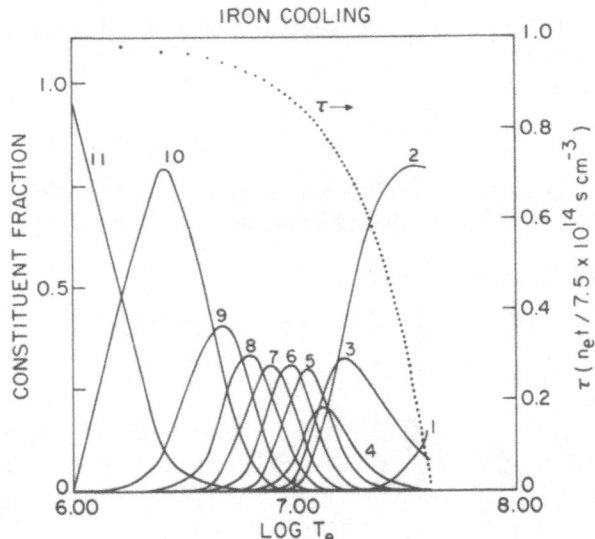

Figure 1a. Ionization fractions of iron (left hand scale) vs. temperature for a plasma cooling radiatively from an initial temperature of 3×10^7 K to 10^6 K. The solid curves give the fraction of iron atoms which have the number of electrons given by the labels. The initial ionization fractions are the equilibrium values for the initial temperature. The dotted curve gives the product $\tau = tn$ (right hand scale) vs. temperature.

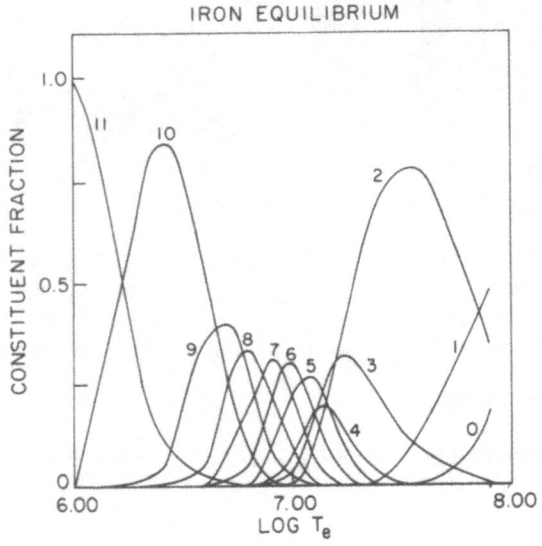

Figure 1b. Ionization fractions of iron vs. temperature for plasma in statistical equilibrium.

Therefore, if the plasma cools completely from an initial temperature T_{max}, then the total luminosity in the line is

$$L_i = \dot{M} \times (5k/2\mu m) \times \int_0^{T_{max}} [\varepsilon_i(T)/\Lambda(T)] \, dT, \qquad (9)$$

which can be inverted to give

$$\dot{M} = L_i/G(T_{max}), \qquad (10)$$

where $G(T_{max})$ is everything that multiplies $\dot{M}$ on the right hand side of eq. 9.

This derivation neglects any work done on the gas by the gravitational potential or by any other energy sources such as heating or conduction. All these would appear as extra positive terms on the right hand side of eq. 8, reducing the value of $\dot{M}$ required to produce a given line luminosity. Therefore, if eq. 10 is applied to a real cluster, it gives an upper limit to the true cooling rate.

The factor $G(T_{max})$ is quite insensitive to the overall metal abundance of the cooling gas. That is because the cooling function $\Lambda(T)$ has a large contribution from line emission and from free-bound continuum so that it, as well as ε_i, is roughly proportional to the overall metal abundance. Thus, the ratio ε_i/Λ, which appears in eq. 9, is nearly independent of overall abundance. However, relative abundance enhancements of particular elements can affect $G(T_{max})$; the size of the effect is less than proportional for a relative enhancement of iron, whose emission is a major contributor to $\Lambda(T)$, but it is roughly proportional for enhancements of other elements.

Figure 2 shows $G(T_{max})$ for two important lines. These computations use the emissivities of Raymond and Smith (1977 and private communication) and the cooling function of Mewe and Gronenschild (1981). Although in each case there is a single dominant line, for ε_i we take the sum of emissivities of all lines and continuum within the indicated pass band. For the Fe XVII line $dG(T)/dT$ is reasonably peaked around $\log T = 6.8$, so $G(T_{max})$ is rather insensitive to T_{max} above this value (the slow rise at higher T is due mainly to contributions from the continuum). Thus the use of eq. 10 does not require very precise knowledge of T_{max}. For O VIII the emissivity, and hence $dG(T)/dT$ has a long tail to high temperatures, so $G(T_{max})$ continues to rise.

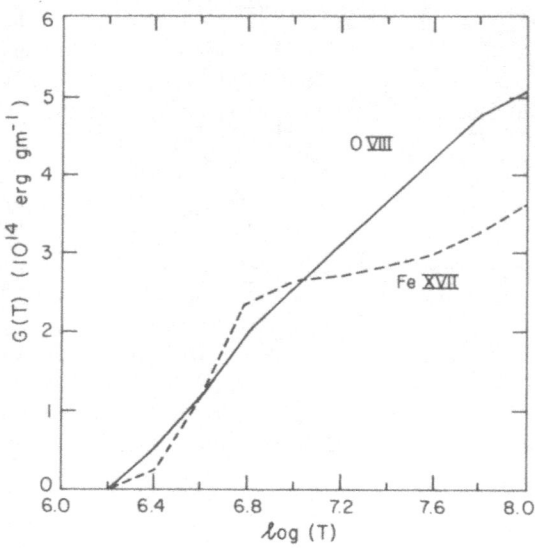

Figure 2. The function G(T) defined in the text vs. temperature: $\dot{M} = L_i/G(T)$, where L_i is the total luminosity in the selected band. The curve labeled O VIII is for the band 647-656 eV in which the most prominent line is O VIII Lyα. The Fe XVII curve is for 820-931 eV, which is dominated by an Fe XVII 3p-2s line. The computation of G(T) accounts for all line and continuum emission in each band.

4. FPCS OBSERVATIONS OF COOLING FLOW CLUSTER

4.1 The Centaurus Cluster

We observed three regions of the spectrum of the Cen cluster. The derived luminosities corrected for Galactic absorption are listed in Table 1 (we assume $H_0 = 50$ km s^{-1} Mpc^{-1} throughout). The statistical and systematic uncertainties are typically $\approx 30\%$. Using eq. 10 to estimate $\dot{M}$ from the two detected lines gives 20-30 $M_\odot$ yr^{-1} for $\log T_{max} = 7.4$-7.8 (Mushotzky 1983; the higher $\dot{M}$ corresponds to the lower T_{max}). The upper limit to the O VIII line gives no additional information; it implies an upper limit to $\dot{M}$ of ≈ 150-200 $M_\odot$ yr^{-1}.

TABLE 1. Centaurus Cluster

E(eV)	Lines	L(10^{42}erg s^{-1})
647-656	0 VIII Lα	<4
1006-1046	Ne X + Fe XXI	1.6 ± 0.8
1087-1153	Fe XXIII, XXIV	1.6 ± 0.4

We can compare our estimate with the results obtained from a deconvolution of the imaging data. Arnaud (private communication) obtains ≈ 30 $M_\odot$ yr^{-1} within ≈ 100 kpc, which is in excellent agreement with our value. Thomas, Fabian, and Nulsen (1987), who use a multi-phase analysis, find that $\dot{M}$ increases roughly linearly with radius to a value of ≈ 90 $M_\odot$ yr^{-1} at ≈ 180 kpc, assuming that the effective cooling radius does extend this far. Because our finite aperture subtends 60×600 kpc at the source, we must apply an aperture correction of ≈ 2-2.5 to our value for comparison. This gives 40-75 $M_\odot$ yr^{-1}, which is in reasonable agreement with the values derived from the imaging analysis.

4.2 The Perseus Cluster

For the Perseus cluster we observed a 45 eV region centered near 825 eV in the frame of the cluster. This scan includes a strong 3p-2s line of Fe XVII, two weaker Fe XVII lines and the Lyman γ and Lyman δ lines of O VIII (see Figure 3). An additional observation of the O VIII Lyman β line gave only an uninteresting upper limit. Table 2 lists the derived luminosities assuming absorption by a column $N_H = 2 \times 10^{21}$ cm^{-2}; there is some uncertainty in the value of N_H which introduces a 40% systematic uncertainty in the line fluxes in addition to the statistical and instrumental uncertainties, which are smaller.

The estimates of $\dot{M}$ obtained from eq. 10 are listed in Table 2. We use $\log T_{max} = 7.9$ (Mushotzky 1983). As discussed above, $\dot{M}$ deduced from the Fe XVII line flux is relatively independent of the assumed T_{max}, and it is only weakly dependent on the Fe abundance. Therefore, the estimate of 200 $M_\odot$ yr^{-1} derived from Fe XVII is reasonably secure. Note that both oxygen lines give $\dot{M}$ estimates that agree with each other but are significantly larger than the estimate from Fe XVII. Possible uncertainties in T_{max} cannot explain such a large discrepancy; to lower sufficiently the $\dot{M}$ estimates derived from the oxygen lines requires values of $G(T_{max})$ several times larger than we have used, yet $G(T_{max})$ increases too slowly near the assumed temperature to permit that (the $G(T)$ distributions for O VIII Lyξ and Lyδ are similar in shape to that for Lyα plotted in Figure 2). This suggests a relative abundance enhancement of O to Fe. We discuss this further below.

The estimate of $\dot{M}$ we derive from the Fe XVII line is within the range of values found independently from imaging and other spectral observations. Earlier analyses gave $\approx$300-400 $M_\odot$ yr^{-1} (Mushotzky *et al.* 1981, Fabian, Nulsen and Canizares 1984), although more recent work favors values of $\approx$100-150 $M_\odot$ yr^{-1} (Mushotzky, this meeting; Arnaud, private communication).

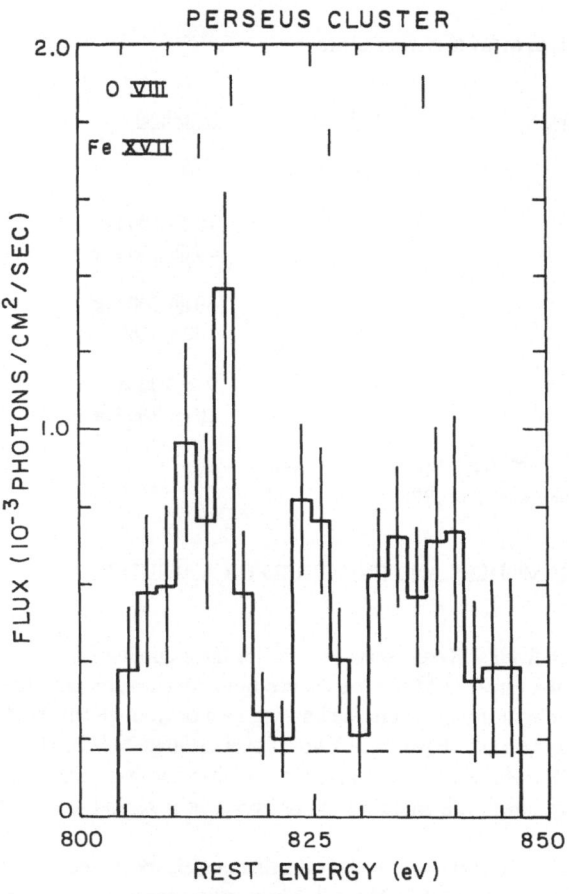

Figure 3. A portion of the spectrum of the Perseus Cluster obtained with the Focal Plane Crystal Spectrometer on the *Einstein* observatory. The energies of prominent emission lines of O VIII and Fe XVII are marked. The horizontal dashed line marks the level of the instrumental background which is measured simultaneously by the position sensitive detector (e.g. see Canizares *et al.* 1982). The error bars are statistical.

TABLE 2. Perseus Cluster

E(eV)	Lines	L(10^{42}erg s^{-1})	$\dot{M}$(M$_\odot$yr^{-1})
804-820	O VIII Lγ	14 ± 2.8	600 ± 120
820-831	Fe XVII	4.8 ± 1.4	220 ± 64
831-847	O VIII Lδ	10 ± 3.1	640 ± 200

4.3 Other Clusters

We obtained upper limits to the line emission from the four cooling flow clusters listed in Table 3. In each case, the deduced upper limit on $\dot{M}$ is consistent with the mass accretion rate found from analyses of the imaging data.

TABLE 3. Upper Limits to M from X-Ray Lines

Cluster	log(Tmax)	$\dot{M}$(lines)	$\dot{M}$(other)
0335+096	7.4	<690	~20
A262	7.4	<40	18 @ 100 kpc 45 @ 200 kpc
A496	7.2	<310	60 @ 100 kpc
	7.9	<200	100 @ 180 kpc
A1060	7.4	<35	9 @ 100 kpc
	7.9	<25	30 @ 200 kpc

References for $\dot{M}$(other): 0335+096, Mushotzky (1983); A262, A496, A1060, Thomas *et al.* (1987).

5. CONSTRAINTS ON CONDUCTION MODELS AND EVIDENCE FOR A MULTI-PHASE INTRACLUSTER MEDIUM

We have measured the strengths of individual X-ray emission lines for three cooling flows: M87, Centaurus and Perseus. For each of these, the values of $\dot{M}$ that we deduce from the emission lines using eq. 10 agree to within a factor of 2 with the values found from the imaging data or from global spectral studies at lower resolution from the SSS. As we note above, eq. 10 would over estimate $\dot{M}$ if conduction (or other heat sources) were significant. Therefore, the rough agreement with other determinations is circumstantial evidence in favor of cooling flows with significant mass accretion rates and against the dominance of conduction.

Bertschinger and Meiksen (1986) have constructed detailed conduction models for Perseus that are roughly consistent with the imaging data but have $\dot{M} \approx$ 20-75 $M_\odot$ yr^{-1}. These values are three to ten times smaller than our estimate using the Fe XVII line. Although we have not yet made precise comparisons, we suspect that these models would be deficient in Fe XVII flux by similar factors.

For the Perseus Cluster, the large luminosity of the Fe XVII line has significant implications for the structure of the emitting plasma: it indicates that the plasma must be a multi-phase medium. To see this, we estimate the minimum volume emission measure required to produce the Fe XVII line by assuming that all the gas in this volume is at $T = 10^{6.6}$ K, the temperature at which the line emissivity is at its maximum. The volume emission measure required is $n^2V = L_i/\varepsilon_i(T) = 1.6\times10^{66}$ cm^{-3}. Analysis of the imaging data gives a density $n \approx 0.1 \, (r/10")^{-0.7}$ cm^{-3} in the central region of Perseus assuming the plasma is in a single phase (Fabian *et al.* 1981). Thus a sphere of outer radius R_{kpc} in kpc has a volume emission measure $n^2V \approx 5.7\times10^{63} \, R_{kpc}^{1.6}$ cm^{-3}. For a single phase plasma, all the gas in the inner 30 kpc of the Perseus Cluster would have to be at $T = 10^{6.6}$ K or kT = 0.3 keV to account for the observed Fe XVII line luminosity. This is highly implausible and is contradicted by other spectral measurements (cf. Mushotzky *et al.* 1981). On the other hand, if the plasma is multi-phase and in pressure equilibrium, the material that has cooled to $10^{6.6}$ K would be ≈ 10 times denser that the surrounding medium and the same volume emission measure would occupy only 1% of the above volume. A multi-phase medium would be expected if matter is cooling and dropping out of the flow over a wide range of radii, as indicated by analyses of the X-ray images (Fabian, Nulsen and Canizares 1984; Thomas, Fabian and Nulsen 1987). The apparent necessity for a multi-phase medium is an additional argument against models in which large scale conduction replenishes the energy lost by radiation in a single phase; however, the possible role of conduction on small scales has not yet been fully explored (e.g. see discussion by Binney and Nulsen in this volume).

6.   LIMIT ON COOLING RATE IN THE COMA CLUSTER

As more and more cooling flow clusters have been identified, it has raised the question of whether or not gas is cooling in all clusters. Cooling might be a ubiquitous phenomenon in a multi-phase intergalactic medium, although an actual cooling flow may form only in clusters with cD galaxies or otherwise centrally concentrated gravitational potentials (e.g. Fabian, Nulsen and Canizares 1984).

We have investigated the possibility that cooling is ubiquitous even in relaxed clusters by searching for X-ray line emission from low temperature gas in Coma. We made two observations of the central regions of the cluster, both of which yielded marginal detections which are summarized in Table 4. In each case the aperture accepted approximately 14% of the total X-ray flux from the source, and it is likely that much or all of the detected flux is simply due to the continuum from plasma at the mean temperature of $\approx 10^8$ K (Mushotzky 1983). For the present, we subtract our estimate for the continuum contribution and treat the remainder as an upper limit to the line flux. Then application of eq. 10 gives upper limits to the amount of material cooling from $T_{max} = 10^8$. The tighter limit, which is from the Fe XXIV line, is $\dot{M} < 17$ $M_\odot$ yr^{-1} for the rate at which gas is cooling within the 120×1200 kpc region viewed by the aperture. If we make the assumption that matter cools over the whole cluster with a distribution similar to the distribution of X-ray surface brightness, then the limit for all of Coma is $\dot{M} < 125$ $M_\odot$ yr^{-1}. Thus we cannot rule out the possibility that gas does cool in Coma (and presumably in other Coma-like clusters), but it must do so at a rate smaller than that in Perseus, and certainly any cooling material must be much less centrally concentrated than it is in the well identified cooling flow clusters.

TABLE 4. Coma Cluster

E(eV)	Lines	L(10^{42}erg s^{-1})
987-1049	Ne X, Fe XVII-XXIII	11 ± 4
1114-1237	Fe XIX-XXIV	8.4 ± 3.5

7.   THE ABUNDANCE OF OXYGEN IN CLUSTER GAS

We detect lines of both oxygen and iron for two of the cooling flows studied with the FPCS, M87 (Canizares *et al.* 1982) and Perseus (section 4.2 above). In both cases the oxygen lines are 3-5 times stronger than one would expect based on the strengths of the iron lines. We can think of no plausible reason for this excess line strength other than an enhancement of the oxygen to iron abundance ratio compared to its solar value. Such an enhancement would not be surprising if the intracluster medium had been enriched primarily by an early population of short-lived, massive stars, since these are thought to have oxygen rich ejecta (e.g. see Tinsley 1979).

ACKNOWLEDGEMENTS

We thank Paula Blizzard, Leah Bateman, Joan Coyne and John Culver for assistance with data analysis, and Andrew Bernoff for computation of non-equilibrium ionization fractions for Fe. We are grateful to Keith Arnaud, Edmund Bertschinger, Andrew Fabian, Richard Mushotzky, and Paul Nulsen for useful conversations.

This work was supported in part by NASA Grant NAG 8-494.

REFERENCES

Bertschinger, E. and Meiksen, A. 1986, *Ap. J. (Letters)*, **306**, L1.

Canizares, C. R., Clark, G. W., Markert, T. H., Berg, C., Smedira, M., Bardas, D., Schnopper, H., and Kalata, K. 1979, *Ap. J. (Letters)*, **234**, L33.

Canizares, C. R., Clark, G. W., Jernigan, J. G. and Markert, T. H. 1982, *Ap. J.*, **262**, 33.

Cowie, L. L. 1981, in Giacconi, R. (ed.) *X-ray Astronomy with the Einstein Satellite*, (D. Reidel), 227.

Edgar, R. J. and Chevalier, R. A. 1986, *Ap. J. (Letters)*, **310**, L27.

Fabian, A. C., Hu, E. M., Cowie, L. L., and Grindlay, J. 1981, *Ap. J.*, **248**, 47.

Fabian, A. C., Itoh, H., Stewart, G. C., Canizares, C. R. and Nulsen, P. E. J. 1984, *Nature*, **307**, 343.

Fabian, A. C., Nulsen, P. E. J., and Canizares, C. R. 1984, *Nature*, **310**, 733.

Mewe, R. and Gronenschild, E. H. B. M. 1981, *Astron. Ap. Suppl.*, **45**, 11.

Mushotzky, R. 1983, *Physica Scripta*, **T7**, 157.

Mushotzky, R. F., Holt, S. S., Smith, B. W., Boldt, E. A. and Serlemitsos, P. J. 1981, *Ap. J. (Letters)*, **244**, L47.

Shapiro, P. R. and Moore, R. T. 1976, *Ap. J.* **207**, 460.

Shull, M. 1983, *Ap. J.*, **262**, 308.

Thomas, P. A., Fabian, A. C., and Nulsen, P. E. J. 1987 *M.N.R.A.S.*, **228**, 973.

Tinsley, B. 1979, *Ap. J.*, **229**, 1046.

OPTICAL, ULTRAVIOLET, AND INFRARED OBSERVATIONS OF COOLING FLOWS IN RICH CLUSTERS

Esther M. Hu
Institute for Astronomy
University of Hawaii
2680 Woodlawn Drive
Honolulu, HI 96822 USA

ABSTRACT. Many rich clusters are now known to possess centrally located optically emitting gas systems, which accompany large ($\sim 100 M_\odot$/year) inflows of cooling cluster gas. We report on recent results of optical, UV, and IR studies on cooling flow clusters. The work centers on three topics: (1) Spectroscopic studies of the velocity structure of the line-emitting gas systems have been used to probe the dynamics of the central cluster regions. (2) Filter imaging studies of cluster emission line systems have yielded line luminosities and characteristic angular scales. We discuss the implications of both data sets for the cosmological evolution of galaxies and gas in clusters. (3) Finally, we report the results of combining our optical survey with similar surveys undertaken by us using ultraviolet and infrared satellite data. Carefully flux-calibrated data from this multi-wavelength data base have been used to show that a small but significant component of dust extinction is associated with the environment of these clusters. The measured IR flux and the extinction values are used to suggest that substantial recent injection of material into the cluster atmospheres has taken place.

1. INTRODUCTION

In this talk I'd like to discuss what we can learn about cooling flow systems from studies of the optical emission systems often found at their cores. Excellent reviews of the general X-ray properties of cooling flow systems are given elsewhere in the present volume, and will not be detailed here. However, in order to maintain a homogeneous sample of data in the hopes of clarifying the relationship between the X-ray and optical properties, the present discussion will be confined to cooling flow systems found in the centers of rich clusters of galaxies.

2. THE EVIDENCE FOR COOLING FLOWS

The first question to pose is whether the emission systems which we see are in fact cooling flows, that is, whether their presence is a consequence of the cooling of the

A. C. Fabian (ed.), Cooling Flows in Clusters and Galaxies, 73–86.

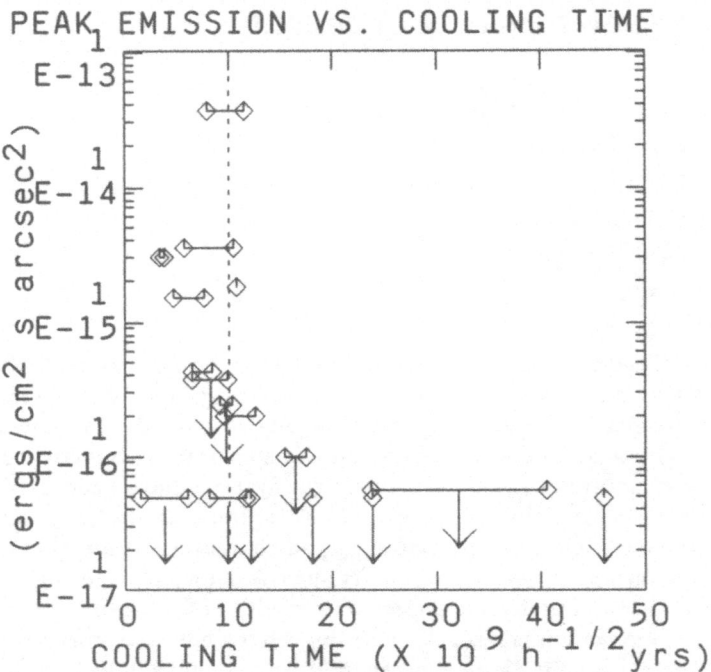

Figure 1. Incidence of optical emission systems in the central core as a function of the time for the extended cluster gas to cool to $10^{4°}$ K. Cooling time determinations are based on isothermal temperature fits and central densities extrapolated from the outer core. The length of the bars give an estimate of the errors (primarily due to uncertainties in the central density determinations). A dashed reference line marks off a cooling time of 10^{10} yrs.

large scale cluster gaseous atmosphere as opposed to being a purely local active nucleus phenomenon. From Figure 1, adapted from Hu, Cowie, and Wang (1985), it can be noted that all cases for which optical emission can be seen in the central core correspond to cluster cooling times less than some critical value which can be empirically determined as $\sim 10^{10} h^{-1/2}$ yrs ($h = H_0/100$ km s^{-1} Mpc^{-1}), and that *no* cases with central emission are seen for clusters with cooling times longer than this critical value. Since the original figure appeared, data have become available for a number of additional clusters, all consistent with the behavior noted above. This data set now includes nearly all rich clusters having published X-ray data for cooling times below the critical value, and a good many clusters with cooling times above this value. The strong correlation between the incidence of optical emission with global cooling properties of the X-ray cluster halo gas provides compelling evidence that these systems are indeed the results of cooling inflows.

Accepting that the emission systems arise from cooling of the cluster gas, the determined cooling times in the left-hand side of Figure 1 may be compared with

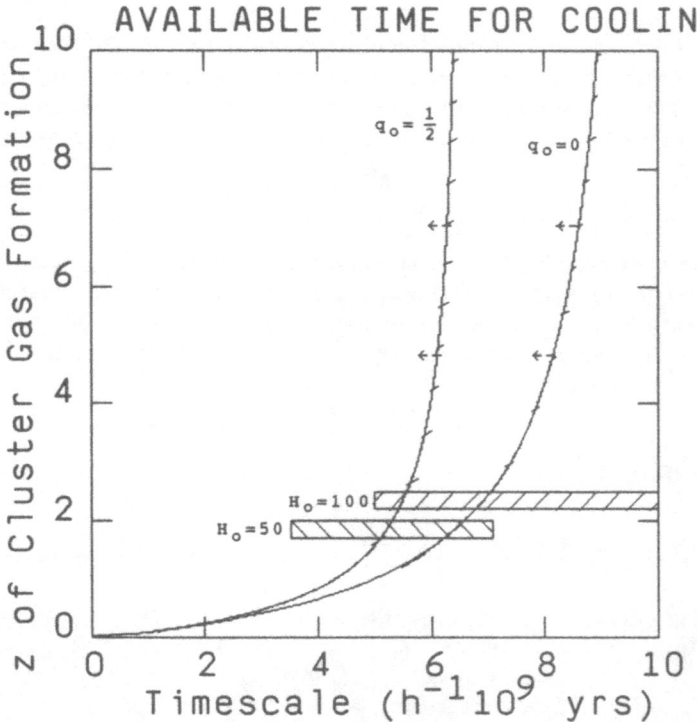

Figure 2. Shown here are plots of the cosmological time between the epoch at which cluster atmospheres began to cool up to the epoch of currently observed central emission systems ($z \sim 0.05$ for the clusters plotted in Figure 1) as a function of z of cluster gas formation. Curves for two different cosmological models ($q_0 = 0$ and $q_0 = \frac{1}{2}$) are given. Also shown (hatched bars) is the range of cooling times found for the optical emission system clusters graphed in Figure 1 for values of the Hubble constant $H_0 = 50$ and $H_0 = 100$. In graphical terms, all points in the bars should lie to the *left* of the available time curve evaluated at the y-value at which cluster gas formed (*i.e.*, the measured cooling times should not exceed the time available.)

the available time (*i.e.*, the time between the epoch at which cluster atmospheres formed and began to cool, and the epoch at which the clusters in our sample are observed to show emission). Figure 2 gives curves of the available times, as a function of the redshift at which cluster halos formed, for $z \sim 0.05$ clusters (representative of the redshifts of clusters in Figure 1) in two different cosmologies ($q_0 = 0$ and $q_0 = \frac{1}{2}$). If the observed optical emission arises from the cooling gas, then the associated measured cooling times, whose range is shown graphically by the heavy bars, must all be less than the time available. Even allowing for errors in the cooling time determinations (primarily due to uncertainties in the density measurements, approximately of order 30%), this would suggest that a value of H_0 as large as 100 would be extremely difficult to accommodate, and that for $H_0 = 50$ a value of q_0 close to 0 is preferred.

The above analysis assumed cooling times based on isothermal temperature fits. If adiabatic models are preferred, as recent analyses (Cowie, Henriksen, and Mushotzky 1987) have indicated, the resulting derived cooling times are increased and the available time constraints for a low value of H_0 are increased.

3. DYNAMICS OF THE CENTRAL CORE

Though the central velocity structures are quite complex in detail, the inflowing gas generally forms a rough disk with apparent shear at the cluster center, and spectroscopy of the central emission can be used to probe the dynamics of the central core. Applying this technique to A1795, assuming a spherical potential for material in a rotating disk, we obtain

$$M \lesssim 10^{11} h^{-1} M_\odot \text{ inside of } 4h^{-1} \, kpc$$

and a mass-to-light ratio

$$\frac{M}{L_v} \lesssim 20$$

which is not significantly different from mass-to-light ratios measured for normal galaxies.

This highlights a well-known problem in cooling flow studies: *If the mass inflow rates are of order 100 $M_\odot$/yr as inferred from the X-ray data, even if only 10% of this material arrives at the center (where the optical emission is seen) this process, if proceeding over a Hubble time, should have produced a sizable disruption of the dynamical mass in the core which is not seen.*

A variety of solutions to this dilemma have been proposed:

- mass condenses out at larger radii (*e.g.*, Fabian, *et al.* 1984)
- the mass inflow rate is reduced by conduction (*e.g.*, Tucker and Rosner 1983; Bertschinger and Meiksen 1986) or reheating (*e.g.*, Silk, *et al.* 1986)
- the large mass inflows have only recently switched on (this paper)

3.1. How Much Gas Flows Into the Center?

Let us consider the energetics of the cooling gas. We can distinguish three regimes: (1) an initial stage when energy is radiated away at constant pressure, where gas temperatures are of order $10^8 \to 10^{7}°$K, (2) a stage where gas cools rapidly at constant density, with corresponding temperatures in the range of $10^7 \to$ less than $10^{4}°$K, and (3) a stage where gas is repressurized by shocks driven by the surrounding hot gas. It is in this final phase that substantial optical and UV emission is produced, whether by direct shocking or by photoionization.

Assuming for the moment that photoionization may be neglected and only dynamical energy is available, an absolute upper bound to the combined optical and UV luminosity is given by the product of the available energy per particle multiplied by the mass inflow rate

$$L_{opt+uv} \leq \frac{3}{2} k \frac{\dot{M}}{m} T_{ISO} \tag{1}$$

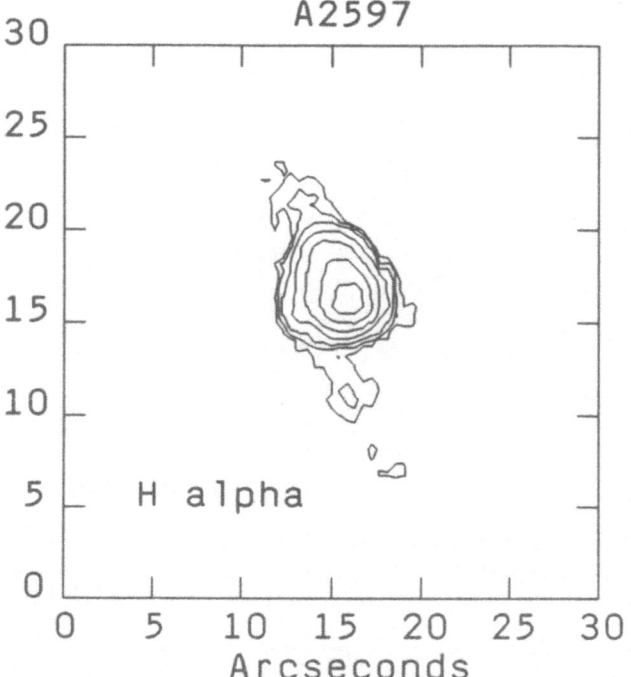

Figure 3. Extended optical emission in the central galaxy of A2597. This subtracted image was created by forming the difference between a 25^m *in* narrow-band exposure centered on Hα + [N II] and an adjacent continuum band exposure. Data taken at the 3.6 m Canada-France-Hawaii Telescope on Mauna Kea under $0\rlap{.}''7$ seeing using the IfA Galileo CCD Camera (Hlivak, *et al.* 1982; 1983). North is up and east is to the left.

where T_{ISO} is the temperature for the onset of constant density cooling

$$T_{ISO} = 1.3 \times 10^7 \left\{ \frac{R}{100~kpc} \right\}^{0.32} \left\{ \frac{nT}{5 \times 10^{5\circ}K~cm^{-3}} \right\}^{0.32} {}^\circ\mathrm{K} \qquad (2)$$

or

$$L_{opt+uv} \leq 1.7 \times 10^{43} \mathrm{ergs~s}^{-1} \left\{ \frac{\dot{M}}{100\,M_\odot\,yr^{-1}} \right\} \qquad (3)$$

If we compare this maximum with the *measured* optical and UV luminosities for the central regions of A1795 and A2597 (cf. Figure 3), we find

$$L_{opt+uv}(A2597) \sim 10^{43}~h^2\mathrm{ergs/s} \Longrightarrow \dot{M} \gg 60M_\odot~yr^{-1} \qquad (4a)$$

$$L_{opt+uv}(A1795) \sim 4 \times 10^{42}~h^2\mathrm{ergs/s} \Longrightarrow \dot{M} \gg 30M_\odot~yr^{-1} \qquad (4b)$$

for lower bounds on the mass inflow *into the central few kiloparsecs*. For A1795 considerably more than 30% of the mass flow ends up in the middle of the galaxy.

Photoionization by a population of young stars has been suggested (Johnstone *et al.* 1987) as an additional process contributing to the optical emission line flux.

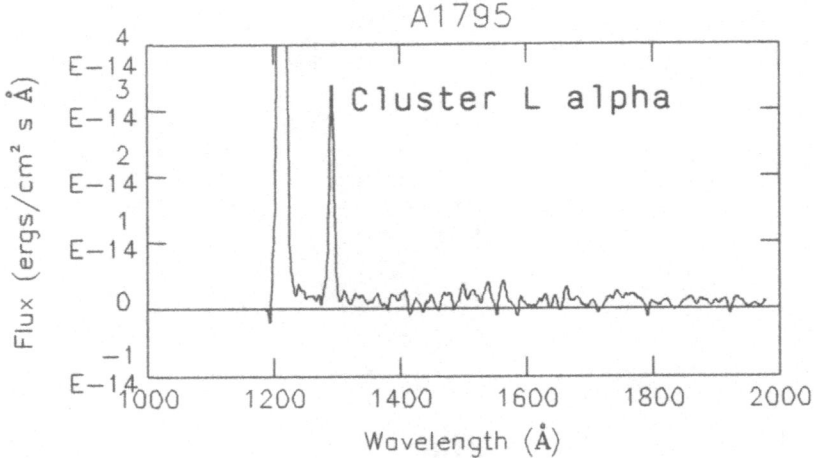

Figure 4. Flux-calibrated spectrum of A1795 from an 820-minute exposure taken with SWP camera aboard *IUE*. Spectrum has been corrected for contamination by residual geocoronal Lα in the region shortwards of the redshifted cluster Lα feature.

The most direct way of testing the validity of this hypothesis is with far UV observations. In Figure 4 we show a flux-calibrated spectrum of A1795 taken with the *IUE* satellite. The integrated Lα flux is 3.6×10^{-13} ergs cm^{-2} s^{-1}, and the measured continuum level near the line is 1.4×10^{-15} ergs cm^{-2} s^{-1} Å^{-1} for an equivalent width $W_\lambda \gtrsim 250$ Å. This is well in excess of the predicted Lα equivalent width from a young stellar population of ~ 90 Å, calculated from Meier (1976) in the limit where *every* Lyman continuum photon produces a Lα photon. (In fact the systems appear density bounded rather than ionization bounded, so this is an upper limit.) Thus photoionization contributions cannot substantially change the energy arguments given above.

We can also translate the UV continuum value into an equivalent young star population if the continuum is interpreted as arising from stars with a specified IMF. (Alternatively, it could originate from a nuclear source.) *This result is highly dependent on the form of the assumed IMF and cutoff masses.* However, if we assume a Salpeter IMF stretching from $0.02\ M_\odot \rightarrow 50\ M_\odot$ then $\sim 5\%$ of the mass is contained in stars 10 $M_\odot$ and larger. The following relation (Cowie 1987; similarly Baron and White 1987) between spectral energy distribution and star formation rate

$$F_\nu = 2.5 \times 10^{29} \text{ergs s}^{-1}\ \text{Hz}^{-1}\ \left(\frac{\dot{M}}{100 M_\odot\ \text{yr}^{-1}}\right) \tag{5}$$

applied to the UV continuum of A1795 translates into $0.2\ h^{-2}\ M_\odot\ \text{yr}^{-1}$ in massive ($> 10\ M_\odot$) stars, scaling to a total star formation rate of $\sim 5\ h^{-2}\ M_\odot\ \text{yr}^{-1}$. The inferred star formation rate is in agreement with the range of values determined by O'Connell (1987) and Romanishin (1987) from photometric observations sensitive formation rates in disk galaxies, Kennicutt (1983) has pointed out the fragility of estimates of total star formation rates, since the bulk of the mass is at the low mass end of the spectrum whereas the estimators of young populations all come from properties of the high mass end. We therefore offer these star formation rates as extremely approximate, model-dependent quantities.

The above estimates assumed no reddening corrections (see, however, discussion in the following section for evidence of low level extinction). It should be noted that the massive star formation rates inferred from the UV continuum measurements and optical photometry of the cluster cores provide a dynamical dilemma analogous to the one posed by the inferred X-ray mass inflow rates, if the measured rates are extrapolated back over a full Hubble time. Regardless of the detailed form of the IMF, most of the mass must be in low mass stars. And if an IMF preferentially weighted towards low mass stars (e.g., Sarazin and O'Connell 1984) is adopted, the dynamical mass of stars formed in the central regions over a Hubble time becomes a severe problem.

3.2. The Case for Recent Onset of Substantial Mass Inflow

The measured optical luminosities applied to the above arguments indicate that substantial present mass inflow rates are occuring within the central few kiloparsecs of the cluster core. To reconcile this with the absence of large dynamical effects at the core, we conclude that sizable mass inflows into the central (galaxy-scale) regions must be a relatively recent phenomenon. A number of other considerations support this conclusion.

First, this behavior is consistent both with the available times argument given in section 2, and with our knowledge of the cooling function. As is well known, thermal bremsstrahlung cooling is relatively inefficient and appreciable cooling does not take place until stage (2) in the above description (i.e., most of the time cooling is spent at temperatures above $10^7 °$K). Consequently, the sizable mass accretions onto the central galaxy may not have begun until recently. The consideration of available times for cooling showed that these times were very closely approached by the actual measured cooling times in systems with optical emission. Furthermore, this would explain the case of A2029, where the cooling time falls below the critical value, but no optical emission is seen, as an instance where we are looking just before the time of 'turn on' for the optical emission.

This is also consistent with the failure to observe substantial amounts of extended emission far away from the core (as one might expect for mass condensations at large radii). Typical imaging exposures cover fields of view of $\sim 5'$ and fail to show such emission structures at sensitivity levels more than an order of magnitude below the observed filament surface brightnesses.

Finally, Silk et al. (1986) have pointed out that the predominently A-type population noted in Perseus would be consistent with a *recent* accretion flow (e.g. persisting only a few billion years). They further consider the possibility that dynamical evolution in clusters up to relatively recent times could also be responsible for a relatively recent onset of flows.

4. ABUNDANCES AND DUST IN CLUSTERS

It is generally believed that the gaseous atmosphere is formed by release of material from the cluster galaxies combined with some primordial gas. From X-ray spectrometry (Mushotzky 1984) abundances in the halo gas are generally about half solar, and observations (Hu, Cowie, and Wang 1985; hereafter, HCW) of the optically emitting filaments also indicate roughly solar abundances. A natural presumption is that a substantial fraction of dust was released with this material. However, the current dust content is clearly nowhere near standard values for the interstellar medium. The reason for this deficiency is of course the destruction of the dust by thermal sputtering in the hot gas, with only the most recently injected material still surviving (Yahil and Ostriker 1973, Silk and Burke 1974). In this section we discuss the UV and IR evidence for the presence of dust in clusters, and consider the theoretical basis for interpreting these observations.

4.1. Ultraviolet Spectra of Cluster Emission Systems

The large lever arm for extinction by dust at ultraviolet wavelengths suggests that a sensitive method for detecting foreground dust is to compare the Lα flux from the emission line systems with Balmer line fluxes at optical wavelengths. Such flux determinations can be made with the *IUE* satellite (cf., Nørgaard-Nielsen *et al.* 1984, Fabian *et al.* 1984, and Bertola *et al.* 1986) and, when combined with accurately fluxed optical data over a comparable aperture (e.g., narrow-band imaging by CHJY) can provide an estimate of the cluster dust content from comparison with the theoretical ratios for the photoionization or shock models which accurately characterize the optical spectra.

The results for nine clusters are summarized in Table 1 (from Hu, Cowie, and Blades 1987), and demonstrate a small but systematic component of extinction above foreground reddening associate with the cluster. The procedure for each cluster (col. 1) was to measure the Lα flux in the *IUE* aperture (col. 2) from the line-by-line spectra, construct the comparable Hα flux over a similarly positioned aperture (col. 3) from flux-calibrated optical imaging data, and compute the Lα:Hα flux ratio (col. 4). The difference between this value and the theoretical ratio (~ 16, with conservative limits taken from 8.3 $\rightarrow$ 27) predicted either by shock (HCW) or photoionization (Ferland and Osterbrock 1986) models capable of accurately describing the optical spectrum was then interpreted using a standard reddening curve (Seaton 1979) to infer a measured E_{B-V} (col. 5). The foreground extinction component (col. 6) was estimated for these generally high galactic latitude sources using the neutral hydrogen data (Burstein and Heiles 1978), and the net cluster component of extinction was calculated (col. 7). The reader is referred to Hu, Cowie, and Blades (1987) for a more detailed description of this procedure and for a log of the observations and archival data used. It should be noted that tabulated fluxes are aperture measurements and not total fluxes.

The average extinction obtained, $\langle E_{B-V} \rangle \sim 0.15$, appears significant. It is in reasonable agreement with the estimates of Zwicky, and of Karachentsev and Lipovetskii (1969) who used the more uncertain method of number counts of background galaxies to estimate cluster extinction. The principal uncertainty in the

TABLE 1. UV/OPTICAL DATA ON CLUSTER EXTINCTION

	$L\alpha$ Flux (ergs cm^{-2} s^{-1})	$H\alpha$ Flux (ergs cm^{-2} s^{-1})	$\frac{F(L\alpha)}{F(H\alpha)}$	Meas'd E_{B-V}	Foreground $E_{B-V}{}^*$	Cluster E_{B-V}
A85	3.0×10^{-14}	2.3×10^{-15}	13.0	$0.03\{{}^{0.11}_{-0.07}$	0.02	$0.01\{{}^{0.09}_{-0.09}$
A426	$\gtrsim 6.6 \times 10^{-13}$	4.4×10^{-13}	$\lesssim 1.5$	$0.33\{{}^{0.42}_{0.24}$	0.19	$0.13\{{}^{0.22}_{0.05}$
A496	7.2×10^{-14}	2.6×10^{-14}	2.8	$0.25\{{}^{0.32}_{0.16}$	0.04	$0.21\{{}^{0.28}_{0.12}$
A1126	$<2.0 \times 10^{-14}$	6.0×10^{-15}	$\lesssim 3.3$	$\gtrsim 0.23\{{}^{0.30}_{0.13}$	0.03	$\gtrsim 0.20\{{}^{0.27}_{0.10}$
A1795	3.6×10^{-13}	5.9×10^{-14}	6.1	$0.14\{{}^{0.22}_{0.04}$	-0.01	$0.15\{{}^{0.23}_{0.03}$
A1991	$:1.7 \times 10^{-14}$	9.2×10^{-15}	$:1.8$	$\gtrsim 0.31\{{}^{0.39}_{0.22}$	0.02	$\gtrsim 0.29\{{}^{0.37}_{0.20}$
A2052	4.9×10^{-14}	8.6×10^{-15}	5.7	$0.15\{{}^{0.23}_{0.05}$	0.03	$0.12\{{}^{0.20}_{0.02}$
A2199	4.4×10^{-14}	1.6×10^{-14}	2.8	$0.25\{{}^{0.32}_{0.16}$	0.03	$0.22\{{}^{0.29}_{0.13}$
A2597	4.0×10^{-13}	5.5×10^{-14}	7.3	$0.11\{{}^{0.19}_{0.02}$	0.02	$0.09\{{}^{0.17}_{0.00}$

*Burstein and Heiles (1978)

$\langle E_{B-V} \rangle = 0.16 \left\{{}^{0.23}_{0.07}\right. $ mag

method applied here lies in the assumption of an intrinsic $L\alpha/H\alpha$ ratio, with the possibility of introducing systematic errors. By comparison, the aperture flux estimates which are generally good to 50%, and often better than 30%, except for the very weakest $L\alpha$ flux measurements, introduce small errors for the large reddening effects being measured. The bracketed subscripts and superscripts give the range from extremal values of $L\alpha/H\alpha$ corresponding to the limits adopted in the above analysis. It may be noted that even the extremal lower limits support the presence of a cluster extinction component of order $E_{B-V} \sim 0.1$ mag. (One caveat here is that the dust may be associated with the cool gas if grains reform in the cooling flow.)

If the measured extinction is cluster wide it can be used to estimate the characteristic timescales for the injection of dust and gas into the cluster atmospheres. In the central regions of the cluster only recently injected grains have not been sputtered since the sputtering time for destruction is (Draine and Salpeter 1979):

$$\tau_{sp} = 2 \times 10^8 n_{-3}^{-1} \text{ yrs} \tag{6}$$

for 0.1μm grains, where n_{-3} is the gas density in units of 10^{-3} cm^{-3}. Now let the rate of current gas injection into the cluster atmosphere be

$$\dot{\rho}_{gas} = \rho_{gas}/\theta \tag{7}$$

where θ is a characteristic timescale. If the gas has a grain abundance of half the interstellar value, and a similar dust-to-gas ratio by mass, we can obtain ρ_{grain} by integrating the above equation over a sputtering time as

$$\rho_{grain} = 8 \times 10^{-33} \left\{ \frac{10^{11} \text{ yr}}{\theta} \right\} \text{ g cm}^{-3} \tag{8}$$

Normalization for the timescale is chosen to match the typical mass loss timescales of a few times 10^{11} yrs from stellar evolution studies. Calculating the column density out to a radius of ~ 6 Mpc (typical for $\tau_{sp} \approx t_{Hubble}$), and again assuming grain properties similar to the interstellar case, we obtain the following relation between extinction and characteristic time:

$$E_{B-V} = 0.10 \left\{ \frac{1.5 \times 10^{10} \text{ yr}}{\theta} \right\} \tag{9}$$

Scaling from the measured extinctions given in Table 1 would suggest characteristic timescales of a few times 10^{10} yrs. This relatively low value of θ argues that much of the gaseous atmosphere is recently formed, in contrast to a scenario where most of the gaseous enrichment is done in a large initial burst (e.g., following stellar evolution models) followed by a long tail-off to a long present mass loss time scale of order 10^{11} yrs.

4.2. Infrared Emission from Dust in Clusters

If the dust seen above is indeed embedded in the hot cluster atmospheres it should produce a substantial amount of IR emission as it is heated by the hot gas. An initial estimate of associated IR flux at the level of a few Janskys was given in HCW, and suggested that although these sources were marginally too faint to be detected in scanning mode on the *IRAS* satellite, they might be detectable in the pointed observations.

Nine clusters were identified with pointed observations in the *IRAS* database, with redshifts $0.032 < z < 0.065$. A major difficulty in working with the 100μm data, which is near the expected peak in the dust emission, is the presence of infrared cirrus as a contaminant. The following methodology was adopted to deal with the problem of cirrus: for each cluster a pattern of five $5'$ circles (size chosen to approximate a cluster core radius at the specified redshift range) was used to define the source (beam centered on cluster) and four background fields (beams tangent to source beam). Each data point specifying cluster flux was given by the difference between the source and average background field. The sampled point was considered 'clean' of cirrus only when the average background was less than 2 Jy, and all background fields had measured fluxes of less than 3 Jy.

By these criteria, only four clusters were classifiable as free of possible cirrus contamination. The histogram of their flux distribution, and a similar distribution from a comparable empty field sample, are shown in Figure 5. The mean flux from the cluster sample is 0.36 Jy, while the mean flux of the empty field sample is -0.03 Jy, with $\sigma_{empty\,field}=0.46$ Jy. Thus the excess of the mean cluster flux is very marginal but the level of detected flux is consistent with the predicted values, corresponding to a flux from a rich cluster at $z \approx 0.02$ of approximately 2 Jy at 100μm.

A direct comparison of this result with the Perseus cluster, which was not included in the above sample by the redshift criterion, shows that this famous case is much brighter than the above prediction. A plot of the radial flux distribution

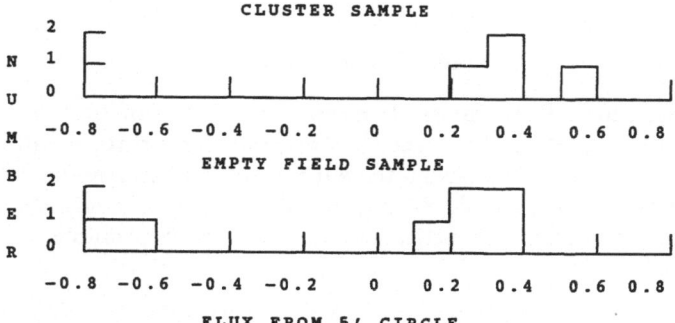

Figure 5. Distribution of 100μm fluxes within 5' circle from *IRAS* pointed mode observations for cluster and empty field samples. Data points represent excess flux of source above average background, with the angular size of the circle approximating a cluster core radius for the redshift range considered ($0.032 < z < 0.065$). The average flux for the cluster sample deemed free of cirrus is 0.36 Jy, while the average flux of the empty field sample is -0.03 Jy. By contrast, the average flux for the complete cluster sample (no selection against cirrus) is 0.49 Jy.

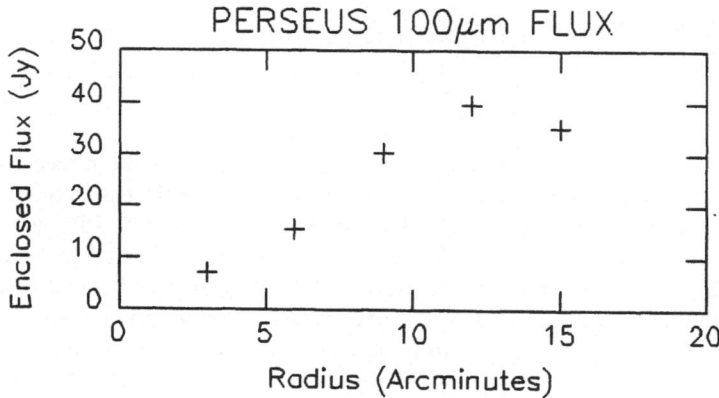

Figure 6. Radial distribution of 100μm flux emission from the Perseus cluster. Points give the flux (in Jy) within the enclosed circle, with a total flux $\approx$ 35 Jy. No correction has been applied here for an extended background component.

for Perseus is shown in Figure 6. Part of the high flux ($\approx$ 35 Jy) may be associated with the foreground impacting galaxy.

Finally, let us consider the theory of IR emission and, as with the UV measurements, use the data to place constraints on the timescales for the recent injection of material into the cluster atmospheres. Grains are heated by thermal deposition

(Draine 1981) at a rate

$$\dot{\epsilon} = 9.5 \times 10^{-12} \, n_{-3} \, T_8^{3/2} \text{ ergs s}^{-1} \tag{10}$$

for 0.1 μm grains, where T_8 is the gas temperature in units of $10^{8\,\circ}$ K and n_{-3} is the hot-gas density in units of 10^{-3} cm^{-3}. The grain temperature will lie around $T = 18^{\circ}$ K $(n_{-3} \, T_8^{3/2})^{1/5}$ (for graphite, with a similar expression for silicates) so that emission will peak around 100–200 μm depending on composition. Again scaling from interstellar grain properties, and integrating over volume, we obtain an IR luminosity

$$L_{\text{IR}} = 3 \times 10^{44} \left\{ \frac{M_{gas}}{10^{13} \, M_\odot} \right\} \left\{ \frac{1.5 \times 10^{10} \text{ yrs}}{\theta} \right\} \text{ ergs s}^{-1} \tag{11}$$

from a volume containing a mass of gas M_{gas} with $T_8 \sim 1$.

Within a $10'$ beam

$$M_{gas} \sim 10^{13} \, h_{50}^{-5/2} \, M_\odot \tag{12}$$

at $z = 0.02$ – roughly the mass of gas in a cluster core, with an assumed Hubble constant of 50 km s^{-1} Mpc^{-1}. The associated flux is

$$F_{\text{IR}} = 1.8 \times 10^{-10} \left\{ \frac{1.5 \times 10^{10} \text{ yrs}}{\theta} \right\} h_{50}^{-1/2} \text{ ergs cm}^{-2} \text{ s}^{-1} \tag{13}$$

for a net source strength

$$\approx 6 \text{Jy} \left\{ \frac{1.5 \times 10^{10} \text{ yrs}}{\theta} \right\} h_{50}^{-1/2} \tag{14}$$

Comparing the characteristic timescales obtained from the measured IR emission with the values from the UV extinction measurements gives reasonable consistency: the mean of the four distant clusters yields $\theta \approx 4 \times 10^{10}$ yrs, while the flux value from Perseus gives $\theta \approx 3 \times 10^9$ yrs, as compared with the range in θ of $6 \times 10^9 \rightarrow 2 \times 10^{10}$ yrs from the extinction data. Again the typical characteristic timescale value of $\theta \approx 2 \times 10^{10}$ yrs is short compared with current stellar mass loss timescales of a few times 10^{11} yrs from stellar evolution calculations.

5. SUMMARY AND CONCLUSIONS

The correlation of optical emission systems with cluster cooling times offers strong evidence that largescale cooling of cluster atmospheres is occuring, and that the emission systems seen at the cluster cores represent such gas which has cooled to temperatures at which optical emission can be seen.

When the measured cooling times are compared with the available cosmological times the data argue for a value of $H_\circ$ near 50 and $q_\circ \ll \frac{1}{2}$.

Spectroscopic data on the emission systems indicate that mass-to-light ratios $M/L_v \stackrel{<}{\sim} 20$ in the central regions, so that cooling flows have not had strong dynamical consequences in the cluster core despite currently measured large mass inflow rates of $\sim 100 M_\odot$ yr^{-1} derived from the X-ray data. We use measured optical luminosities to argue that most of the large accretion inflows are in fact reaching the central cores, and conclude that the sizable accretion flows into the central regions are a recent phenomenon.

Ultraviolet data from the *IUE* satellite in combination with fluxed optical data on nine cluster emission systems are used to demonstrate the presence of extinction associated with clusters. A mean $\langle E_{B-V} \rangle \sim 0.15$ above foreground reddening is obtained. Scaling from the properties of interstellar grains, this result is interpreted in terms of a characteristic timescale roughly an order of magnitude below current stellar mass loss timescales, thus pointing to relatively recent injection of much of the cluster atmosphere if the dust has a cluster wide distribution.

Pointed observations at 100μm from the *IRAS* satellite were examined to search for infrared emission from dust in the hot cluster. Mean cluster fluxes of around 0.4 Jy were found, consistent with the marginally significant detections expected from this data, after correction for cirrus contamination. By comparison, the flux from the Perseus cluster was markedly larger. Interpretation of fluxes in terms of characteristic timescales was in agreement with the extinction measurements. A brief discussion of the theory of dust in clusters is also given.

We would like to acknowledge support for this research through NSF grant AST-8607375 and NASA grant NAGW-959. CCD research at the Institute for Astronomy is partially supported by NSF grant AST-8615631. Satellite data reduction was supported under the *IRAS* extended mission program by JPL contract 957695, and by NASA grant NAG 5-829 covering *IUE* observations. Access to *IUE* archival data was provided through the facilities of the National Space Science Data Center.

REFERENCES

Baron, E. and White, S. D. M. 1987, *preprint.*

Bertola, F., Gregg, M. D., Gunn, J. E., and Oemler, Jr., A. 1986, *Ap. J.,* **303**, 624.

Bertschinger, E. and Meiksen, A. 1986, *Ap. J. (Letters),* **306**, L1.

Burstein, D., and Heiles, C. 1978, *Ap. J.,* **225**, 40.

Cowie, L. L. 1987, *in preparation.*

Cowie, L. L., Henriksen, M., and Mushotzky, R. F. 1987, *Ap. J.,* **317**, 593.

Cowie, L. L., Hu, E. M., Jenkins, E. B., and York, D. G. 1983, *Ap. J.,* **272**, 29 (CHJY).

Draine, B. T. 1981, *Ap. J.,* **245**, 880.

Draine, B. T., and Salpeter, E. E. 1979 *Ap. J.,* **231**, 77.

Fabian, A. C., Nulsen, P. E. J., and Arnaud, K. A., 1984, *M.N.R.A.S.,* **208**, 179.

Fabian, A. C., Nulsen, P. E. J., and Canizares, C. R., 1984, *Nature* **310**, 733.

Ferland, G. J., and Osterbrock, D. E. 1986, *Ap. J.,* **300**, 658.

Hlivak, R. J., Pilcher, C. B., Howell, R. R., Colucci, A. J., and Henry, J. P. 1982, *in Instrumentation in Astronomy IV,* ed. D. L. Crawford, (*SPIE Proc.,* **331**, 96).

86

Hlivak, R. J., Henry, J. P., and Pilcher, C. B. 1983, *in Instrumentation in Astronomy V*, ed. A. Boksenberg and D. L. Crawford, (*SPIE Proc.*, **445**, 122).

Hu, E. M., Cowie, L. L., and Blades, J. C. 1987, *submitted to Ap. J.*

Hu, E. M., Cowie, L. L., and Wang, Z. 1985, *Ap. J. Suppl.*, **59**, 447 (HCW).

Johnstone, R. M., Fabian, A. C., and Nulsen, P. E. J. 1987, *M.N.R.A.S.*, **224**, 75.

Karachentsev, I. D., and Lipovetskii, V. A. 1969, *Soviet Astr.*, **12**, 909.

Kennicutt, R. C. 1983, *Ap. J.*, **272**, 54.

Mushotzky, R. F. 1984, *Phys. Scripta*, **T7**, 157.

Nørgaard-Nielsen, H. U., Jørgensen, H. E., and Hansen, L. 1984, *Astr. Ap.*, **135**, L3.

O'Connell, R. W. 1987, *IAU Symposium No. 127: Structure and Dynamics of Elliptical Galaxies.*

Romanishin, W. 1987, *preprint.*

Sarazin, C. L. and O'Connell, R. W. 1983 *Ap. J.*, **268**, 552.

Seaton, M. J. 1979, *M.N.R.A.S.*, **187**, 73P.

Silk, J., and Burke, J. R. 1974, *Ap. J.*, **190**, 11.

Silk, J., Djorgovski, S., Wyse, R. F. G., and Bruzual, A., G. 1986, *Ap. J.*, **307**, 415.

Tucker, W. H., and Rosner, R. 1983, *Ap. J.*, **267**, 547.

Yahil, A., and Ostriker, J. P. 1973, *Ap. J.*, **185**, 787.

DYNAMICS OF COSMIC RAYS IN COOLING FLOWS

HANS BÖHRINGER AND GREGOR E. MORFILL
Max-Planck-Institut für Physik und Astrophysik
Institut für Extraterrestrische Physik
Garching bei München, W. Germany

ABSTRACT. Galaxies presumably are sources of cosmic rays due to supernovae or active nuclei. In the centre of clusters of galaxies with high gas densities, where cosmic rays cannot easily escape, the cosmic rays may have an influence on the dynamics of the cluster gas. We have investigated the cosmic ray - gas interaction by means of steady models with spherical symmetry. It was found that the cosmic ray pressure may become comparable to the gas pressure in the halo region around a central galaxy with a cooling flow. Rayleigh Taylor instabilities might develop there and set the scale for inhomogeneities leading to filaments or star formation regions.

1. INTRODUCTION

Like the Milky Way galaxy, where cosmic rays are produced at an energy rate of $\sim 3 \cdot 10^{40}$ erg s^{-1}, galaxies in clusters should be sources of cosmic rays due to supernovae and active galactic nuclei. Even though most of the galaxies in the centre of clusters are ellipticals, which show little sign of formation of massive stars and supernovae at present, star formation and supernova activity must have been much larger at earlier times as can be deduced form the large amount of iron observed in the intracluster gas (Mushotzky, 1984). In addition the giant central galaxies in clusters often have active nuclei with a high radio synchrotron luminosity indicating large amounts of relativistic electrons, like for example M87 in the Virgo cluster and NGC 1275 in the Perseus cluster (c.f. Pedlar *et al.*, 1983). Furthermore the lifetime of the nuclear component of the cosmic rays (assuming the major part are protons with an energy of a few GeV) is comparable to the lifetime of the clusters where the major loss rate is due to collisions with the background gas. This implies, that different from our galaxy, "historic" cosmic rays from the early formation times may still play a role in galactic clusters. In the centre of compact clusters cosmic rays cannot easily escape because of the high ambient gas densities; in clusters with cooling flows cosmic rays are even partly convected inwards with

87

A. C. Fabian (ed.), Cooling Flows in Clusters and Galaxies, 87–91.

the gas flow. Thus cosmic ray pressures pile up that might become comparable to the thermal gas pressure. In this case the influence of cosmic rays on the dynamics of the gas can no longer be neglected.

To gain some first understanding of the dynamics of cosmic rays and thermal gas we constructed simple steady, spherically symmetric models which approximately apply to the case of M87 and to the Perseus cluster (Böhringer and Morfill, 1988). The results show that Rayleigh Taylor instabilities may develop, which has important effects on the cooling flow gas.

2. THE MODEL

The description of cosmic ray propagation in the intracluster medium applied here is similar to that used for the cosmic rays in the halo of our galaxy (e.g. Wentzel, 1974; Jones, 1979). Cosmic rays propagate along the magnetic field lines where they are scattered by field irregularities (hydromagnetic waves) which themselves travel outwards with Alfvén velocity. We further assume that enough field lines are connecting the galaxies and the intergalactic medium such that on a large scale the cosmic rays can be viewed as propagating radially outward from sources in the core of the cluster of galaxies. Hydromagnetic waves are generated by cosmic ray streaming instabilities and the waves are damped by the thermal gas. The cosmic rays are thus convected with the waves and in addition are spread by diffusion.

The dynamics of the thermal gas, the cosmic rays, and the hydromagnetic waves can be approximately described by a three-fluid-model with spherical symmetry. As a first approach we studied this system by means of a steady model involving the following hydrodynamic equations.

$$\frac{1}{r^2}\frac{d}{dr}\left(r^2\rho u\right) = Q_S - Q_L \tag{1}$$

$$u\frac{du}{dr} + \frac{1}{\rho}\frac{d}{dr}\left(P_G + P_C + P_W + P_B\right) = -\frac{d}{dr}\Phi - \frac{Q_S u}{\rho} \tag{2}$$

$$\frac{1}{r^2}\frac{d}{dr}\left\{r^2 u\left(E_G + P_G\right)\right\} - u\frac{dP_G}{dr} = -L_R + \gamma_C n_H E_C$$
$$+\gamma_W E_W + \left(q_S - q_L + \frac{Q_S u^2}{2}\right) \tag{3}$$

$$\frac{1}{r^2}\frac{d}{dr}\left\{r^2\left(u+v\right)\left(E_C + P_C\right) - r^2\kappa_C\frac{dE_C}{dr}\right\} - \left(u+v\right)\frac{dP_C}{dr}$$
$$= -\gamma_C n_H E_C + Q_{CR} \tag{4}$$

$$\frac{1}{r^2}\frac{d}{dr}\left\{r^2\left(\left(u+v\right)E_W - uP_W\right)\right\} - u\frac{dP_W}{dr} = -v\frac{dP_C}{dr} - \gamma_W E_W \tag{5}$$

The equations are the continuum and momentum equations for the gas, the equation for the internal energy of the gas, and the energy equations for the cosmic rays and the hydromagnetic waves, respectively. r, ρ, n_H, u, and v are the radius, the gas density, the number density of hydrogen atoms, the flow velocity of the gas, and the propagation velocity of the hydromagnetic waves. P_G, P_C, P_W, and P_B are the pressure of the thermal gas, the cosmic rays, the waves, and the magnetic field and E_G, E_C, and E_W are the corresponding energy densities. Q_S (Q_L) and q_S (q_L) are source (loss) functions for the mass and the internal energy of the gas. Φ is the function for the gravitational potential. L_R, γ_C, and γ_W are the energy loss due to radiative cooling, the rate coefficient for collisional energy losses of the cosmic rays, and the damping rate of the waves, respectively. For L_R we used an approximation to the cooling function of Gaetz and Salpeter (1983). κ_C is the diffusion coefficient for the cosmic rays and Q_{CR} a cosmic ray source function. Heat conduction has been neglected here. (Effects of heat conduction have been discussed recently by Rosner et al. (1987) and Bertschinger and Meiksin (1986)).

For the following investigation we make some further simplifications. The source and sink terms for the thermal gas are set to zero which is equivalent to a constant mass flow rate, $\dot{M} = 4\pi r^2 \rho u$. This does not correspond to a realistic cooling flow where $\dot{M}$ is found to decrease with decreasing radius indicating condensation of hot gas over a larger radial zone (Fabian et al., 1981: Fabian et al., 1984). For this first approach to study the cosmic ray - gas interaction the radial dependence of $\dot{M}$ is of secondary importance, however. The pressure of the magnetic field and the hydromagnetic waves is neglected which is justified for the expected magnetic field strength of the order of $1\mu G$ (Sarazin, 1986). The vdP -work done by the cosmic rays generating hydromagnetic waves is directly converted to internal energy of the gas. The system can then be reduced to a two-fliud model with the role of the hydromagnetic waves implicitly involved. Energy density and pressure for the remaining two fluids are related by the following equations of state.

$$E_G = 3/2 P_G \; ; \; E_C = 3P_C \tag{6}$$

For the model applicable to the halo of M87 the inte:gration of the equations was started at an inner radius of $R_o = 5 \text{kpc}$ with $n_H(R_o) = 3 - 5 \cdot 10^{-2} \text{cm}^{-3}$, $T(R_o)$ $= 1.5 \cdot 10^7 K$ and a mass flow rate of $\dot{M} = 10 M_\odot \text{year}^{-1}$. For Φ we use a King function with a core radius $a = 20 \text{kpc}$ and a central mass $M_c = 4\pi \rho_o a^3 = 7.5 \cdot 10^{12} M_\odot$ (Fabricant and Gorenstein, 1983; Stewart et al., 1984). Fig. 1 shows a typical result with the following parameters: $\kappa_C = 10^{28} \text{cm}^2 \text{ s}^{-1}$, $v = 3 \text{km s}^{-1}$, and a source flux of cosmic ray energy $F_C = 4.5 \cdot 10^{41} \text{erg s}^{-1}$. For comparison the same model without cosmic rays is also shown in dashed lines. The results show a sharp decrease of the cosmic ray pressure with increasing radius and a corresponding inversion layer in the gas density profile at small radii. Such an inversion layer occurs in the steady models for cosmic ray source fluxes of at least a few $10^{41} \text{erg s}^{-1}$ and diffusion coefficients, κ_C, not much larger than $10^{28} \text{cm}^2 \text{ s}^{-1}$.

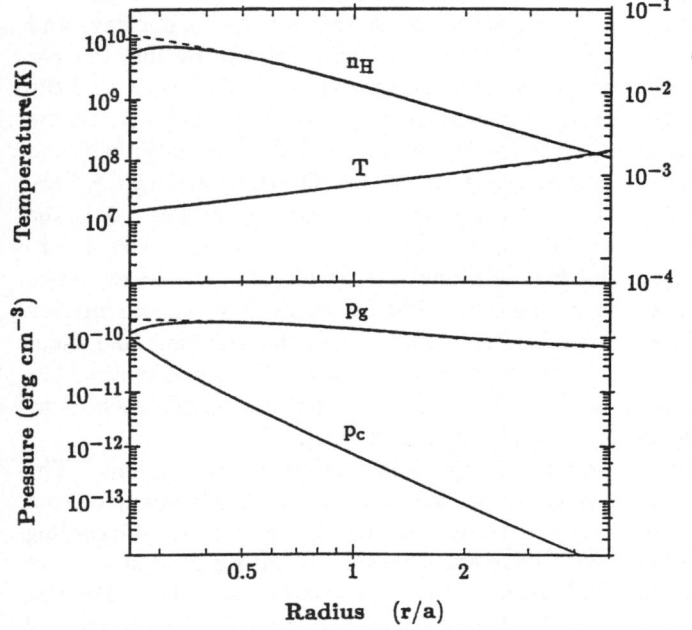

Fig. 1. Calculated radial profiles for temperature, T, gas density, n_H, cosmic ray and gas pressure ,P_c, P_g, for the model applicable to halo of M87. The radius is given in units of the core radius a = 20 kpc. The density profile shows an inversion at small radii.

Similar models were calculated for Perseus cluster parameters: $a = 250$kpc, $M_c = 3.5 \cdot 10^{14} M_\odot$, $R_o = 25$kpc, $n_H(R_o) = 2 - 3 \cdot 10^{-2} \text{cm}^{-3}$, $T(R_o) = 2 \cdot 10^7$K, and $\dot{M} = 300 M_\odot \text{year}^{-1}$. For energy source fluxes in cosmic rays larger than a few 10^{42}erg s^{-1} and diffusion coefficients not lager than used above we again obtain an inversion layer in the density profile.

3. RAYLEIGH-TAYLOR INSTABILITIES

The steady models showed that for larger fluxes of cosmic rays from a central source, an inversion layer in the gas density profile can occur. Such a configuration is very likely to be Rayleigh Taylor unstable. The growth time of Rayleigh Taylor instabilities in the approximation of the inversion layer as a discrete boundary is given by

$$\tau_{RT} = \left\{ \frac{\rho_2 + \rho_1}{\rho_2 - \rho_1} \frac{\lambda}{2\pi g} \right\}^{1/2} \tag{7}$$

where index 2 refers to the denser fluid, g is the gravitational acceleration, and λ is the length scale of the instability. Cosmic ray diffusion will damp the fluctuations unless the growth of the instabilities is faster than diffusion of particles through a distance λ, i.e.:

$$\tau_{RT} < \tau_{diff} \sim \frac{\lambda^2}{\kappa_C} \qquad (8)$$

Application to the above model results gives minimum values for the growth time, $\tau_{RT} \geq 4 \cdot 10^6 (1.6 \cdot 10^7)$years, legth scale $\lambda \geq 0.4$ (0.7) kpc, and the masses $M_b \geq 2 \cdot 10^6 (6 \cdot 10^6)$ $M_\odot$ for unstable regions in M87 and Perseus, respectively. These instabilities should develop into blobs of material with a topologically separated magnetic field, which shields them from outside heat conduction. This is a favourable condition for the sinking gas blobs to become thermally unstable and subsequently form filaments or star forming regions.

4. IMPLICATIONS AND CONCLUSIONS

The above models show that for a cluster of galaxies with a cooling flow and a strong cosmic ray source in the centre, propagation of the cosmic rays will not always be smooth but may give rise to instabilities causing inhomogeneities in the structure of the cooling flow region, which might be an important trigger for thermal condensation. Radio observation of synchrotron radiation in many clusters indicate that cosmic ray fluxes large enough to produce these effects might well exist (see also discussion by Pedlar in this volume). The necessary fluxes are still orders of magnitude smaller than the power needed to reheat the X-ray gas. Cosmic ray and hydromagnetic wave heating may, however, be important for the cooler gas phases like the filaments.

To observe γ-rays from the cosmic ray halos with present day experiments like the COS B satellite, a flux of cosmic rays of at least one or two orders of magnitude larger than that used in the models would be needed (c.f. Strong and Bignami, 1983).

REFERENCES

Bertschinger, E. and Meiksin, A., 1986, *Astrophys.J.*, **306**, L1.
Böhringer, H. and Morfill, G.E., 1988, *in preparation*.
Fabian, A.C., Hu, E.E., Cowie, L.,L., and Grindlay, J., 1981, *Astrophys.J.*, **248**, 47.
Fabian, A.C., Nulsen, P.E.J., and Canizares, C.R., 1984, *Nature*, **310**, 733.
Fabricant, D. and Gorenstein, P., 1983, *Astrophys. J.*, **267**, 535.
Gaetz, T.J. and Salpeter, E.E., 1983, *Astrophys. J. Suppl.*, **52**, 155.
Jones, F.C., 1979, *Astrophys. J.*, **229**, 747.
Mushotzky, R.F., 1984, *Physica Scripta*, **T7**, 157.
Pedlar, A., Booler, R.V., and Davies, R.D., 1983, *Mon. Not. R. astr. Soc.*, **203**, 66.
Rosner, R., Tucker, W.H., and Nijita, J., 1987, *preprint*.
Sarazin, C.L., 1986, *Rev. Mod. Phys.*, **58**, 1.
Stewart, G.C., Canizares, C.R., Fabian, A.C., and Nulsen, P.E.J.,1984, *Astrophys. J.*, **278**, 536.
Strong, A.W. and Bignami, G.F., 1983, *Astrophys. J.*, **274**, 549.
Wentzel, D.G., 1974, *Ann. Rev. Astr. Astrophys.*, **12**, 71.

HOT AND COLD GAS IN EARLY-TYPE GALAXIES: A COMPARISON OF X-RAY, HI AND
FAR INFRARED EMISSION

G. R. Knapp
Department of Astrophysical Sciences
Princeton University
Princeton, NJ 08544, U.S.A.

ABSTRACT. The relationship between the hot and cold phases of the
interstellar medium in nearby elliptical and S0 galaxies is examined by
a comparison of X-ray, HI and $100\,\mu m$ emission. The data suggest that
there is little relationship between the presence and amount of hot and
cold gas in these galaxies. The X-ray emission is more closely related
to the stellar content of Es than are the HI or infrared emission,
suggesting that the hot gas originates in mass loss from stars while the
cold gas is the remains of the original interstellar medium plus
accretion from outside the galaxy.

I. INTRODUCTION

It has recently become possible to study the interstellar medium in
elliptical galaxies by a variety of observational techniques, allowing
access to both the hot and cold phases. Nearby bright ellipticals show
an excess of X-ray emission which likely arises from hot coronal gas in
these galaxies (Fabricant and Gorenstein 1983; Forman et al. 1979;
Biermann and Kronberg 1983; Nulsen, Stewart and Fabian 1984; Dressel
and Wilson 1985; Forman, Jones and Tucker 1985; Fabbiano 1986;
Canizares 1986; Canizares, Fabbiano and Trinchieri 1986). The mass of
gas detected in X-ray emission is typically $\sim 10^{10}\ M_{\odot}$, and this
material may well be the "missing gas", i.e. the gas lost by evolved
stars in the galaxies (Forman et al. 1985). The role of this gas both
in forming stars and in fuelling the central active regions often seen
in those galaxies has been discussed extensively (Nulsen et al. 1984;
Fabian et al. 1986; Sarazin 1986; Fabbiano et al. 1987). If the gas is
cooling and forming stars, it may at some point be in the form of
atomic or molecular hydrogen. The presence of atomic hydrogen in
several ellipticals has now reliably been demonstrated via 21 cm line
emission or absorption, but the origin of the cold gas is an open
question. This review compares observations of hot and cold gas in
nearby ellipticals in loose groups or the Virgo Cluster. The great
clusters, with their massive amounts of X-ray emitting gas, are not
discussed because there are almost no HI observations of these clusters.

A. C. Fabian (ed.), Cooling Flows in Clusters and Galaxies, 93–102.

In the next section the available information on cold gas in ellipticals and SOs is summarized. Since much of the information on HI is already in the literature, this part of the discussion is kept brief. A more extensive description is given of another manifestation of cold interstellar matter, i.e. long wavelength emission measured by the IRAS satellite which is plausibly arising from cool dust.

Canizares et al. (1987) have recently summarized and discussed the X-ray data for some eighty nearby E and SO galaxies. Using this sample, the X-ray characteristics are compared with the available information from HI and infrared in §III.

II. HI AND INFRARED EMISSION IN EARLY-TYPE GALAXIES

A. HI About 10% of well studied ellipticals and ~ 25% of well studied SOs contain detectable amounts of HI, and the available data show that in both cases the relative HI content covers a very wide range (M_{HI}/L_B ~ 10^{-3} to 1 $M_\odot/L_\odot$ for Es) (Knapp, Turner and Cunniffe 1985; Wardle and Knapp 1986), unlike the situation for spiral galaxies. A close connection has also been found between the presence of HI and that of radio continuum emission (Knapp and Wardle 1987). These distributions in global content, plus kinematic and structural misalignments in the stellar and gaseous components (e.g. NGC1052, van Gorkom et al. 1986) strongly suggest that, at least for the ellipticals, much of the HI does not originate from the stars in the galaxy.

A small number of ellipticals with strong nuclear radio point sources show HI absorption (Shostak et al. 1983; NGC1052, van Gorkom et al. 1986; NGC5128, van der Hulst et al. 1983; and NGC5363, van Gorkom and Laing 1987). As discussed by van Gorkom et al. (1986), all show HI absorption redshifted with respect to the galaxy's systemic velocity and all except NGC315 also show absorption near the galaxy's systemic velocity. Three of the galaxies are known to be X-ray sources (I know of no data for NGC5363). These galaxies do have properties which correspond to those of cooling flows, although other explanations are possible. Gunn (1979) has discussed the infall of clouds from an extended system of cold gas slowed down by frictional drag in the hot gas, while van Gorkom et al. (1986) have discussed elliptical orbits in triaxial potentials. The sample size needs to be increased before these possibilities can be further investigated.

B. Infrared A recent very promising development has been made possible by the great success of the IRAS mission (Neugebauer et al. 1984). About 10% of bright ellipticals ($m_B < 11^m$) are detected at long wavelengths by the satellite and appear in the Point Source Catalogue. The infrared colors of the galaxies strongly suggest that the emission is from dust, and also that some of the galaxies are undergoing star formation at a low level (Jura 1986; Jura et al. 1987; Tytler 1987; Thronson and Bally 1987). The IRAS data thus provide another probe of cold interstellar matter in early-type galaxies; observations of spirals show that long-wavelength emission is closely correlated with both the atomic and molecular gas in galaxies (e.g. Helou 1987).

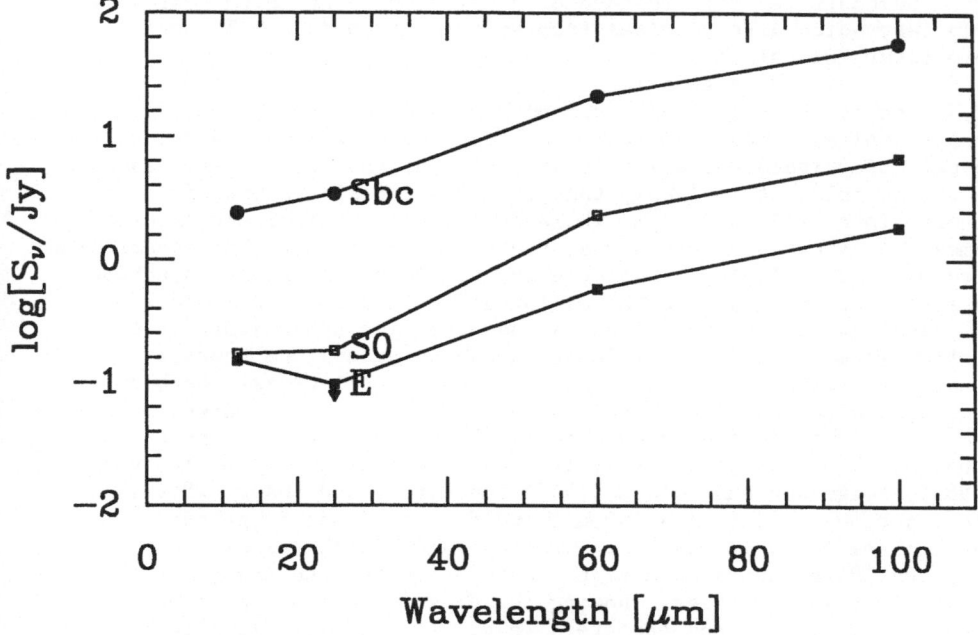

FIGURE 1. IRAS spectra for NGC4501, type Sbc (filled circles), NGC1387, type SO (open squares) and NGC4278, type E (filled squares).

Jura et al. (1987) demonstrated that averaging of the IRAS data results in about a factor of three sensitivity increase over the Point Source Catalogue, with a corresponding increase in the detection rate of the weak sources associated with early-type galaxies. A large project is underway to obtain IRAS flux densities for some thousand E and SO galaxies. Some preliminary results (Guhathakurta et al. 1987, and Kim et al. 1987) are summarized below.

The detection rates at each of the IRAS bands for the galaxies brighter than ~ 13^m are shown in Table 1. The limiting flux densities are typically ~ 0.3 Jy at $\lambda12\mu$m and ~ 0.8 Jy at $\lambda100\mu$m, though at the long wavelengths the sensitivity is sometimes compromised by Galactic cirrus emission. Examination of 'blank sky' data shows that the confusion rate is less than 2%. For comparison, data for a sample of spirals from the Virgo cluster study of Helou (1987) are listed. IRAS spectra are compared in Figure 1, which shows the data for NGC4501 (type Sbc), NGC1387 (SO) and NGC4278 (E).

These results show the following:

(1). The similarity of the spectra of all three galaxies at 60 μm and

$100\,\mu m$ demonstrates the presence of a cold component in the early-type
galaxies, which likely arises from emission from dust. The mean
temperature of this component, assuming a $Q_\nu \sim \nu$ emissivity law, is
~ 30 K for all three galaxies.
(2). The bright ellipticals are much more often detected at $\lambda 12\,\mu$ than
are the spirals. As examination of Figure 1 shows, the shapes of the
spectra of spirals and ellipticals differ at the shorter wavelengths.
In the spirals, it is likely that the $12\,\mu m$ emission arises from the
general interstellar medium, while in the ellipticals, the emission
comes from the stars themselves, via a combination of photospheric and
circumstellar emission. This can be used to estimate the injection
rate of dust and gas into the elliptical's interstellar medium.
    What about the undetected Es and SOs? The upper limits are
similar in magnitude to the lowest detected values, so there is no
clear indication of dust-rich and dust-poor populations. Methods of
finding the maximum likelihood value of Φ (L_{IR}/L_B), the distribution of
the flux ratios in the IR and visible bands for data sets containing a
substantial number of upper limits, have been discussed by Knapp et al.
(1985), Feigelson and Nelson (1985) and Wardle and Knapp (1986). The
distributions at $12\,\mu m$ and $100\,\mu m$ found for the E and SO galaxies in
Table 1 are shown in Figure 2. For spirals, which are all detected at
$60\,\mu m$ and $100\,\mu m$, there is a tight correlation between the far infrared
and visible luminosities, showing the close relationship between the
interstellar medium and the stars (e.g. Helou 1987). In diagrams such
as those in Figure 2, such a distribution would have a sharp peak and a
small dispersion. For ellipticals, Figure 2 shows that there is a
close correspondence between L_{IR} and L_B (and this correspondence
improves when correction is made for the relative size of the IRAS $12\,\mu m$
beam to the galactic light). There is a small population of galaxies
with peculiar colors (e.g. NGC1275, Perseus A) resembling those of
starburst galaxies. These galaxies also have high values of L_{IR}/L_B.
Otherwise, the close correspondence between L_{IR} and L_B for ellipticals
confirms that we are seeing radiation from stars at $\lambda 12\,\mu m$. Plausible
calibrations of the mass loss rate corresponding to the $12\,\mu m$ excess
lead to mass injection rates of ~ 1 M_Θ yr^{-1}.
    The distributions again show tails towards high values of L_{IR}/L_B
due to peculiar galaxies such as NGC1275. The distributions for
L_{100}/L_B for both ellipticals and SOs show that, while the dispersion is
very large, a mean value of L_{100}/L_B can be defined. The data support
the hypothesis that essentially all ellipticals and SOs contain cold
interstellar matter. With reasonable dust models and a Galactic gas to
dust ratio, we find $\langle M(H)/L_B \rangle \sim 6 \times 10^{-3}$ M_Θ/L_Θ for Es and $\sim 2 \times 10^{-2}$
M_0/L_0 for SOs. For SO/a galaxies, Wardle and Knapp (1986) found
$\langle M(HI)/L_B \rangle \sim 0.04$ M_Θ/L_Θ. Further, the infrared colors (Thronson and
Bally 1987) and the uv excesses (Burstein et al. 1987) are consistent
with continuing star formation at a rate of ~ 1 M_Θ yr^{-1} for some
galaxies. The amounts of gas and star formation show much wider
variation for ellipticals than for spirals, and probably reflect the
much greater impact of accretion from an external source on the
properties of an elliptical.

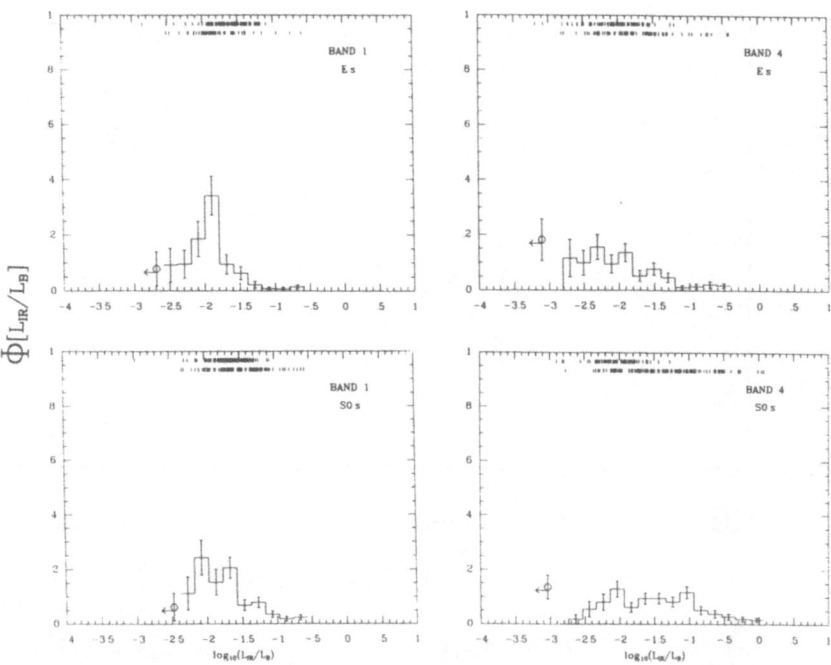

FIGURE 2. Fractional distribution $\Phi = N/N_0$ (N_0=sample size) of L_{IR}/L_B for Es and SOs for the 12 μm and 100 μm IRAS bands, where L_{IR} and L_B are in $L_\odot$. In each case 1σ error bars are shown and the lowest point corresponds to the likely fraction of the galaxies with values of L_{IR}/L_B lower than the lowest detected values. The upper row of tick marks shows the values of the individual upper limits; the lower row shows the individual detected values.

TABLE 1. Detection Rates with IRAS for Galaxies as a Function of Morphological Type (Guhathakurta et al. 1987).

Galaxy	12 μm	25 μm	60 μm	100 μm
Es	47/148	22/148	57/148	74/148
SOs	86/207	73/209	135/207	145/207
Sas+Sbs	2/29	5/29	28/29	29/29
Scs+Sds	10/29	11/29	29/29	29/29

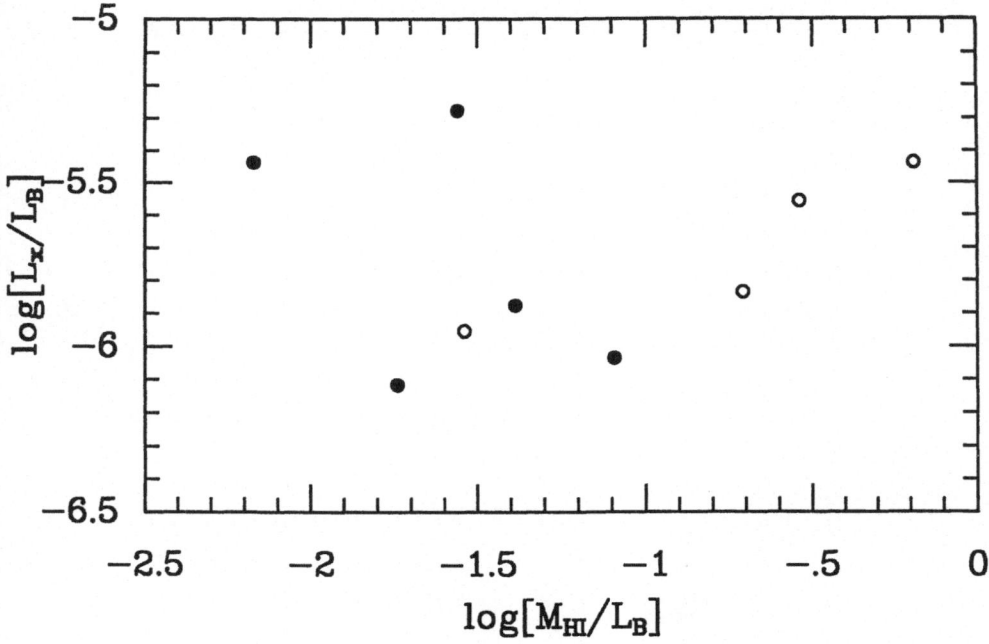

FIGURE 3. Plot of M_{HI}/L_B in M_Θ/L_Θ versus L_x/L_B for Es (filled dots) and SOs (open dots) detected in both HI and X-rays.

III. COMPARISON OF X-RAYS, DUST AND HI

Canizares, Fabbiano and Trinchieri (1987) have recently analyzed X-ray data for a sample of 81 nearby E and SO galaxies. I have used this sample to compare X-ray, HI and IR data with two changes. NGC1052, which has an X-ray observation by Helfand (1983) has been added, and IC989 has been removed because of position ambiguities. The comparison has been made with the total X-ray fluxes, uncorrected for the contribution of discrete sources. All except five of these 81 galaxies have been observed by IRAS, in HI, or both. Using the morphological type information in the Revised Shapley–Ames Catalogue (Sandage and Tammann 1981) and the Second Reference Catalogue (de Vaucouleurs, de Vaucouleurs and Corwin 1976), the galaxies can be grouped into 42 ellipticals and E/SOs (hereafter Es) and 34 SOs and SO/as (hereafter SOs). The detection rates break down as shown in Table 2. Since the HI detection rate is so low, the comparison between hot and cold gas could better be made if we can establish a connection between the HI

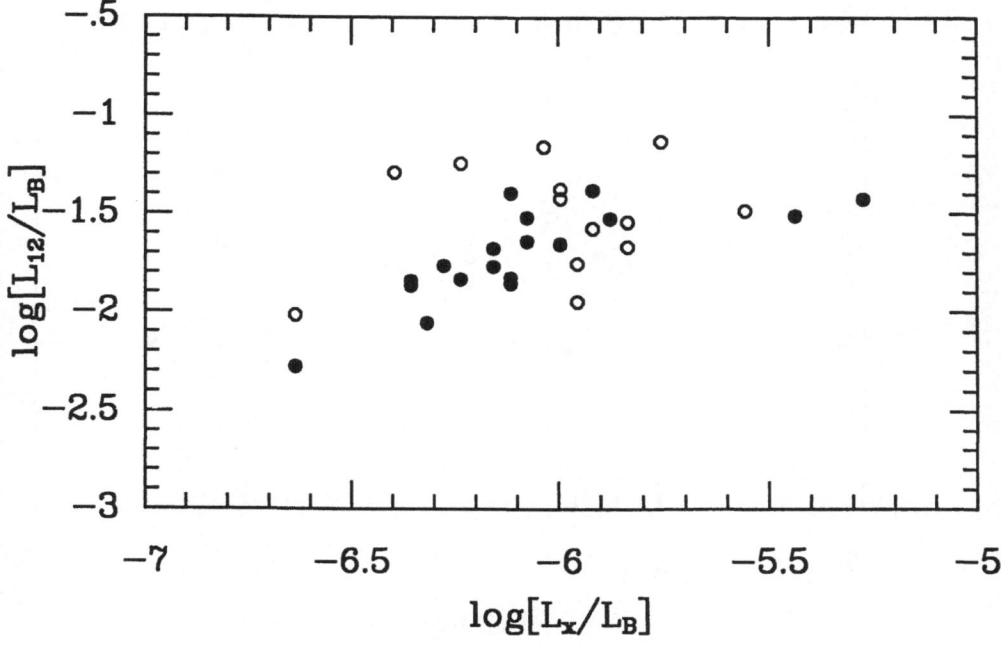

<u>FIGURE 4.</u> L_{12}/L_B versus L_x/L_B for SOs (open circles) and ellipticals (filled circles).

<u>TABLE 2.</u> Detection Rates in X-ray, HI and Far-infrared Emission for 76 nearby E and SO galaxies.

Type	X-ray det.		HI det.	12μ det.	100μ det.
E	31/42	74%	6/33	24/42	27/42
SO	24/34	70%	7/29	16/29	20/29

and 100 μm (dust) emission. Table 3 shows the breakdown between the detection rates for HI and 100μm emission. There is a very strong correlation between the detections in HI and at 100μm, and this correspondence is even more striking in our larger sample, where all 26 HI-detected ellipticals are IRAS sources at 100μm, and comprise most of the strong 100μm sources. The value of L_{100}/M_{HI} varies widely in ellipticals, and this is at least partly attributable to the large extent of the HI relative to that of the visible galaxy (e.g. NGC1052,

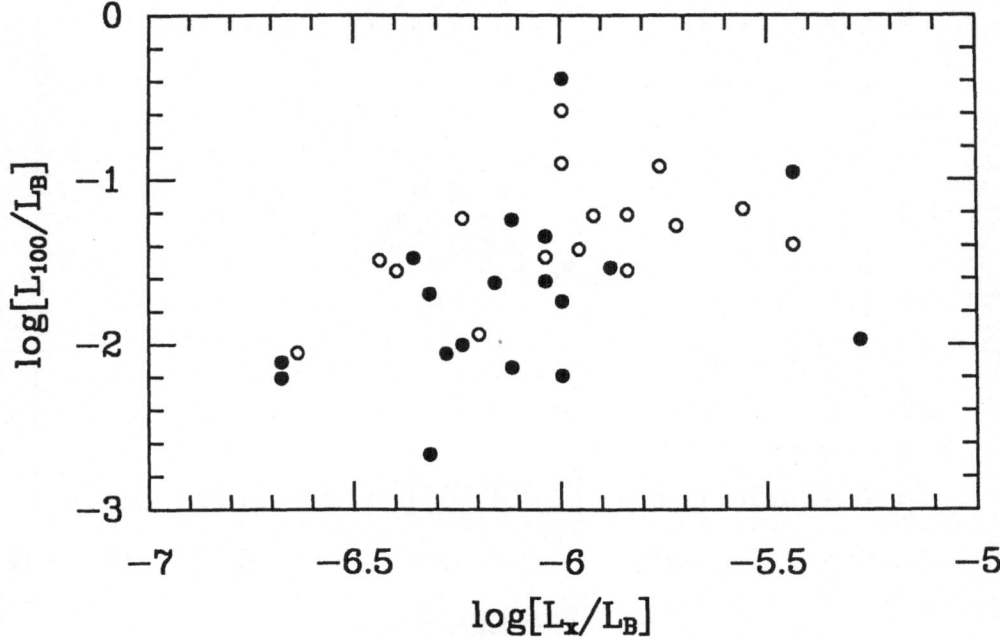

FIGURE 5. L_{100}/L_B versus L_x/L_B for SOs (open circles) and ellipticals (filled circles).

TABLE 3. Comparison between HI and 100 μ Data for Nearby E and SO Galaxies.

	HI Detections		HI Upper Limits	
	100μ detected	100μ limit	100μ detected	100μ limit
E	6	0	14	13
SO	6	0	13	6

van Gorkom et al. 1986). Emission from dust is only likely to be seen where the dust is warm, i.e. within the stellar body of the galaxy.

We now look at the relationship between X-ray and infrared emission. The detection rates are summarized in Table 4. Figure 3 shows $M(HI)/L_B$ versus L_x/L_B, and Figures 4 and 5 show, respectively, L_{12}/L_B and L_{100}/L_B versus L_x/L_B. A strong caveat about all of these plots is that upper limits are not included. These plots and Tables 1 and 4 show that there is essentially no relationship between cold gas and X-rays in E and SO galaxies (cf. Canizares et al. 1987). However,

TABLE 4. Comparison of X-ray and IRAS Detection Rates

	X-ray Detections		X-ray Limits	
	12μ detection	12μ limit	12μ detection	12μ limit
E	20	11	4	6
S0	13	10	3	4
	100μ detection	100μ limit	100μ detection	100μ limit
E	18	13	8	3
S0	16	6	4	3

(Figure 4) there is a proportionality between 12μ emission and X-rays for Es, though not for S0s. For the latter galaxies, some of the emission at short IRAS wavelengths may come from interstellar dust as is the case for spirals. For the ellipticals, though, the data suggest a correlation between starlight, mass injection and hot gas.

The above preliminary analyses lend to the following conclusions about the interstellar medium in ellipticals and S0s.
1. Most or all elliptical galaxies contain both hot and cold gas, and may be undergoing continuing star formation. Unlike spirals, though, the amounts of hot gas exceed the amounts of cold gas.
2. Cold gas and dust are correlated. Thus the cold gas is not primordial.
3. Cold gas and hot gas are little correlated. There are many observational difficulties in the way of establishing any relationship, but a tentative conclusion is that it is unlikely that the hot gas originates in evaporated cold gas or, conversely, that the cold gas is due to thermal instabilities in the hot gas.
4. There are suggestions that the hot gas is proportional to the evolved star population while the cold gas is not. Thus the hot and cold gas may have different origins.

In preparing for this review, I have made extensive use of unpublished work done in collaboration with others, and would like to thank Bruce Draine, Jacqueline van Gorkom, Raja Guhathakurta, Jim Gunn, George Helou, Mike Jura, Don-Woo Kim and David Weinberg for the use of these results. George Helou, Tom Soifer and Walter Rice have been extremely helpful with the IRAS data. This research is supported by N.A.S.A. under the Extended IRAS Mission and by the N.S.F. via grant AST86-02698 to Princeton University.

REFERENCES

Biermann, P., and Kronberg, P.P. 1983, Ap.J. (Letters) 268, L69.
Burstein, D. Bertola, F., Buson, L.M., Faber, S.M., and Lauer, T.R. 1987, in preparation.
Canizares, C.R. 1986, in 'Dark Matter in the Universe", ed. J. Kormendy and G.R. Knapp, D. Reidel Co., p.165.
Canizares, C.R., Fabbiano, G., and Trinchieri, G. 1987, Ap.J. 312, 503.
Dressel, L.L., and Wilson, A.S. 1985, Ap.J. 291, 668.
Fabian, A.C., Arnaud, K.A., and Thomas, P.A. 1986, in 'Dark Matter in the Universe', ed. J. Kormendy and G.R. Knapp, D. Reidel Co., p201.
Fabbiano, G. 1986, P.A.S.P. 98, 525.
Fabbiano, G., Klein, U., Trinchieri, G., and Wielebinski, R. 1987, Ap.J. 312, 111.
Fabricant, D., and Gorenstein, P. 1983, Ap.J. 267, 535.
Feigelson, E.D., and Nelson, P.I. 1985, Ap.J. 293, 192.
Forman, W. Schwarz, J., Jones, C., Liller, W., and Fabian, A.C. 1979, Ap.J. (Letters) 234, L27.
Forman, W., Jones, C., and Tucker, W.H. 1985, Ap.J. 293, 102.
van Gorkom, J., Knapp, G.R., Raimond, E., Faber, S.M., and Gallagher, J.S. 1986, A.J. 91, 791.
van Gorkom, J., and Laing, R. 1987, in preparation.
Guhathakurta, P., Knapp, G.R., Kim, D.-W., and Jura, M. 1987, B.A.A.S. 18, 926.
Gunn, J.E. 1979, in 'Active Galactic Nuclei', ed. C. Hazard and S. Mitton, Cambridge University Press, p.213.
Helfand, D. 1983, unpublished.
Helou, G. 1987, in preparation.
van der Hulst, J.M., Golisch, W.F., and Haschick, A.D. 1983, Ap.J. (Letters) 264, L37.
Jura, M. 1986, Ap.J. 306, 483.
Jura, M., Kim, D.-W., Knapp, G.R., and Guhathakurta, P. 1987, Ap.J. (Letters) 312, L11.
Kim, D.-W., Jura, M., Guhahakurta, P., and Knapp, G.R. 1987, B.A.A.S. 18, 926.
Knapp, G.R., Turner, E.L., and Cunniffe, P.E. 1985, A.J. 90, 454.
Knapp, G.R., and Wardle, M. 1987, in preparation.
Neugebauer, G., et al. 1984, Ap.J. (Letters) 278, L1.
Nulsen, P.G.J., Stewart, G.C., and Fabian, A.C. 1984, M.N.R.A.S. 208, 185.
Sandage, A., and Tammann, G.A. 1981, 'A Revised Shapley-Ames Catalogue of Galaxies (Washington: Carnegie Institute of Washington).
Sarazin, C.L. 1986, in 'Dark Matter in the Universe', ed. J. Kormendy and G.R. Knapp, D. Reidel Co., p.183.
Shostak, G.S., van Gorkom, J.H., Ekers, R.D., Sanders, R.H., Goss, W.M., and Cornwell, T.J. 1983, Astron. Astrophys. 119, L3.
Thronson, H.A., and Bally, J. 1987, Ap.J. (Letters) (in press).
Tytler, D. 1987, Ap.J. (in press).
de Vaucouleurs, G., de Vaucouleurs, A., and Corwin, H.C. 1976, 'Second Reference Catalogue of Bright Galaxies', University of Texas Press.
Wardle, M., and Knapp, G.R. 1986, A.J. 91, 23.

EVIDENCE FOR STAR FORMATION IN COOLING FLOWS

Robert W. O'Connell and Brian R. McNamara
Astronomy Department
University of Virginia
Charlottesville, VA 22903 USA

ABSTRACT. There is good evidence that star formation occurs in cluster cooling flows. In no case, however, are the observed effects large enough to be consistent with the inflow rates predicted by X-ray data if the initial mass function resembles that in our Galaxy. It seems most likely that the bulk of the inflow in the clusters as well as in normal gE galaxies is deposited in the form of very low mass stars with $\langle m \rangle << 0.7$ $M_\odot$. This efficient, low mass mode of star formation is a plausible candidate for the dark matter.

INTRODUCTION

The subject of cooling flows is a surprisingly controversial one. Much of the blame for this probably falls on the fact that the final, low temperature repository for the very large amounts of cooling gas indicated by the X-ray data has not yet been identified observationally.

For reasons reviewed in Sarazin (1986) and O'Connell (1987), it is hard to imagine a suitable final repository *other* than stars for the $\sim 10^{12}$ $M_\odot$ which would be deposited by a typical cluster cooling flow over a Hubble time . So most workers, by default, have assumed that star formation is the sink for the accreted mass. The difficulty is that ordinary star formation at a rate equal to the mass flux for a typical cluster flow, ~ 100 $M_\odot$ yr^{-1}, would be spectacular. Such rates correspond to those in "starbursts" (*e.g.* Montmerle & Thuan 1987), which have unmistakable signatures at UV, optical, and IR wavelengths—signatures which are normally *not* found to be associated with cooling flows.

This basic contradiction seems to have been the impetus for critics to search for ways to suppress cooling flows and for some even to propose that the flows may not exist at all. However, a consensus has already emerged from this meeting that none of the mechanisms studied to date is likely to be effective in suppressing cooling flows and that the mass fluxes estimated from simple models are correct to factors of 2-3. Thus, the dilemma posed by the absence of an observed final state for this material remains to be resolved.

A. C. Fabian (ed.), Cooling Flows in Clusters and Galaxies, 103–113.
© 1988 by Kluwer Academic Publishers.

When the signature of vigorous star formation is discussed what is usually meant is one or more of the following: *(i)* "blueness" in colors or absorption lines; *(ii)* spatial structure, particularly spiral arms, clumps, or strong gradients in color; *(iii)* emission lines; *(iv)* absorption or emission from dust clouds, especially infrared emission; or *(v)* large quantities of molecular or atomic gas.

But this familiar list of expectations is conditioned by what one might call "disk prejudice"—*i.e.* an implicit assumption that active star forming regions will always resemble those in the well-studied disks of our own and other nearby spiral galaxies. This will be true *only* if the environment is sufficiently similar to that in spiral disks and the star forming regions have a *disk-like* chemical and thermal history, time scale ($\sim 10^8$ yrs), velocity field, and, most importantly, initial mass function (IMF). In Tinsley's (1980) notation, the local IMF is

$$\psi(m) \sim m^{-(1+x)}, \text{ for } m_L \leq m \leq m_U, \text{ where } \begin{cases} x \sim 1 \\ m_U \sim 100 \text{ M}_\odot \\ m_L \sim 0.1 \text{ M}_\odot \end{cases}$$

(*e.g.* Scalo's 1986 review), with a mean mass, $\langle m \rangle$, of ~ 0.7 M$_\odot$. It is the large value of m_U which leads to items *(i)-(iii)* in the list above (and also to strong far-IR dust emission).

Star formation in cooling flows, where the environment is very different from the disks of spirals, could easily be characterized by a very different IMF. Several groups (Cowie & Binney 1977, Fabian *et al.* 1982, Sarazin & O'Connell 1983, Jura 1986) have argued that the higher gas pressures and perhaps the higher shear or lower dust density (owing to high initial temperatures) present in cooling flows can induce significant reductions in m_U or $\langle m \rangle$. This could render vigorous star formation much less conspicuous than for the local IMF. Searches for evidence of such star formation must be correspondingly more careful. They must include a proper control sample and must consider a wider range of signatures—*e.g. redder* colors than normal as well as bluer.

The following sections review results from the searches to date for star formation in cluster and normal gE cooling flows, starting with a discussion of the detectability of such "accretion populations". We want to emphasize that these searches *have* been successful—*i.e. detectable star formation does occur in cooling flows*, at least in the case of clusters. While a completely satisfactory understanding of this phenomenon is not yet at hand, we think it is probably correct to assume that stars are the final repository for the mass inflow.

DETECTABILITY OF ACCRETION POPULATIONS

It is unclear what to expect for the cool gaseous or dusty components which accumulate just prior to star formation in a cooling flow, and the distance of most massive flows is so large that direct detection of cool gas would be difficult. So, most searches have involved attempts to detect accretion population (AP) starlight directly in the UV-optical-IR region through spectroscopy, photometry, or imaging.

Consider the problem of doing this against the background light of a normal elliptical galaxy stellar population (*cf.* Sarazin & O'Connell 1983). The V-band luminosity of the accretion population will be $L_{AP} = \dot{s}\, t_{AP}(M/L_v)_{AP}^{-1}$, where $\dot{s}$ is the star formation rate, t_{AP} is the lifetime of the cooling flow, and $(M/L_v)_{AP}$ is the mass-to-light ratio of the accretion population, which is determined by its IMF. We assume a constant $\dot{s}$ throughout t_{AP}.

The resulting distortion in the spectral energy distribution (SED) of the accreting galaxy is proportional to the *fraction* of the total light contributed by the AP, or L_{AP}/L_{tot}. The distortion will therefore be a function of the IMF, t_{AP}, and $\beta \equiv \dot{s}/L_{tot}$, which is the star formation rate per unit V-luminosity for the region observed. (Note that if, as the X-ray observations suggest, $\dot{s} \sim M(r)$, where $M(r)$ is the mass enclosed within r, then $\beta \sim$ const., and any region will be representative.) As a convenient unit for β, we define $\beta_0 \equiv \dot{m}_0/L_v$, where $\dot{m}_0$ is the mass loss rate from a normal gE population giant branch. Estimates of this rate are good to about a factor of 3; we adopt $\beta_0 = 10^{-11}$ M$_\odot$ yr^{-1} L$_{v,\odot}^{-1}$ (*e.g.* Faber & Gallagher 1976).

For reference, typical gE galaxy cooling flows would yield $\beta/\beta_0 \lesssim 1$ if all the mass were converted to stars (Thomas *et al.* 1986, Canizares *et al.* 1987), and typical cluster cooling flows (where $\dot{m}_{CF} \sim 100$ M$_\odot$ yr^{-1} and $L_{tot} \sim 2 \times 10^{11}$ L$_{v,\odot}$) have $\beta/\beta_0 \sim 50$.

As a fiducial threshold change in a flux ratio or color we take $|\delta| = 0.1$ mag. An SED distortion of this size would represent a 3σ effect for high quality nuclear photometry, for example. Based on spectral synthesis models for the local IMF by Larson & Tinsley (1978) and for truncated IMF's by Sarazin & O'Connell (1983), we have estimated the $(\beta/\beta_0)_{min}$ values required to produce 0.1 mag changes in the $m_\lambda(1500 \text{ Å}) - V$, $U - V$, and $V - K$ colors in the case of matter accreting onto a galaxy with normal gE colors. More details will be given elsewhere, but the following items are worth noting here:

- For the local IMF with $t_{AP} = 10$ Gyr, $(\beta/\beta_0)_{min} \sim 1$ for UVK colors but, because of the rich population of hot massive stars in this IMF, is only 0.003 for $m_\lambda(1500 \text{ Å}) - V$. The UV is clearly the region of choice to search for accretion populations with a normal IMF.

- For the local IMF, $(\beta/\beta_0)_{min}$ is *insensitive to* t_{AP}, and decreases only a factor of 2 between 0.1 Gyr and 10 Gyr. This is because $(M/L_v)_{AP}$ increases as t_{AP} increases, keeping their ratio roughly constant in the expression for L_{AP} above. The increase in $(M/L_v)_{AP}$ occurs because of evolution of the most luminous stars to remnants during t_{AP}. (At short wavelengths, the spectrum is nearly independent of the flow duration since it is dominated by massive OB stars, whose number depends on $\dot{s}$ only during their main sequence lifetimes, $\sim$ few $\times 10^7$ yrs.) The implication is that above the threshold β/β_0's quoted above, accretion populations with local IMF's will be readily detectable *regardless of the age of the cooling flow*.

- Truncated IMF's are much harder to detect. For $t_{AP} = 10$ Gyr and an IMF in which m_U is reduced to 1 $M_\odot$, $(\beta/\beta_0)_{min}$ becomes 8.7 for $U - V$ and 3.0 for $V - K$. These values rapidly increase as $\langle m \rangle$ becomes smaller, reaching > 200 for $m_U = 0.45$ if $x = 2$.

- Because no evolution occurs off the main sequence if $m_U \lesssim 1$ $M_\odot$, $(M/L_v)_{AP}$ does not depend on t_{AP}. Hence, $(\beta/\beta_0)_{min} \propto t_{AP}^{-1}$, and a change of the assumed t_{AP} from 10 Gyr to 1 Gyr increases the values quoted above by a factor of 10. The detectability of accretion populations with truncated IMF's is therefore *greatly reduced if cooling flow lifetimes are reduced*.

- As an extreme example of a non-local IMF, consider a population consisting entirely of objects below the hydrogen burning main sequence ($m \lesssim 0.085$ $M_\odot$). We can use the "brown dwarf" models of Nelson *et al.* (1986) here. If we take $t_{AP} = 10$ Gyr, $x = 1.5$, $m_U = 0.1$ $M_\odot$, and $m_L = 0.01$ $M_\odot$, for example, we find that $L_{bol} \sim 2 \times 10^6 \, \dot{s} \, L_{bol,\odot}$. Most of this light will emerge in the K-band, where the fraction of the light contributed by the accretion population will be only $\sim 3 \times 10^{-5} \, \beta/\beta_0$. Cooling flows which deposit most of their mass in this form would evidently be *undetectable* by current methods.

The implications of these results are clear. In normal E galaxies, where $\beta/\beta_0 \lesssim 1$, star formation in cooling flows will *not* be detectable in photometry or spectroscopy *unless* the IMF is $\sim$ local. The local IMF would be readily detectable in the vacuum UV. For $\lambda > 3300$ Å, however, effects would be marginal, and the resulting limits on $\dot{s}$ will be strongly dependent on modeling of the background galaxy light.

In clusters, where $\beta/\beta_0 \sim 50$, star formation with a local IMF will be readily detectable throughout the UV-optical-IR *independent of the lifetime of the cooling flow*. Star formation with a local IMF would remain detectable in many clusters *even if their $\dot{m}_{CF}$'s have been overestimated by factors of 10-100*. On the other hand, truncated IMF's are undetectable if $\langle m \rangle$ is small enough—*e.g.* if $m_U \lesssim 0.5$ $M_\odot$ or $m_L << 0.1$ $M_\odot$—or if the lifetime of the flow is $<< 10$ Gyr.

STAR FORMATION IN gE COOLING FLOWS

The rate of ongoing star formation in normal early-type galaxies was recently reviewed in O'Connell (1986), so we will cover this only briefly. From optical data, one finds that massive stars contribute $\lesssim 2\%$ of the V light (*e.g.* Rose 1985), which implies $\beta/\beta_0 \lesssim 0.4$ for the local IMF. Considering the uncertainties, this would be consistent with the X-ray estimates of mass deposition ($\dot{m}_{CF} \sim 0.1$-3 $M_\odot$ yr^{-1}, Thomas *et al.* 1986) from gE cooling flows.

As noted above, effects in the vacuum UV should be much larger. Now, it is well established that normal early-type systems exhibit far-UV fluxes well in excess of those expected from the main sequence turnoff of an old population (Bertola, this conference). These amount to $\delta[m_\lambda(1500 \text{ Å}) - V] \sim - (2\text{-}4)$ mags. If these were interpreted as due to star formation with a local IMF, then the estimated β/β_0's would be as high as 0.7—again consistent with the optical and X-ray estimates.

However, we believe that the UV excesses are probably *not* produced by massive stars for reasons outlined in O'Connell (1986) and Bertola (this conference), one of the most persuasive of which is that the expected OB stars are simply not seen individually in the bulge of M31. Instead, we think the UV light is produced by low mass, post-asymptotic giant branch stars in rapid collapse to the white dwarf phase. If this PAGB interpretation is correct, then $\beta/\beta_0 \lesssim 0.1$ for the local IMF. Such a rate is *not* consistent with X-ray estimates of $\dot{m}_{CF}$ in many cooling flows.

If the IMF is truncated, then there is little hope of detecting effects in the UV-optical-near IR at the $\dot{m}_{CF}$'s estimated for gE's. An argument by Jura (1986) based on stellar binding energies indicates that low mass star formation at these rates will not be detectable even in the far-IR ($\sim$ 60-100 μ) against the thermal background of the thin ISM of gE's.

Overall, we think it is most likely that gE cooling flows deposit the bulk of their mass in the form of stars with a non-local IMF.

STAR FORMATION IN CLUSTER COOLING FLOWS

In clusters, mass fluxes estimated from X-ray data cover the range $\dot{m}_{CF} \sim$ 10-1000 $M_\odot$ yr^{-1}, with corresponding $\beta/\beta_0 \sim$ 5-500 expected if the mass is deposited in the form of stars. There is already a substantial literature concerning the interpretation of $\sim$ 20 individual central dominant galaxies in cluster cooling flows ("CFD" objects hereafter) in the context of accretion populations (Wirth *et al.* 1983, Sarazin & O'Connell 1983, Hu *et al.* 1985, Bertola *et al.* 1986, Arnaud & Gilmore 1986, Romanishin 1986, Silk *et al.* 1986, Johnstone *et al.* 1987). Some of this was reviewed last year in O'Connell (1987), which also summarized our own spectroscopic survey of CFD's (O'Connell & McNamara, in preparation, = OM). Rather than describing the individual studies, we will instead try to consolidate them concerning the points of agreement which have emerged.

 • About 50% of the CFD sample exhibits significant color/spectral anomalies ($|\delta| \gtrsim 0.1$ mag) when compared to samples of non-accreting cD galaxies. Effects take the form of excess light at $\lambda \lesssim 5000$ Å, and no infrared anomalies have been found except in the case of NGC 1275. The most dramatic examples, with $|\delta| > 0.5$ mag, are NGC 1275, PKS 0745-191, and A 1795. Romanishin's imaging program (this conference) has strongly confirmed earlier spectroscopic conclusions in these and other cases that the effects are *spatially* extended, which eliminates the possibility that nuclear nonthermal radiation is responsible for the anomalies.

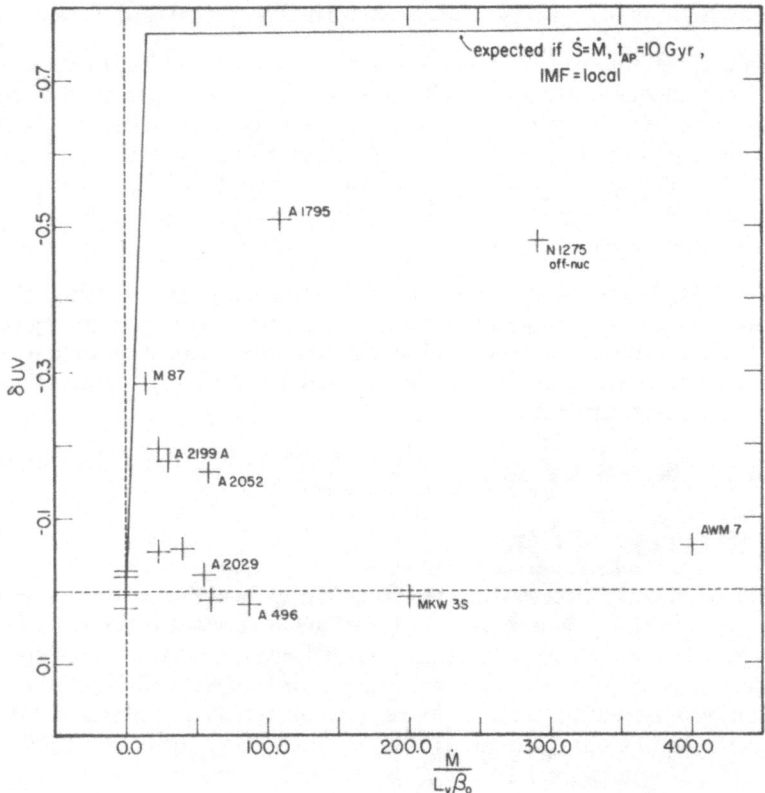

FIGURE 1: Nuclear color excesses *vs.* normalized mass accretion rates for the OM cD sample. $\delta UV = \Delta[m_\lambda(3600\,\text{Å}) - m_\lambda(4500\,\text{Å})]$, where the difference is with respect to a non-accretor cD template. Galaxies without cooling flows are plotted at $\dot{m}_{CF} = 0$. The solid line indicates the δUV resulting if 100% of the inflow formed stars with the local IMF. The excess reaches saturation at $\beta/\beta_0 = 20$. Although many objects have significant excesses, all have $\dot{s}(\text{local IMF})/\dot{m}_{CF} \ll 1$.

• Such anomalies are extremely rare among normal luminous galaxies, including non-accreting cD's, and their occurrence in this relatively small sample of clusters is *almost certainly due to the presence of the cooling flows.*

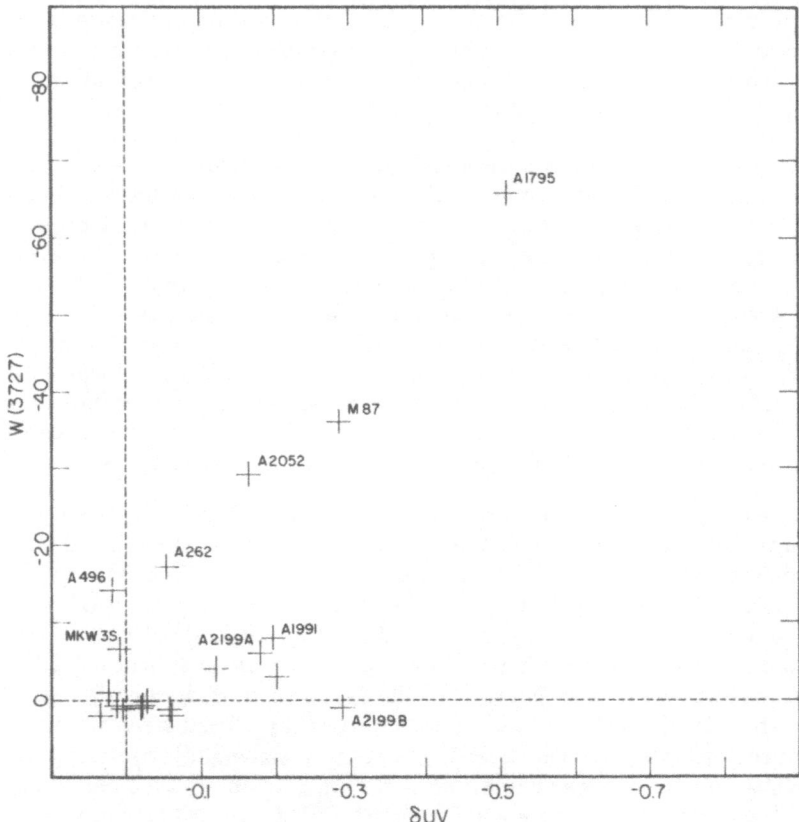

FIGURE 2: The equivalent width of nuclear [O II] λ3727 emission *vs.* the color excess defined in Figure 1 in the OM cD sample. Objects with large continuum excesses have stronger [O II], but there is considerable scatter, and some source of ionization other than hot stars is present in some objects. The continuum excess and line emission for M87 are probably produced by its nonthermal nuclear source.

• The remaining CFD objects have no significant anomalies *despite having predicted* β/β_0's $\gtrsim$ 30. Examples include A 2029, MKW 3s, and AWM 7. For such objects $\dot{s}$(local IMF)/$\dot{m}_{CF} \lesssim 0.01$.

• There seems to be *no correlation* between color/spectral anomalies and the X-ray estimates of $\dot{m}_{CF}$ or β/β_0. Results from the OM spectral survey are illustrated in Figure 1.

• Spectral analysis of A 1795, A 2052, A 2199, and PKS 0745-191 by OM, Bertola *et al.* (1986), and Johnstone *et al.* (1987) indicates that massive OB stars are present. But even here, conversion of the full $\dot{m}_{CF}$'s into stars with a local IMF would yield much larger effects, and in all cases it appears that $\dot{s}$(local IMF)$/\dot{m}_{CF} \lesssim 0.15$.

• The correlation between emission lines and spectral/color anomalies has been studied by Johnstone *et al.* (1987) and OM; the latter results are plotted in Figure 2. There is a correlation in the sense that objects with large UV excesses have strong [O II] $\lambda3727$ emission. However, there are objects—*e.g.* A 262, A 496, and MKW 3s—with [O II] but without significant continuum anomalies. In cases such as A 1795 or A 2052 the emission line fluxes are comparable to those expected for photoionization-bounded H II regions forming stars with the local IMF at the rates derived from the UV excesses. However, the scatter indicates that the situation is complicated and, in particular, that a source of ionization other than hot stars is important in some objects.

• In the case of NGC 1275 (Perseus), it appears that *O and early B stars are absent from the IMF*. Sarazin & O'Connell (1983) showed that the unusual off-nuclear spectrum (characterized by A-type absorption lines) was consistent with $t_{AP} = 10\,\mathrm{Gyr}$ and a truncated IMF with $m_U \sim 3\,\mathrm{M}_\odot$, $m_L \sim 0.001\,\mathrm{M}_\odot$, and $x \sim 1.5$. Romanishin's (1986) $V - K$ observations are consistent with this model. Population synthesis of the nuclear spectrum by Wirth *et al.* (1983) showed no evidence for stars hotter than B5 ($\sim 5\,\mathrm{M}_\odot$), and contamination by the nuclear nonthermal source probably caused them to overestimate the B5-9 contribution. Other UV and IR studies (Fabian *et al.* 1984, Gear *et al.* 1985) also suggest the absence of massive stars. Hence the data cannot, as claimed by Silk *et al.* (1986), be consistent with the local IMF *unless* star formation has been strongly suppressed during the last $\sim$ few $\times 10^7$ yrs. (Silk *et al.* suggested a young age of ~ 3 Gyr for the cooling flow, but the preceding statement is independent of the starting time of the flow.) This point is also evident from NGC 1275's position in Figure 1. NGC 1275 appears to be the best and most easily studied candidate for a truncated IMF, and a careful analysis of high-quality off-nuclear spectrometry in the UV-optical-IR should be a high priority.

We conclude that star formation is definitely occurring in some CFD galaxies and that this has almost certainly been induced by the cooling flows. Except for NGC 1275, where the IMF appears truncated, the detected accretion populations contain massive OB stars and may be consistent with the local IMF. However, if the local IMF is adopted, in no case does the $\dot{s}$ estimated from continuum or emission lines amount to more than $\sim$ 15% of the mass flux estimated from X-ray data. 50% of the CFD sample exhibits no evidence of massive star formation. Therefore, in all cases, if the X-ray analyses are valid and if stars are the final repository, *the bulk of the mass inflow must be deposited in objects with* $\langle m \rangle << 0.7\,\mathrm{M}_\odot$, which implies $m_U \ll 10$ $M_\odot$ and probably $m_L \ll 0.1$ $M_\odot$.

The situation is perhaps best interpreted in the context of multimode star formation in which a very low mass mode is the norm but where a high mass mode is occasionally triggered if conditions are right. The efficiency ratio of the low mass to the high mass mode must be $> 10{:}1$. One immediately thinks of bimodal star formation of the type reviewed by Larson (1986) with an enhanced low mass mode. Whether the kind of low mass mode he describes would be quantitatively consistent with the accretion flow data remains to be examined. Johnstone et al. (1987) have sketched a possible mechanism whereby cooling flows could drive two modes of star formation. Note that if conditions changed such that all modes were inhibited, cool gas would accumulate rapidly in the galaxy centers at the $\dot{m}_{CF}$'s typical of clusters, and this would quickly become observable. Similarly, the efficiency of the low mass mode must be high to prevent observable accumulations of residual gas.

High S/N photometry in the 0.7-10 μ region could presumably place improved limits on the character of the low mass mode, but the available data suggests very small masses, $\langle m \rangle < 0.1$ $M_\odot$—i.e. brown dwarfs. As pointed out above, the stellar continuum from accreting matter in this form would be nearly undetectable with current techniques. Non-uniformities in the flow which produce spatial concentrations might give some hope of detection, but it seems preferable to search for other kinds of signatures, perhaps emission lines associated with the early collapse of the stars.

If the assumed age of the cooling flows is reduced, then the spectral effects predicted for truncated IMF's become smaller (see above), and less extreme parameters can be assigned to the low mass mode. An attractive compromise might be to assume a small enough t_{AP} that the Galactic low mass mode of Larson (1986) satisfies the observational constraints. As Hu (this conference) points out, there are also other attractive features of assuming flows are young. However, Nulsen, Sarazin, and others (this conference) argue that young flows would not yield the observed consistency in the X-ray imaging and spectroscopic results and also that it is statistically unlikely that all flows could have begun within only the last Gyr or so. It is not possible from the stellar observations alone to place good upper limits on the age of the flows.

Finally, the low mass mode does seem, as Fabian et al. (1986) point out, to be a plausible candidate for the dark matter in E galaxies, particularly now that constraints on the baryonic component of the universe have been relaxed by newer big bang nucleosynthesis studies.

SUMMARY

Do stars form in cooling flows? The answer is definitely yes in about 50% of the cluster flows. Concerning the IMF, we find that $\dot{s}(\text{local IMF})/\dot{m}_{CF}$ could be as large as $\sim 15\%$ in some cases but must be $\lesssim 1\%$ in many cases and that $\dot{s}(\text{local IMF})/\dot{m}_{CF} = 1$ is strongly excluded for any cooling flow lifetime.

In normal gE cooling flows, there is little unambiguous evidence of star formation. If the far-UV excesses are due to post-asymptotic giant branch stars, as seems likely, then $\beta/\beta_0 \lesssim 0.1$ for the local IMF.

Unless the cluster $\dot{m}_{CF}$'s have been overestimated by factors $\gtrsim 10$, the best interpretation seems to be that cooling flows efficiently deposit most of their mass in the form of very low mass stars with $\langle m \rangle << 0.7$ M$_\odot$. Direct detection of star formation in this low mass mode will be very difficult. A high mass mode, perhaps with the local IMF but encompassing only a small fraction of $\dot{m}_{CF}$, can be triggered under some circumstances and contributes significantly to the ionization of the emission line filaments. NGC 1275 remains a good candidate for a truncated IMF with $m_U \sim 3$ M$_\odot$ and deserves careful study.

One can make the parameters of the low mass mode less extreme by assuming smaller cooling flow lifetimes, but this appears to conflict with other properties of cluster flows. The low mass mode appears to be a plausible candidate for the dark matter.

This research has been supported in part by NASA through NAG 5-700.

REFERENCES

Arnaud, K.A., and Gilmore, G. 1986. *M.N.R.A.S.*, **220**, 759.

Bertola, F., Gregg, M.D., Gunn, J.E., and Oemler, A. 1986. *Ap.J.*, **303**, 624.

Canizares, C.R., Fabbiano, G., and Trinchieri, G. 1987. *Ap.J.*, **312**, 503.

Cowie, L.L. and Binney, J. 1977. *Ap.J.*, **215**, 723.

Faber, S.M., and Gallagher, J.S. 1976. *Ap.J.*, **204**, 365.

Fabian, A.C., Arnaud, K.A., Nulsen, P.E.J., and Mushotsky, R.F. 1986. *Ap.J.*, **305**, 9.

Fabian, A.C., Nulsen, P.E.J., and Arnaud, K.A. 1984. *M.N.R.A.S.*, **208**, 179.

Fabian, A.C., Nulsen, P.E.J., and Canizares, C.R. 1982. *M.N.R.A.S.*, **201**, 933.

Gear, W.K., Gee, G., Robson, E.I., and Nolt, I.G. 1985. *M.N.R.A.S.*, **217**, 281.

Hu, E.M., Cowie, L.L., and Wang, Z. 1985. *Ap.J. Suppl.*, **59**, 447.

Johnstone, R.M., Fabian, A.C., and Nulsen, P.E.J. 1987. *M.N.R.A.S.*, **224** , 75.

Jura, M. 1986. *Ap.J.*, **306**, 483.

Larson, R.B. 1986. In *Stellar Populations*, eds. C.A. Norman, A. Renzini, and M. Tosi (Cambridge: Cambridge University Press), p. 101.

Larson, R.B., and Tinsley, B.M. 1978. *Ap.J.*, **219**, 46.

Montmerle, T., and Thuan, T.X. 1987. *Starbursts and Galaxy Evolution* (Paris: Editions Frontieres).

Nelson, L.A., Rappaport, S.A., and Joss, P.C. 1986. In *Astrophysics of Brown Dwarfs*, eds. M.S. Kafatos, R.S. Harrington, and S.P. Maran (Cambridge: Cambridge University Press), p. 177.

O'Connell, R.W. 1986. In *Stellar Populations*, eds. C.A. Norman, A. Renzini, and M. Tosi (Cambridge: Cambridge University Press), p. 167.

O'Connell, R.W. 1987. In *Structure and Dynamics of Elliptical Galaxies*, ed. T. de Zeeuw (Dordrecht: Reidel).

Romanishin, W. 1986. *Ap.J.*, **301**, 675.

Rose, J.A. 1985. *A.J.*, **90**, 1927.
Sarazin, C.L. 1986. *Rev. Mod. Phys.*, **58**, 1.
Sarazin, C.L., and O'Connell, R.W. 1983. *Ap.J.*, **268**, 552.
Scalo, J. 1986. *Fund. Cosmic Phys.*, **11**, 1.
Silk, J., Djorgovski, S., Wyse, R.F.G., and Bruzual, A. 1986. *Ap.J.*, **307**, 415.
Thomas, P.A., Fabian, A.C., Arnaud, K.A., Forman, W., and Jones, C. 1986. *M.N.R.A.S.*, **222**, 655.
Tinsley, B.M. 1980. *Fund. Cosmic Phys.*, **5**, 287.
Wirth, A., Kenyon, S.J., and Hunter, D.A. 1983. *Ap.J.*, **269**, 102.

DYNAMICAL FRICTION IN THE RICH CLUSTER A2029

R. G. Bower[1], R. S. Ellis[1] and G. F. Efstathiou[2]

1) Department of Physics, 2) Institute of Astronomy,
   University of Durham,   Madingley Road,
   Durham,   DH1 3LE,   Cambridge,   CB3 0HA,
   UK.   UK.

ABSTRACT: Redshifts have been obtained for galaxies close to the cD in the cluster A2029 in order to determine the fraction that are bound to the cD. No difference in the velocity dispersion of 'core' and 'mid-cluster' galaxies was found, however, suggesting that the bound fraction is small. Intriguing structure in the velocity field leads us to ponder the possibility that the present-day cluster formed from the decay of a binary system.

1. Introduction

Dynamical friction of cluster galaxies against the dark-matter causes them to lose energy and spiral in towards the cluster centre. The rate of 'in-fall' is uncertain, but Ostriker and Hausman (1978) suggested the galaxies that reach the centre might merge to form a giant cD type galaxy there.

An overdensity of galaxies around the cD is indeed observed (cf. Schneider, Gunn and Hoessel (1983), Beers & Tonry (1986)); but Merritt (1984a,b) has argued that this is not conclusive evidence in favour of Ostriker and Haussman's scenario as galaxies that are initially on radial orbits will be 'focussed' so as to crowd the centre without slowing down sufficiently to be cannibalised.

Distinction between these two theories can be made by measuring the velocities of objects close to the cD. Studies of 'multiple nuclei' have been made by Tonry (1985) and Smith *et al.* (1985), and, for a much fainter sample of serendipitous objects, by Cowie & Hu (1986). The dispersions derived are much higher than would be expected for a set of objects than were soon to merge with the cD (eg., Smith *et al.* obtain a dispersion of 833 $km\,s^{-1}$ for multiple nuclei with separations less than 50 Kpc.); but this is partly or wholly due to the contamination of the sample by *mid-cluster* galaxies that are chance projections onto the core. Cowie and Hu were able to disentangle the two populations by combining all the available data to create a sample of 75 objects. With such a large sample, they were able to demonstrate that a single gaussian was not a good fit to the data, and that two gaussians with dispersions of 250 and 1400 $km\,s^{-1}$ were required in the ratio 6 : 4.

Doubt has been cast on this result, however, because the combined sample is very inhomogeneous both in the richness and global velocity dispersion of clusters included, and in the magnitude and separation of the galaxies. We have attempted to confirm the result using a homogeneous sample drawn from close to the cD in a *single* cluster.

115

A. C. Fabian (ed.), Cooling Flows in Clusters and Galaxies, 115–119.

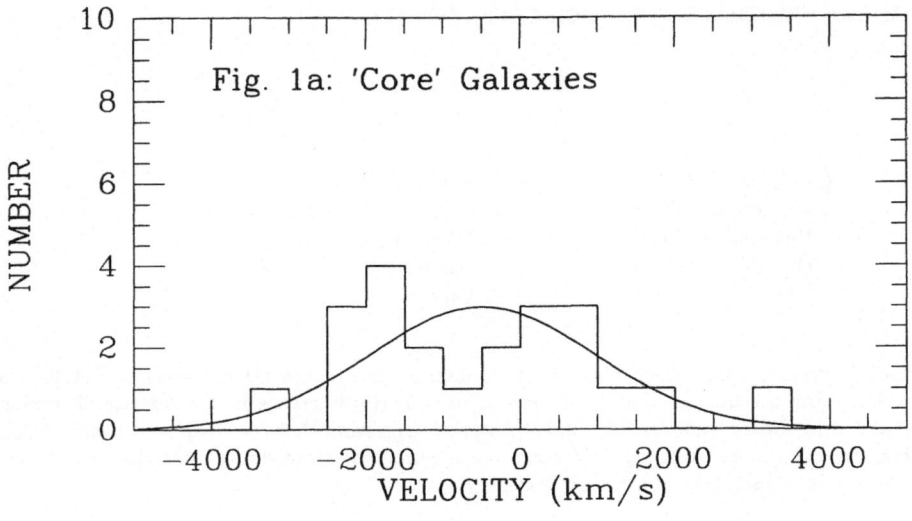

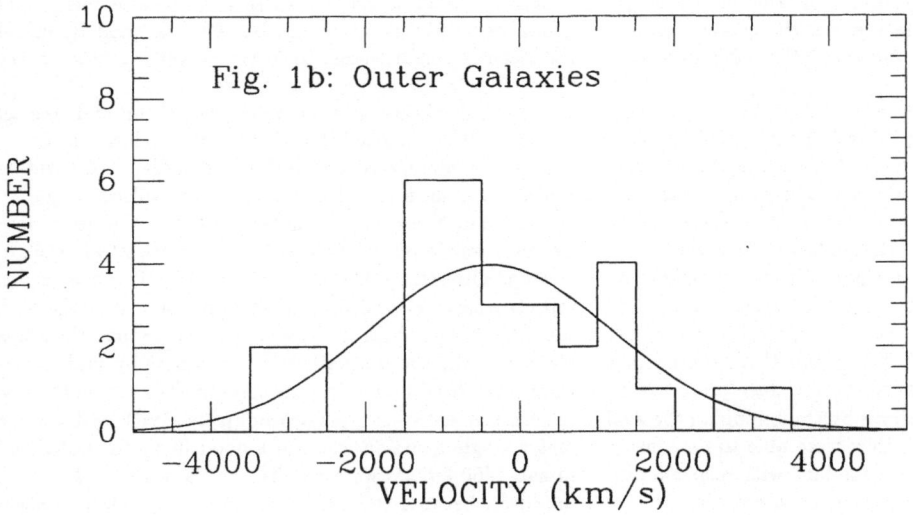

Figure 1. Histograms showing the velocity distributions of 'core' and general cluster galaxies. The fitted gaussians are also shown.

2. Observations

The cluster A2029 was chosen for this project because of its richness and regularity, and because of the large volume of previously published data—most notably Dressler's measurement of the velocity dispersion of the cD halo (1979) and of the general cluster (1981). Target galaxies were chosen randomly from within 150 Kpc of the cD. We were unable to complete a magnitude limited survey of the region because of the shortage of telescope time—our final sample is $\sim 70\%$ complete.

Observations were made on the Issac Newton telescope, La Palma, using long- and multi-slit spectrographs with the IPCS as detector, and on the AAO's 3.9m telescope using LDSS in high dispersion single-slit mode, with the RCA CCD as detector. The INT observations have a resolution of 4Å FWHM and dispersion 60Å/mm over the wavelength range 3700–6000Å, while the AAT data have lower resolution (10Å FWHM) and dispersion (165Å/mm) over the same wavelength range. Redshifts were obtained from the spectra by cross-correlation with stellar and elliptical galaxy templates as detailed by Efstathiou, Ellis and Carter (1981).

We were able to estimate our measuring error by comparison of redshifts measured for a few galaxies with differing slit orientations, and by comparison with a few galaxies that had previously been measured by Dressler. The error from internal checks is $\sim 130 \ \mathrm{km\,s^{-1}}$. Comparison with Dressler's measurements suggests that there is no zero-point error.

3. Results

Our final 'core' sample consists of 17 new redshifts plus 4 velocities taken from Dressler (1981) and 1 from Cowie and Hu (1986). We do not have sufficient data to test for non-normality of the distribution, but we compare the sample's dispersion with that of a control sample drawn from further out in the cluster (Dressler 1981).

If a fraction f of our core sample galaxies are genuine satellites, we expect

$$\sigma^2_{\mathrm{sample}} = f.\sigma^2_{\mathrm{core}} + (1-f).\sigma^2_{\mathrm{cluster}}$$

but since $\sigma_{\mathrm{core}} \sim \sigma_{\mathrm{halo}} \sim 300 \ \mathrm{km\,s^{-1}}$ (ie., much less than the global dispersion), the formula reduces to

$$\sigma_{\mathrm{sample}} \simeq \sqrt{1-f}.\sigma_{\mathrm{cluster}}$$

Figures 1a and 1b show the histograms of the 'core' and general cluster velocities (velocities are quoted relative to the cD). The distributions have means of -500 (± 310) $\mathrm{km\,s^{-1}}$ and -410 (± 260) $\mathrm{km\,s^{-1}}$ respectively (1σ errors) and dispersions of $1470^{1770}_{1290} \ \mathrm{km\,s^{-1}}$ and $1450^{1680}_{1290} \ \mathrm{km\,s^{-1}}$ (66% t-statistic confidence limits).

We see that there is no evidence for *any* difference between the two populations. Small-number uncertainties, however, admit the possibility that the bound fraction could be up to 40% (ie., $p(f \geq 0.4) = 0.03$).

4. Discussion

Our *upper limit* of a 40% contribution to our core sample from bound satellites contrasts somewhat with the 60% *detection* of Cowie and Hu. The discrepancy might be resolved in three ways:

1. If the mid-cluster galaxies were on predominantly radial orbits, when seen in projection against the core their line-of-sight dispersion would be increased above the average cluster value, and so might cancel out the reduction induced by any bound satellite population.

118

2. The satellite population might extend out to less than 100 kpc so that our 150 kpc sample region introduces many more mid-cluster galaxies than necessary.

3. There is a chance that we might be viewing the cluster core through a clump of mid-cluster galaxies.

As outlined below, it seems most likely that (3) is the case; there is no evidence to suggest that (1) or (2) is a problem, though they should be noted for future studies.

5. Structure in the Velocity Field

So far in our analysis, it has been assumed that the cluster is well relaxed and homogeneous. However, if we examine the spatial distribution of the velocities (figure 2), we note that (with a few exceptions) galaxies with large negative velocities are concentrated to the South West of the cD, where as the galaxies with small, mostly positive, velocities are offset to the North East. The two spatial groups correspond to the peaks in the 'core' histogram. Monte Carlo simulations suggest that such a pattern is unlikely to occur by chance in a homogeneous cluster. We see no such structure in the outer sample.

We have considered various models for this phenomenon, the two which we consider most likely are:

1. We are viewing the cluster core through a dense group infalling along the line of sight. This would account for the negative velocity group—the remaining galaxies are mostly core members. We must appeal to small number effects to explain why they appear offset from the cD.

2. The present-day A2029 cluster formed from the collision of two smaller clusters. Originally, a binary system was formed. In time, the orbit decays and the sub-cluster cores merge, but the satellite populations continue to rotate.

The second model is appealing because it forms a natural continuation from the sequence of binary and multiple nuclei clusters. However, in the absence of comparable data for other clusters, we cannot consider this phenomenon further.

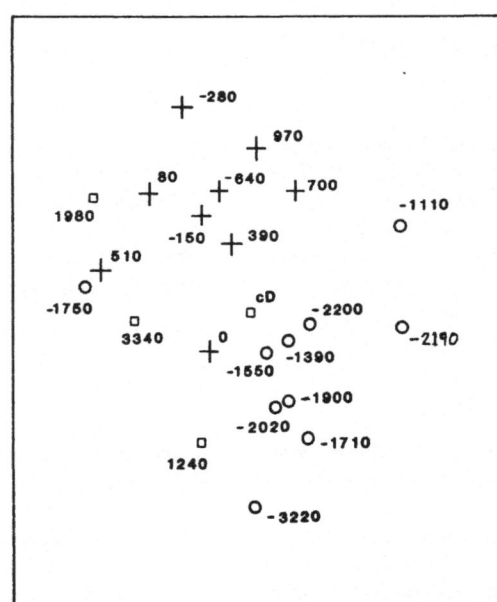

Figure 2. The spatial distribution of the 'core' galaxies. Galaxies in the negative velocity clump are marked by circles, and those in the positive velocity clump by crosses. Galaxies marked with boxes do not belong to either group.

6. Conclusions

We have attempted to demonstrate that the cores of rich clusters contain a population of galaxies that are bound to the dominant central galaxy and may soon be cannibalised by it. Our method is to compare the dispersion of galaxies selected from within 150 kpc of the cD of A2029, with that of galaxies drawn from the general cluster.

We find no evidence for a fall in the velocity dispersion and conclude that bound satellite galaxies make up, *at very most*, 40% of our sample. This contrasts with Cowie and Hu's (1986) *detection* of a 60% contribution. However, there is evidence to suggest that our sample includes an infalling subclump, or that the cluster core has formed the decay of a binary cluster and that the satellite population is rotating about the cD.

7. Acknowledgements

We wish to thank John Lucey and Carlos Frenk for helpful discussions of substructure in clusters, John Lucey for assistance with statistical aspects of our argument, and J. Binney for suggesting the second explanation of the velocity field.

8. References

Cowie L. L. & Hu E. M. (1986) *Astrophys. J.*, **305**, L39.
Dressler A. (1979) *Astrophys. J.*, **231**, 659.
Dressler A. (1981) *Astrophys. J.*, **243**, 26.
Efstathiou G., Ellis R. S. & Carter D. (1980) *Mon. Not. R. astr. Soc.*, **193**, 931.
Merritt D. (1984a) *Astrophys. J.*, **276**, 26.
Merritt D. (1984b) *Astrophys. J.*, **280**, L5.
Ostriker J. P. & Hausman M. (1978) *Astrophys. J.*, **224**, 320.
Smith R. M., Efstathiou G., Ellis R. S., Frenk C. S. & Valentijn A. E.
    (1985) *Mon. Not. R. astr. Soc.*, **216**, 71p.
Tonry J. (1985) *Astrophys. J.*, **291**, 45.

COLOR GRADIENTS IN COOLING FLOW CLUSTER CENTRAL GALAXIES and the IONIZATION OF CLUSTER EMISSION LINE SYSTEMS

W. Romanishin
Physics Department
Arizona State University
Tempe, AZ 85287-1504

ABSTRACT. Preliminary results are given for a program to measure color gradients in the central galaxies in clusters with a variety of cooling flow rates. The objectives are to search for extended blue continuum regions indicative of star formation, to study the spatial distribution of star formation, and to make a quantitative measure of the amount of light from young stars, which can lead to a measure of the star formation rate (for an assumed intial mass function). Four clusters with large $\dot{M}$s and large cluster $H\alpha$ emission fluxes are found to have an excess of blue light concentrated to the centers of the cluster central galaxy. Assumption of a disk IMF leads to the conclusion that the starlight might play a major role in ionizing the emission line gas in these clusters.

1. INTRODUCTION

What happens to the large amounts of gas which x-ray observations tell us may be accreting onto the central galaxies in many galaxy clusters? At least one galaxy, NGC 1275 in Perseus, shows obvious evidence of ongoing star formation which might mark the fate of some of the gas in this cluster. We are making a two-color imaging survey of this cluster and a number of other cooling flow clusters to find more examples of star forming systems and study them in detail. We are looking for extended blue continuum light which would indicate star formation. By taking into account the background light from the old galaxy presumed to lie "underneath" the cooling flow induced star formation, we can make a quantitative estimate of the amount of light associated with the ongoing star formation , and, after assuming an IMF, estimate the total star formation rate implied by the excess light. We comment on the possible role of starlight in providing ionization for the emission line filaments seen in many cooling flow clusters.

2. OBSERVATIONS

The observations were designed to measure the continuum light from stars over a large color baseline, while avoiding emission lines which are strong in cooling flows. In particular, the standard B filter was not used, because it is contaminated by the [OII] 3727 line. Instead, a filter (referred to as b_{45}) was used, which has a central wavelength of 4506 Å,and a FWHM of 355 Å. This avoids the [OII] line up to a redshift of 0.15. The red filter was similar to a standard Mould I filter, with $\lambda_0 = 8200$ Å, FWHM = 2000 Å. These filters span the largest practical optical baseline

A. C. Fabian (ed.), Cooling Flows in Clusters and Galaxies, 121–125.
© 1988 by Kluwer Academic Publishers.

that avoids the [OII] line. It would be useful to go blueward of the [OII] line, but this is difficult due to rising atmospheric extinction and the poor transmission of available UV interference filters.

The sample consists of about 15 clusters chosen from the list of Arnaud (1987). We tried to observe most of the clusters with $\dot{M} \gtrsim 100 \ M_\odot \ \mathrm{yr}^{-1}$, and a number with $\dot{M} = 0$ for comparison. Observations were made with TI CCDs on the Kitt Peak National Observatory 4-m telescope and the Steward Observatory 2.3-m telescope.

The four clusters discussed here form a sample defined by their high cluster emission line luminosities. As far as the author is aware, these are the four cooling flow clusters with the highest L(Hα) presently known. Data on the full sample is in preparation.

3. ANALYSIS

3.1. Excess Blue Light

Our analysis assumes there is an "old" giant galaxy underlying whatever recent star formation has been induced by the cooling flow. Such luminous ellipticals are known to have color gradients, becoming redder with decreasing radius due to increasing metallicity. This color gradient must be carefully taken into account when interpreting any observed colors or color gradients of a composite system of old galaxy plus newly formed stars. We define a "zero $\dot{M}$ " color profile using four clusters without cooling flows. To estimate the amount of blue light in excess of the old galaxy light for a galaxy with some young star contribution, we assume the I profile is unaffected by the light from young stars and use the zero $\dot{M}$ color profile, plus the observed I profile, to predict the old b_{45} profile of the *old population alone*. Any light from young blue stars will show up as a excess in the observed b_{45} profile over the predicted old light profile.

Absolute B magnitudes were derived for the excess blue light in the following central galaxies (assuming $H_0 = 50 \ \mathrm{km \ sec}^{-1} \ \mathrm{Mpc}^{-1}$): -21.8 for NGC 1275, -21.5 for PKS 0745−191 , ~-20.2 for A 1795 , and ~-20.0 for 2A 0335+096. In all clusters, the blue light is concentrated within 10-15 kpc of the nucleus. The outer regions (> 25 kpc radius) have colors approximately appropriate for old elliptical galaxies , but a small amount of extra blue light at these radii cannot be ruled out with the present data. In the Figure, we show the ($b_{45} - I$) color profile for each of the four galaxies. The solid line in each plot is the "zero $\dot{M}$ " color profile. The profiles are normalized to zero color for r > 25 kpc.

3.2. PKS 0745−191 and 2A 0335+096 Hα Luminosities

PKS 0745−191 , discussed by Fabian *et al.* 1985, was observed through an 80 Å wide filter passing Hα+ [NII] at the redshift of the source. A new Galactic absorption value was derived by comparing the colors of PKS 0745−191 cluster ellipticals with the colors of similar galaxies in other, lightly reddened clusters. The final Hα luminosity, corrected for Galactic extinction, and the presence of the [NII] lines in the filter, is $4.8 \times 10^{42} \ \mathrm{erg \ s}^{-1}$ ($H_0 = 50 \ \mathrm{km \ sec}^{-1} \ \mathrm{Mpc}^{-1}$). This is the same H$\alpha$ luminosity as found by Cowie, Hu, Jenkins, and York (1983) for NGC 1275/Perseus. Hu *et al.* (1983) attribute the large Hα luminosity of

Perseus to a galaxy collision responsible for the unique "high-velocity" emission line system found in this cluster. The similarities of NGC 1275 and PKS 0745−191 argue against the "high velocity" system accounting for the high Hα luminosity. Further comparison of NGC 1275 and PKS 0745−191 is found in Romanishin (1987).

We also observed the cooling flow cluster associated with 2A 0335+096 in narrow band filters to derive an Hα luminosity. We find L(Hα)= 1.1 × 10^{42} erg s^{-1}. Details are in Romanishin and Hintzen (1987).

3.3. Star Formation Rate and Hα Luminosity

If the stars producing the excess blue light also produce ionizing radiation, they can account for at least some of the ionization for the emission line systems seen in many cooling flow clusters. The biggest uncertainty is, of course, the IMF and the number of very massive young stars present. For this discussion, we ASSUME that some portion of the accreting gas forms stars with a normal disk IMF, and ignore the rest (Johnstone, Fabian, and Nulsen 1987). For any given age of the cooling flow induced star formation event we can use simple galaxy models with a constant SFR (Struck-Marcell and Tinsley 1978) to relate the blue light and SFR. We can also relate the SFR and the Hα flux directly, as in Johnstone, Fabian, and Nulsen (1987), again with the critical assumption of a disk IMF.

A simple calculation shows that the UV light from the stars producing the excess blue light can account for much of the ionization of the emission line systems. As an example, if we take an excess blue light luminosity of $M_B=-21.7$ (average of NGC 1275 and PKS 0745−191) and assume a star formation event lasting 1 × 10^9 yr, we derive a constant SFR of ∼30 $M_\odot$ yr^{-1} . This SFR, if it produces stars with a disk IMF, could provide ∼80% of the ionization needed for the emission systems of either NGC 1275 or PKS 0745−191 . The clusters A 1795 and 2A 0335+096 scale approximately from these numbers, with both excess blue light and Hα luminosity ∼20 - 25 % of those for NGC 1275 and PKS 0745−191. Obviously, there are many unknowns, such covering factors, internal dust, and lifetime of the star formtion events. However, unless the stars producing the excess blue light do not produce massive stars, stars must be considered as a significant source of ionization.

The observational evidence for the presence or absence of massive stars in cooling flow clusters with star formation is presently very confused. Wirth, Kenyon, and Hunter (1983) find that the optical spectra of the central regions of NGC 1275 requires B stars. The UV situation for NGC 1275 is confused (see Fabian, Nulsen, and Arnaud 1984, particularly the note added in proof). We note that Norgaard- Nielsen, Jorgensen, and Hansen (1984) find a low UV continuum flux from the A 1795 central galaxy. Silk *et al.* (1986) argue that the available data on NGC 1275 does not *require* an upper mass cutoff. We feel that further UV observations are of crucial importance to constrain the high mass stellar population in cooling flow galaxies.

3.4. Optically Dull Clusters

Not all clusters with large x-ray derived $\dot{M}$s are found to have evidence of star formation. One example in our sample is A 644, which has an x-ray $\dot{M}$

of about 300 $M_\odot$ yr^{-1} . The central galaxy has a color profile indistinquisable from the average of the four zero $\dot{M}$ clusters. Narrow- band Hα imaging shows no evidence for emission. This is, of course, consistent with the idea of starlight ionization of the cluster emission line systems.

4. Conclusions and Problems

We offer the following preliminary conclusions:
(1) At least for high Hα luminosity cooling flow clusters, the amount of excess blue light from young stars indicates that these stars might provide significant ionization for cluster emission line systems, PROVIDED a disk IMF holds.
(2) The blue light in these clusters is concentrated to the centers of the dominant galaxies on a ~10 kpc scale, even though the cooling radii are 100 or more kpc.
(3) NGC 1275 is not a uniquely luminous cooling flow cluster in Hα - PKS 0745$-$191 has a very similar Hα luminosity and color profile. This argues that the large Hα luminosity in NGC 1275 is not due to the unique high-velocity interaction observed in that system.

One possible problem with the starlight ionization scenario is the emission line ratios, which are more like those of shocked regions than of HII regions. We feel the ionization situation, and the line ratios, are very complex in cooling flow clusters and that realistic models have not yet been made. Several ionization sources might be present- starlight, shocks, x-ray emitting gas, and nuclear sources.

This research has been partially supported by the HEAO Guest Investigator Program under NASA grant NAG8-576.

REFERENCES

Arnaud, K. A. 1987, preprint.
Cowie, L. L., Hu, E. M., Jenkins, E. B., and York, D. G. 1983, *Ap.J.* **272**,29.
Fabian, A. C., Nulsen, P. E. J., and Arnaud, K. A. 1984, *M.N.R.A.S.* **208**, 179.
Fabian, A. C. , et al. 1985, *M.N.R.A.S.* **216**, 923.
Hu, E. M., Cowie, L. L. , Kaaret, P. , Jenkins, E. B. , York, D. G. , and Roesler, F. L. 1983 , *Ap.J. (Letters)* **275**, L27.
Johnstone, R. M., Fabian, A. C., and Nulsen, P. E. J. 1987, *M.N.R.A.S.* **224**, 75.
Norgaard- Nielsen, H., Jorgensen, H., and Hansen L. 1984, *Astr. Ap.* **135**, L3.
Romanishin, W. 1987. *Ap.J. (Letters)* (in press).
Romanishin, W., and Hintzen, P. 1987. *Ap.J. (Letters)* (submitted).
Silk, J., Djorgovski, S., Wyse, R. , and Bruzual, A.,G. 1986, *Ap.J.* , **307**, 415.
Struck-Marcell, C., and Tinsley, B. M. 1978, *Ap.J.* **221**, 562.
Wirth, A., Kenyon, S.J., and Hunter, D.A. 1983, *Ap.J.* , **269**, 102.

Figure Caption
The figure shows the color profile for the four clusters discussed in detail in the text. The x's mark the observed points, while the solid line is the "zero $\dot{M}$ " color profile.

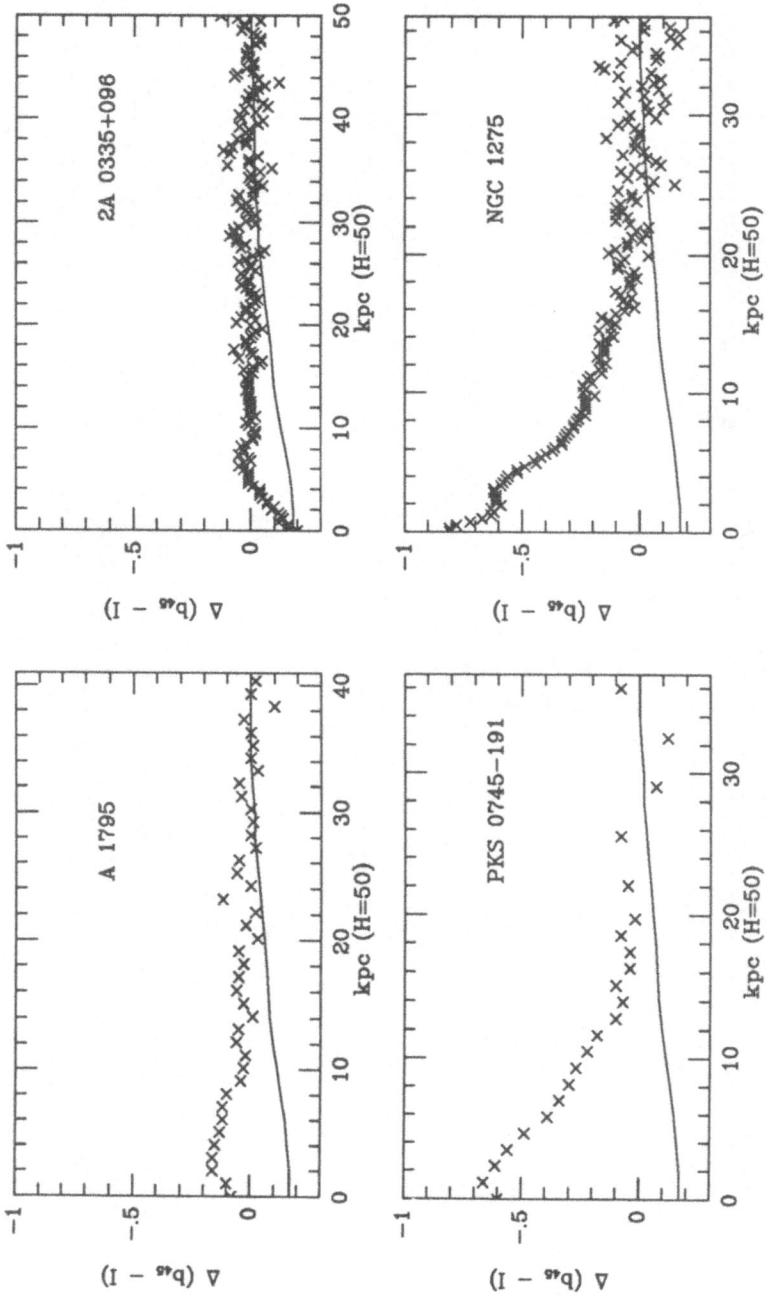

ULTRAVIOLET EVIDENCE FOR STAR FORMATION IN ELLIPTICAL GALAXIES

Francesco Bertola
Department of Astronomy
University of Padova
Vicolo dell'Osservatorio 5
35122 PADOVA, Italy

Abstract. The stellar populations responsible for the UV rising branch observed in elliptical galaxies are discussed. The nuclei of planetary nebulae seem to be the best candidates, due to the presence of a relationship between metallicity and the level of the UV rising branch, which can be understood on the basis of evolutionary properties of the old population stars. On the other hand, the observed energy distribution can be accounted for by models in which continuing star formation takes place. Since there are several forms of evidence that star formation occurs at least in few cases, it is suggested that the observed UV flux could be due to the combined effect of both the old and the young stars. The relationship between the mass infall rate in galaxies with cooling flows and the (1550–V) color is presented.

I. Introduction

One of the important discoveries in the study of the energy distribution of the galaxies has been the detection by telescopes operating outside our atmosphere (especially the IUE) of the rising branch in the UV flux of elliptical galaxies. After reaching a minimum in the region 2000–2500 Å the flux rises again in the region 1000–2000 Å, with a slope typical of a black body at about 30.000 K. This phenomenon is present in large aperture spectra obtained with the OAO2 satellite as well as in the small aperture (10 × 20 arcsec) data obtained with IUE, indicating that it is not characteristic only of the nuclear region. Luminosity profiles perpendicular to the dispersion on IUE spectra are very similar to the profiles in the visual. This suggests that the UV source giving origin to the rising branch is not point like. The similarity of the visual and UV profiles suggests that the UV light originates

A. C. Fabian (ed.), Cooling Flows in Clusters and Galaxies, 127–132.
© 1988 by Kluwer Academic Publishers.

from a stellar population distributed in the same way as the cold one. The most remarkable fact about the UV rising branch is that its level, relative to the visual spectrum, is variable from one object to another, while the remaining spectrum down to the infrared is exactly superimposable, apart from some effects shortward of 5000 Å due to blanketing. The rising branch has a minimum level in M32 and reaches intensities which are almost one order of magnitude higher in other galaxies like NGC 4486 and NGC 4649, with all the intermediate values present (Code and Welch 1979, Bertola, Capaccioli, Holm and Oke 1980, Oke, Bertola and Capaccioli 1981, Bertola, Capaccioli and Oke 1982). At the beginning it seemed that no correlation of the level of the rising branch with some other physical parameter was present. The apparently random variations suggested that the responsibility for the hot component could be ascribed to young upper main sequence stars. Statistical fluctuations in the number of these short–lived stars can easily account for the observed variation in the UV flux shortward of 2000 Å. Faber (1983) has subsequently, however, suggested a correlation between the level of the UV flux and the metallicity parameter Mg_2 in the sense that higher metallicity implies higher UV flux level. This is just the contrary of the effect produced by blanketing at longer wavelength. Since Mg_2 is measured in the cold component of the stellar population, the above relationship could imply that the UV flux shortward of 2000 Å is mainly produced by the old component. Therefore a very thorough analysis of the data, coupled to results from models of stellar evolution and chemical evolution of galaxies is needed in order to establish the presence of star formation phenomena in elliptical galaxies.

II. The old population

A comprehensive study of about 30 elliptical and S0 galaxies (Burstein, Bertola, Buson, Faber and Lauer 1987) has allowed us to establish in detail the relationship between the Mg_2 parameter, which is linked to the metallicity (Terlevich, Davies, Faber and Burstein 1981) and the level of the UV flux, measured as (1550–V) color. This well defined relation is shown in Fig. 1, where it is possible to see that for galaxies with high relative UV level the variations of Mg_2 are small. They become much more conspicuous, however, for the redder galaxies. Due to the correlation of Mg_2 with the central velocity dispersion σ, an attempt has also been made to plot σ versus (1550–V) color. The resulting relationship is much more scattered than the previous one, suggesting that the most direct physical correlation of (1550–V) color is with metallicity rather than with central mass, for which σ can be considered an indicator. The above correlation tends to suggest that an important component of the UV light is contributed by the old stellar population. There are mainly two types of old stars that reach temperatures capable of reproducing the observed rising branch in elliptical galaxies: horizontal branch and PAGB stars. There are several reasons to disregard horizontal branch stars. The peculiar bimodal temperature distribution needed to match the observed spectral energy distribution suggests

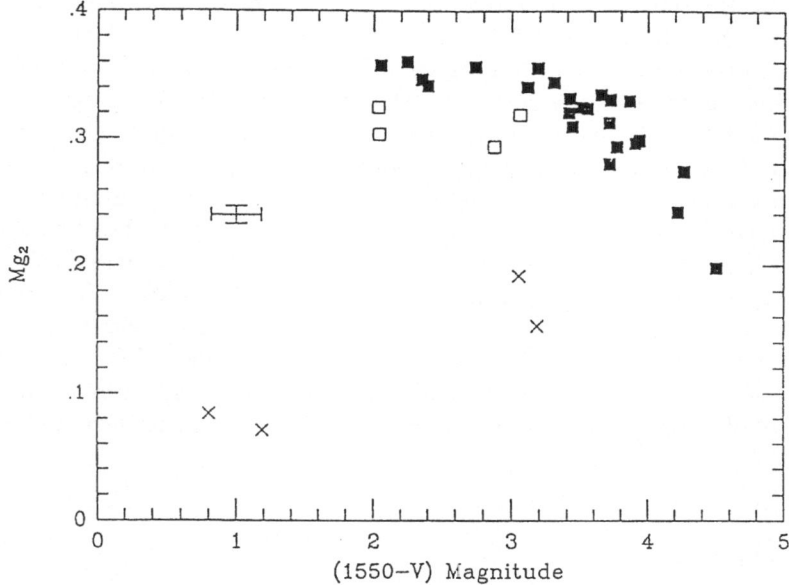

Fig. 1: The optical absorption line index Mg₂ plotted vs. the UV/optical (1550–V) color, expressed in magnitudes. Open squares represent galaxies with emission lines; crosses represent galaxies with known ongoing star formation.

that they are not a good candidate (Nesci and Perola 1985). In addition, the extension of the horizontal branch toward the UV in globular clusters decreases with metallicity, producing an anticorrelation with respect to what is found in ellipticals, although different ranges of metallicity characterize globular clusters and elliptical galaxies. This fact is confirmed by Bica and Alloin (1987) who demonstrate how difficult it is to produce energy distributions steep enough to match the observed one using metal poor globular clusters. It has been demonstrated by Burstein et al. (1987), using the evolutionary tracks by Schoenberner (1983), that nuclei of planetary nebulae can give rise to a composite spectrum which nicely matches the observed slope in the 1000-2000 Å region of the spectrum of the elliptical galaxies. They were also able to show that if a core mass of Mc $= 0.55 \pm 0.01$ is assumed for nuclei of planetary nebulae in M32, the amount of the UV light predicted closely matches that observed. M32 is the galaxy with the lowest UV flux. In order to match the spectrum of NGC 4649, which is one of the galaxies with the highest UV flux (seven times larger than in M32) one needs a core mass for nuclei of planetary nebulae at Mc $= 0.53$. With the theoretical models then available Burstein et al. were unable to find a decrease in the core mass accompanied by an increase in metallicity, as suggested by the observations. A step forward has been made by Bertelli, Chiosi and Bertola (1987), using new models for low and intermediate mass stars, incorporating convection overshooting from central cores, evolved from zero age main sequence to the beginning of the thermally pulsing regime of the asymptotic giant branch phase (Chiosi, Bressan and Bertelli 1986). These models predict a novel calibration between initial and core mass and its dependence on the chemical composition and the mass loss rate. In this way, assuming that the relation between the helium and metal enrichment is $\Delta Y = 3\Delta Z$, they find that an increase in the metal content corresponds to a decrease in the core mass, in the right amount to explain the observed correlation between color and Mg_2. Furthermore, they found that this galaxies have the same age (15–17 Gyr) without invoking a recent episode of star formation in M 32 as in previous models. With this result, PAGB stars seem to be the best and perhaps the unique candidates, at least in the context of old population objects, to explain the UV rising branch in elliptical galaxies.

III. Star formation

The above result does not rule out the possibility that star formation is occurring in elliptical galaxies. Burstein et al.(1987) have shown that continuing star formation models can easily account for the observed slope and flux into the far UV. The only difference is that the contribution to the visual spectrum is much higher than that due to PAGB stars. It should be also mentioned that Bica and Alloin (1987) have shown that the rising branch is easily matched by the spectrum of the young clusters, like those in LMC. However, given the very satisfactory result with PAGB stars, it seems very difficult to consider continuing star formation as the only agent responsible for the UV rising branch in elliptical galaxies. On the other hand there are suggestions, (e.g. presence of cooling flows) that star formation is occurring in elliptical galaxies. In this case, one would expect that the level of the UV rising branch is produced by PAGB stars and also by young stars. The Mg_2 (1550–V) relation defines the level due to the old population, while any excess over this level can be considered as produced by star formation. This approach has been successfully applied to the well known cD galaxy NGC 6166 by Bertola, Gregg, Gunn and Oemler (1986). They carried out a differential spectral synthesis, trying to reproduce the observed spectrum of NGC 6166. They started with the average spectrum of NGC 4472 and NGC 4486, two galaxies with the same metallicity as in NGC 6166, but with a lower level of the UV rising branch. The best fit to the observed fluxes was obtained by adding to the above average spectrum, the energy distribution typical of continuing star formation up to 4% at 5550 Å plus an additional 0.075% contribution of O6 stars. The latter addition was required in order to match the steep rising branch since the assumed young population spectrum was too flat in the 1000–3000 Å region. The rate of star formation implied by this population has been estimated to be 0.1 h^{-2} $M_\odot$ yr^{-1} within the IUE aperture of 10×20 arcsec. It should be noted that the same amount of young stars was deduced in NGC 6166 from analysis of the optical spectrum alone. Similar conclusion about the presence of star formation in a sample of ellipticals was reached also by Gunn, Stryker and Tinsley (1981).

The above differential analysis can be easily understood looking at the Mg_2-(1550-V) color diagram (Fig. 1) where the points representative of ongoing star formation (crosses) are well displaced from the main principal relation and the points representative of galaxies with emission lines (including NGC 6166), and therefore with possible star formation, are also slightly displaced. Both color and Mg_2 are expected to change with increasing amounts of star formation so that a representative point describes in such a diagram a path determined by the color of the star forming population, the percentage of contamination in the visual band and the mean metallicity of the underlying old stars.

IV. Cooling flows

In the previous paragraph we have seen that in NGC 6166, a cD galaxy which is

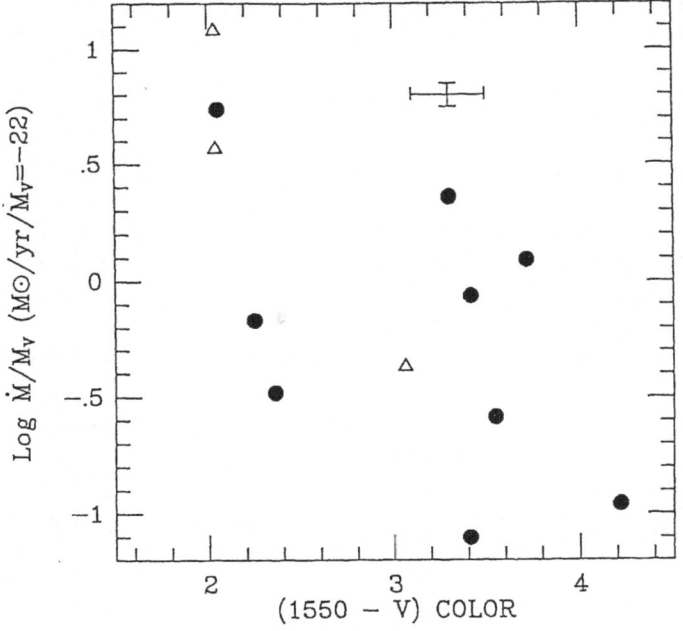

Fig. 2: The logarithm of the ratio of mass cooling flow rate $\dot{M}$ to optical luminosity, normalized to $M_V = -22$ and measured in units of $M_\odot$ yr^{-1} ($M_V = -22$), plotted vs. 1550-V color. Open triangles represent emission line galaxies.

an X-ray emitter, and therefore a galaxy to which the idea of cooling flows can be applied, there is both optical and ultraviolet evidence of star formation. It is therefore interesting to sort out the sample of ellipticals with X–ray emission and with IUE data in order to see whether a correlation between X and UV–flux exists. This was done by Burstein et al. (1987) and is shown in Fig. 2, where the (1550–V) color has been plotted versus the X–ray cooling flow rate dM/dt, normalized to $M_V = -22$. The observational data show the possibility of a trend,although with considerable scatter, of increasing infall rate with increasing level of the UV flux. Taking into account that these galaxies span a relatively narrow range of metallicity where the relation Mg_2 versus (1550–V) color becomes flat, it is not unreasonable to assume that in this metallicity range, variations in color can be due to different rates of star formation. As future work, it should be interesting to compare the data of Fig. 2 with models of elliptical galaxies in which effects of gas infall, metal enrichment and existence of PAGB stars are taken into account.

References

Bertelli G., Chiosi C., Bertola F., 1987, in preparation.

Bertola, F., Capaccioli, M. and Oke, J.B. 1982a, *Ap.J.* **254**, 494.

Bertola, F., Capaccioli,M., Holm, A.V. and Oke, J.B. 1980, *Ap. J. (Letters)* **237**, L65.

Bertola, F., Gregg, M.D., Gunn, J.E., Oemler, A., Jr., 1986, *Ap. J.* **303**, 624.

Bica, E.,Alloin D., 1987 preprint

Burstein, D., Bertola F., Buson L.M., Faber S.M., Lauer T.R., 1987, *Ap. J.*, in press.

Chiosi, C., Bressan, A., Bertelli, G. 1986, in *Advances in Nuclear Astrophysics*, 187, ed. E. Vangioni–Flam, J. Audouze, M. Casse, J.P. Chieze, J. Tran Thanh Van (Editions Frontiers, Paris).

Code, A.D. and Welch, G.A. 1979, *Ap. J.* **228**, 95.

Faber, S.M., 1983, *Highlights of Astronomy* **6**, 165, ed. R.M. West (Dordrechtt: Reidel).

Gunn, J.E., Stryker, L.L. and Tinsley,B.M., 1981, *Ap. J.* **249**, 48.

Nesci, R. and Perola, G.C. 1985, *Astr. Ap.* **145**, 296.

Oke, J.B., Bertola, F. and Capaccioli, M. 1981, *Ap. J.* **243**, 453.

Terlevich, R.J., Davies, R.L., Faber, S.M. and Burstein, D. 1981, *MNRAS* **196**, 381.

CHARACTERISTICS OF EXTENDED GAS AROUND RADIO GALAXIES

I.J. Danziger
European Southern Observatory, Karl-Schwarzschild-Str. 2,
D-8046 Garching b. München, FRG

P. Focardi
Dipartimento di Astronomia, Via Zamboni, 33,
I-40100 Bologna, Italy

ABSTRACT. A qualitative description is given of spectral properties, dynamical behaviour and morphology of a selection of mainly southern radio galaxies which have been observed to have extended ionized gas in their vicinity. The question of the significance of this gas in relation to cooling flows is still very much an open one.

INTRODUCTION

In this presentation a review is given of results (published and unpublished) on radio galaxies which have been accumulating over the past few years through work involving various groupings of collaborators acknowledged at the end. It is worth recalling in brief some of the most significant general results that are emerging from studies of extended emission radio galaxies by other groups.

Recently Heckman et al. (1986) have shown that 30 percent of powerful radio galaxies have peculiar optical morphologies resulting from collisions or mergers. These same galaxies have stronger optical emission, are less luminous optically and occupy regions of lower galaxy density. Van Breugel and Heckman (1982) have noted the following characteristics among radio galaxies that they have studied in detail:

a) Emission line gas tends to have a shell structure morphology tending to lie near the boundaries of the radio structure.

b) The brightness distribution of the optical emission tends to correlate with radio hot spots, but usually with an offset.

c) Observed radio de-polarization seems to be associated with optical ionized gas.

d) Bulk velocities of gas with an amplitude of 200-300 km/sec and asymmetric distributions are common, while velocity widths in the range 300-500 km/sec are frequently observed.

e) Optical line emission occurs mostly within the galaxies.

A. C. Fabian (ed.), Cooling Flows in Clusters and Galaxies, 133–144.
© 1988 by Kluwer Academic Publishers.

In the context of the subject of this meeting it is worthwhile making the following points:

i)   There is no <u>irrefutable</u> evidence for shock heating of extended gas in radio galaxies, although 3C 277.3 (van Breugel et al., 1985) and Cen A (Phillips, 1981) inner regions appear to be the two most likely known candidates for the presence of such gas.

ii)   Photoionization by a power law spectrum has been the most favoured mechanism as in the case of PKS 2158-38 (Fosbury et al., 1982); photoionization by a nuclear black body with a temperature near 130,000°K provides equally satisfactory diagnostic indicators (Robinson et al., 1987; Robinson, this volume) including (importantly) the correct Hβ flux.

iii) In situ photoionization also now seems to be a phenomenon possibly associated with the influence of the radio jet that deserves more study. Three such cases for which supporting quantitative arguments have been given are PKS 0349-27 (Danziger et al., 1984), 3C 171 (Heckman et al., 1984) and PKS 2152-69 (Tadhunter et al., 1987).

iv) Chemical abundances in the extended gas are generally high.

v)   The specific angular momentum content can be as high as for gas in spiral galaxies.

Evidence for most of the characteristics discussed above is present in the group of radio galaxies that are discusssed below. Nevertheless in our selection one is dealing with gas extended on a scale greater on the average than hitherto discussed, and extending up to 90 kpc in the case of PKS 2356-61. This scale is of the same order as that observed for extended emission line gas associated with some quasars.

With the presentation of the following examples of extended emission line gas around radio galaxies it is intended to emphasize the rich variety of phenomena associated with these objects. Which parts, if any, of all of this points an unerring finger in the direction of cooling flows, I leave to the audience to decide.

THE GALAXIES

(a) PKS 0349-27 has been studied in some detail by Danziger et al. (1984) who demonstrated that a photoionizing central source could not easily explain the excitation and flux in some regions of the outlying gas. In the direction of the "jet" towards the north-east, one has a promising candidate example of jet interaction and excitation. The presence of other neighbouring galaxies was considered evidence for a tidal interaction and subsequent presence of extended gas in the form of a ring, arms or a disk; more direct evidence was not then apparent. A recent broad band CCD image (Fig. 1) shows what appear to be two tidal tails extending westward a distance of 47 kpc

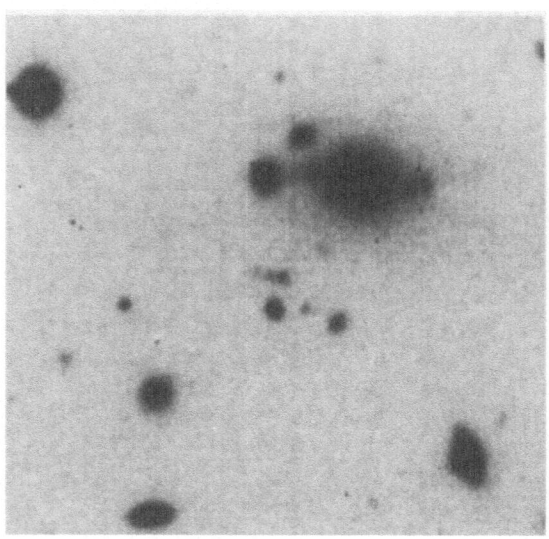

Fig. 1: A continuum image of PKS 0349-27. In all images north is at the top and east to the left. Linear distance information is in the text.

from the radio galaxy, i.e. on the opposite side of the galaxy to the most obvious candidate galaxy for a tidal interaction. There may be other small patches of emission line gas to the south of the main galaxy not hitherto reported. The image suggests that this object is part of a group.

   (b)   The radio and optical properties of PKS 0521-36 have been described elsewhere by Danziger et al. (1983). While the presence of an authentic non-thermal jet and a BL Lac object at the centre of this radio galaxy are probably relevant to many of the other properties of this object, of direct relevance to this symposium is the presence of extended ionized gas and its relationship to the radio structure. This is an example of a system where optical line emitting gas with a fila-mentary or shell structure is found near (see Fig. 2) but not coinci-dent with the radio contours defining the south-east radio lobe and lying at a distance up to 24 kpc from the galaxy. There is evidence at very low surface brightness of ionized gas filling roughly the same volume as the spherically distributed radio emission which is also a characteristic of this source. We see from long slit optical spectra taken on the eastern side that the gas is relatively quiescent with bulk motions no greater than 250 km/sec and without a discernible pattern of motion. All of this gas is of very low excitation according to the small [O III]5007/Hβ ratio ($\sim$ 1).

   Recent radio polarization measurements of this source with the VLA (Goss et al., 1987, in preparation) show a pattern of depolariza-tion that suggests that the ionized gas in the eastern and south-eastern sections is acting as a Faraday screen either because it lies in front of or is mixed with the radio emitting plasma. Because of the extreme faintness of the optical emission in other parts of this object, this work is preliminary and could in principle be further extended to elucidate the radio-optical structure in greater detail.

136

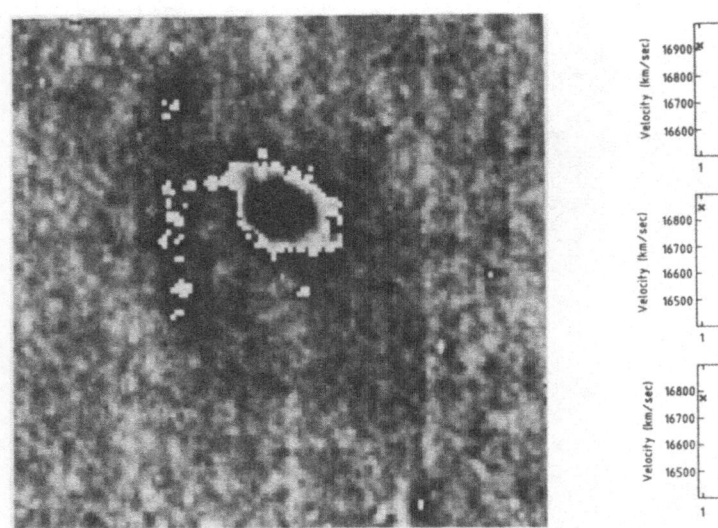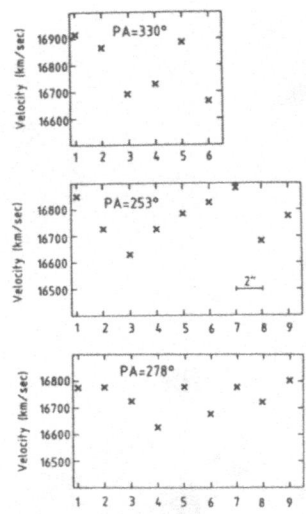

Fig. 2: An Hα image of PKS 0521-36, with velocity data below. Slit PA
330° was centred on filament 10" east of galaxy.

(c) The presence of extended gas around PKS 0634-20 has been
reported by Fosbury et al. (1984). Direct Hα images of this object
(Fig. 3) show well defined arcs of material extending over 70 kpc.
Perhaps the most striking aspect of their morphology is the symmetrical disposition relative to the centre of the galaxy. There is gas also
extending from the centre in a bar-
like or s-shaped structure whose
dynamical behaviour is complex, although very large velocities are not
observed. Although this galaxy has a
well collimated radio jet in the
north-south direction, the lack of
close correspondence between the
gaseous structure and the radio
contours suggests that the morphology
certainly and the excitation probably
are resulting from some other physical processes, such as gravitational
effects and photoionization from a
central source. High excitation
spectra are characteristic of the
central component Danziger et al.,
1978) and the extended gas.

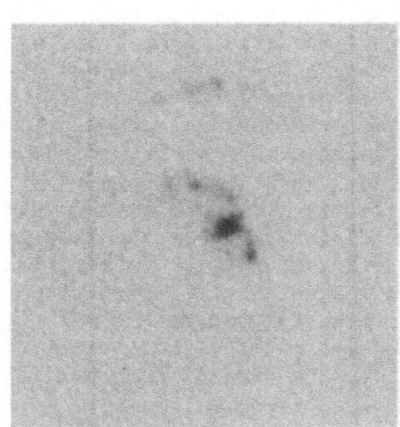

Fig. 3: An Hα image of
PKS 0634-20.

(d) PKS 1216-10 is another high excitation radio galaxy which
remains to be mapped at high resolution at radio wavelengths. The

optical morphology (Fig. 4) provides the most outstanding example of a
complete gaseous ring structure surrounding the optical galaxy with a
diameter of approximately 30 kpc. Because of its distance it has not
been possible to examine the morphology in great detail. However high
quality long-slit spectra at different position angles show systematic
differential velocities that mimic rotation about an axis with a position angle of approximately 0° and a total amplitude of 250 km/sec.
Since the morphology suggests either a disk and ring system viewed
face-on or a spherical system projected on the plane of the sky, it is
not obvious what this motion really means. What little is known about
the radio structure does not suggest an intimate relationship between
radio and optical structure.

If the gaseous structures observed in this galaxy and in PKS
0634-20 result from infall of material resulting from cooling flows or
other disrupted systems, a strong requirement of any modelling is the
very symmetric ring structure. At the moment one cannot say with
certainty whether stars are present in the ring structure. The excitation of the emission line spectrum is too high to be due to photoionization by normal O-B stars.

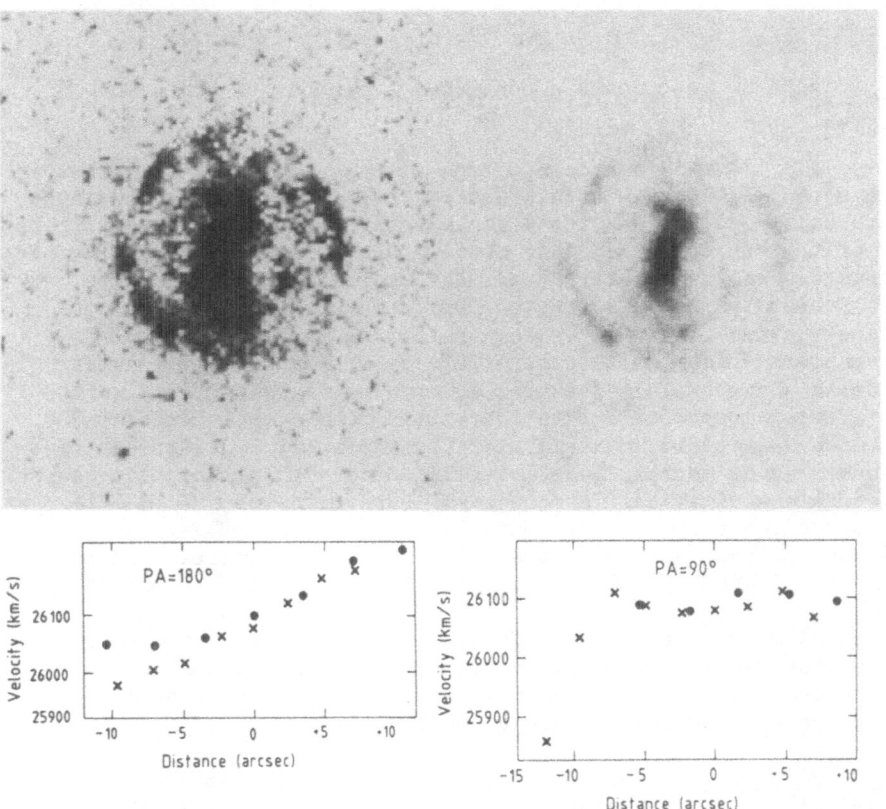

Fig. 4: Hα images of PKS 1216-10 with two cut levels. Velocity data
is below.

(e) NGC 4696 in the Centaurus cluster is an X-ray source. The
filamentary structure (Fig. 5) is very complex and extends more than
7 kpc from the centre. Our unpublished stellar velocity data suggests
a 2-component system, and therefore a possible 2-galaxy interaction.

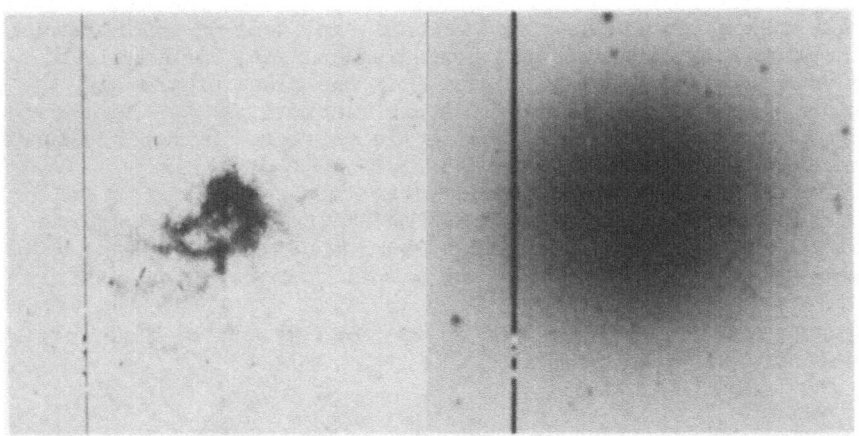

Fig. 5: An Hα image (L) and a continuum image (R) of NGC 4696

(f) PKS 1404-267 stands out from other galaxies in this list in
being a strong extended Einstein Observatory X-ray source, and therefore presumably a good candidate for a cooling flow environment. First
images of this object did indeed seem to show a random scattering of
filaments around the galaxy extending more than 25 kpc from the
centre. Deeper imaging and spectroscopy however reveal a much more
ordered structure (Fig. 6). There does appear to be an irregular
spiral pattern which implies the existence of significant angular
momentum in the gas. This is borne out by the velocity field which
suggests ordered symmetric rotation with a total amplitude of ∼500
km/sec seen at a position angle of 245 degrees. At 270 degrees the
amplitude is considerably less but still ordered. The spectrum shows
only moderate excitation. In many ways (even with respect to being an
X-ray source) this object resembles NGC 5077 in which Bertola et al.
(1987) have proposed that the gaseous ionized disk is misaligned with
both major and minor axes and the stellar configuration is a tri-axial
one.
    A point to be made with this similarity is that if one wants to
conclude that cooling flows are a significant constitutent of the
situation observed in PKS 1404-267, one cannot easily dismiss them in
more mundane galaxies where gaseous disks exist.

(g) PKS 2014-55 provides an example of ionized gas extended
asymmetrically along the major axis of the optical galaxy out to
distances of 24 kpc (Fig. 7). It is not clear whether other galaxies
in the field are interacting with the radio galaxy.

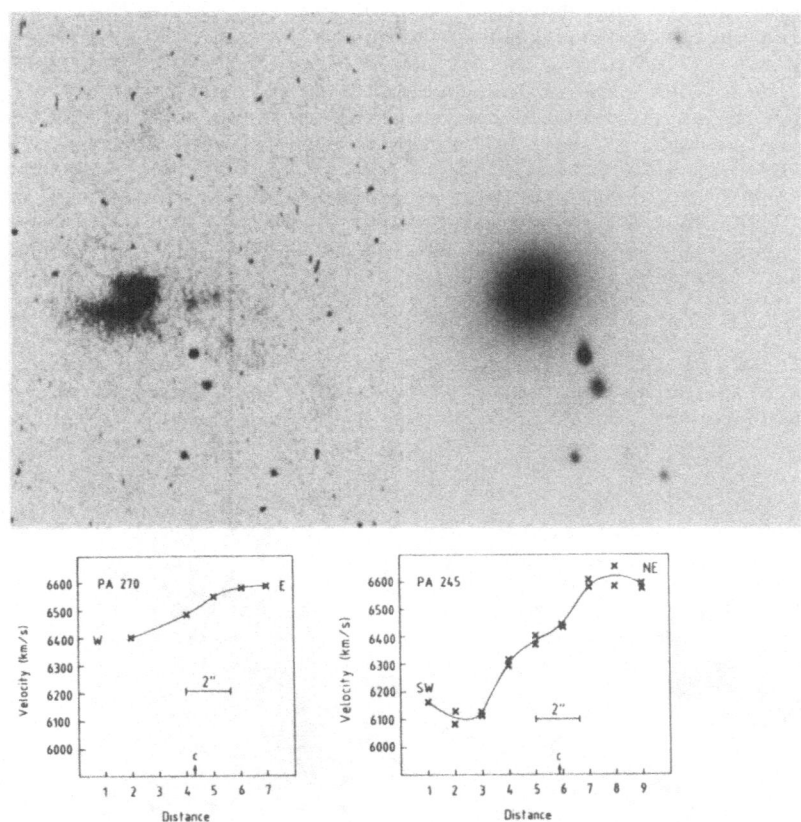

Fig. 6: An Hα (L) and continuum (R) image of PKS 1404-267. Velocity data is below.

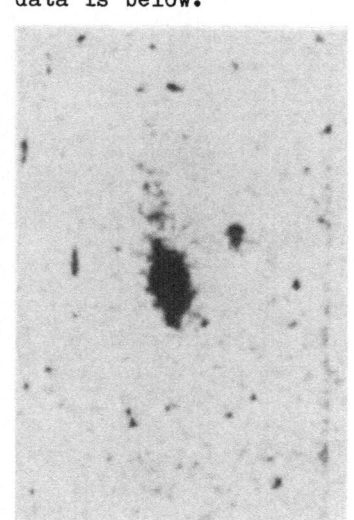

Fig. 7: An Hα image of PKS 2014-55.

(h)　The most striking characteristic of PKS 2152-69 is the
detached nuclear activity described by Tadhunter et al. (1987). A high
surface brightness cloud of gas radiating emission lines of [Ca V],
[Fe VII] and [Fe X], as well as the usual, lower excitation lines,
lies at a projected distance of 8 kpc from the centre of the galaxy.
Although the emission line spectrum of the gas at the centre is of
high excitation with a large [O III]/Hβ ratio it is noticeably less
excited than the detached cloud. This suggests that in situ ionization
is responsible for the remarkable spectrum of the detached cloud. It
seems to lie along the radio axis and therefore suggests that a radio
jet may be responsible for energizing the cloud. Unfortunately, high
resolution radio maps of this object are not yet possible in order to
test this idea. A rather remarkable feature of the cloud which can be
seen in continuum CCD images (Fig. 8) is that a blue continuum seems
to lie on the side facing the centre of the galaxy and therefore in
the direction where the impact of a jet might be felt most.

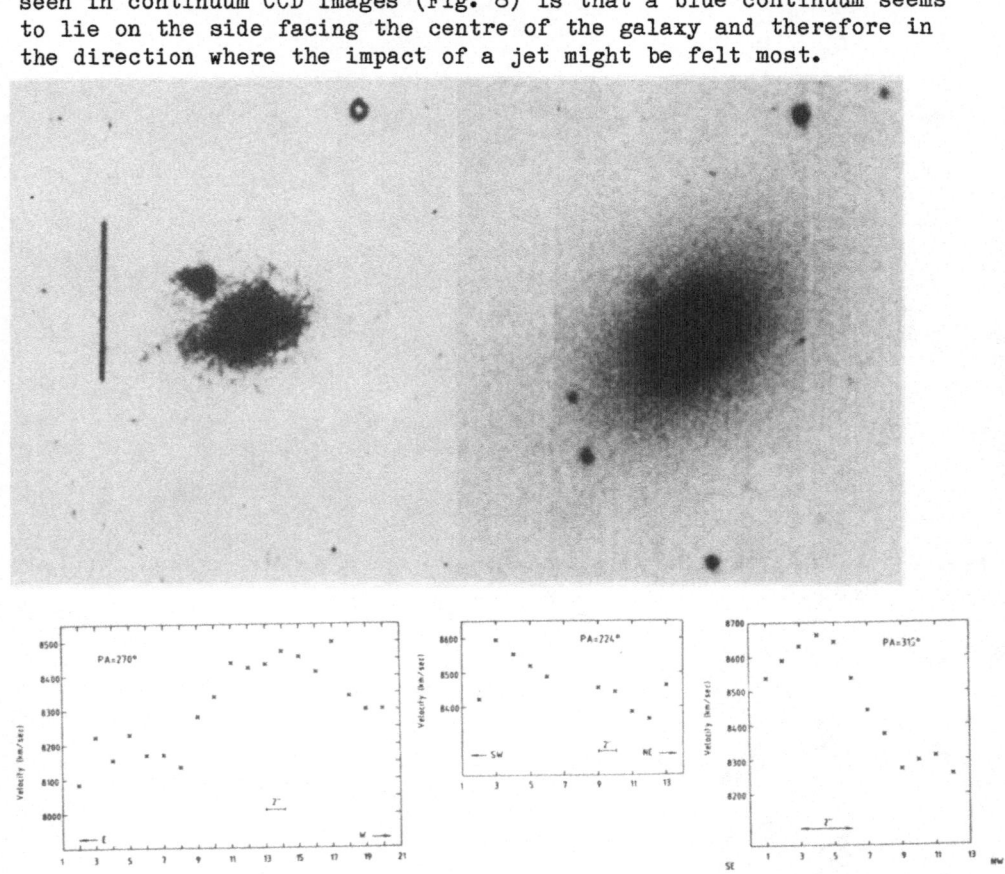

Fig. 8:　An Hα (L) and continuum (R) image of PKS 2152-69. Velocity
data is below. Slit PA 315° was centred on compact object to north-
east.

The dynamical situation of the gas in PKS 2152-69 is not without interest. The detached blob lies approximately on the projected minor axis of the stellar distribution of the main galaxy. There is a significant velocity gradient in the diffuse gas in this direction - a total amplitude of > 250 km/sec. In addition, the detached blob itself has a significant velocity gradient, > 400 km/sec over a distance of 2.3 kpc. In the inner regions of the main galaxy the ionized gas has an insipient spiral structure with filaments arranged in a way that suggests the gas, although having significant angular momentum, is not in an equilibrium situation. Gas extends to distances in excess of 19 kpc from the centre.

(i)   The extension or tail seen in the image of PKS 2206-237 (Fig. 9) clearly points to a galaxy 150 kpc to the south-east as the culprit for interaction, although relative velocities are not known. Whether material has been interchanged is not easy to know. This represents one further variation on the morphology associated with radio galaxies. The tail seen here may have a stellar component.

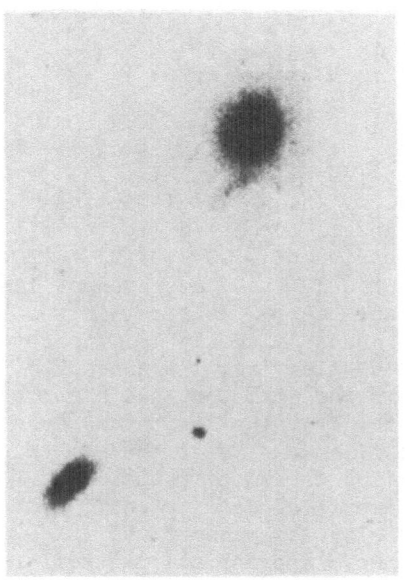

Fig. 9: Continuum image of PKS 2206-237

(j)   PKS 2356-61 is another example of an incomplete shell structure which is more amorphous than the other objects previously mentioned. This galaxy is characterized by asymmetry in morphology and asymmetry in velocity structure. The direct Hα image (Fig. 10) shows a shell reasonably close to the galaxy in the south-east (19 kpc) with another patch much further away (90 kpc) to the north. A long slit spectrum at PA 290° reveals bulk differential motions with an amplitude of 200 km/sec, but a spatial variation difficult to interpret in terms of simple rotation or expansion. The degree of excitation of the

gas in the south-east shell is very high as is the gas at the centre.
Nothing is known about the environment of this galaxy in terms of
interacting neighbouring galaxies.

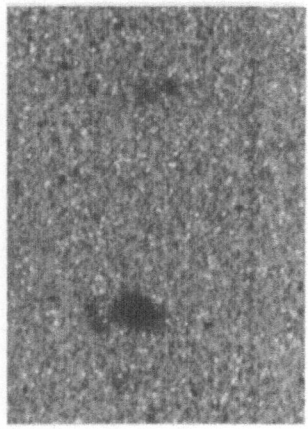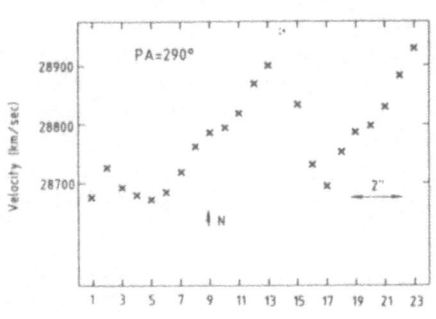

<u>Fig. 10</u>:   An Hα image of PKS 2356-61. Velocity data is next to it.

(k)   PKS 0812+02 focuses our attention on a radio quasar in a
cluster environment (Guzzo et al., in press; Danziger et al., in
preparation). This quasar resides in a galaxy which is interacting
with other smaller galaxies belonging to a well populated cluster. It
has extended gaseous filaments with a dimension of ∼100 kpc (H_o = 100
km/sec/Mpc) that could be gravitationally bound. These are amongst the
largest filamentary structures of this kind observed so far (Hintzen
et al., 1986). Another filament links the quasar directly to a nearby
radio galaxy. The relative velocity of this galaxy and the quasar is
∼1100 km/sec in the rest frame, which seems too high for this extended
structure to result from tidal interactions. From the few parameters
we can measure for these extended gaseous filaments, viz. the
[O II]3727/[O III]4959,5007 ratio, excitation caused by photoionization by the quasar seems possible. Measurements of density would be
invaluable for deciding this question and also the mass of the filaments. Other excitation mechanisms may be more appropriate particularly since this may be an authentic case of a cooling flow triggered by
the massive quasar/galaxy lying in the potential well of a populous
cluster. It was not observed by the Einstein Observatory but will
certainly be on the next X-ray Observatory program.

CONCLUSION

One has seen that our sample of radio galaxies embodies a full
range of properties of the associated extended gas. A complete range
of levels of excitation of the gas is present. There is also a variety

of observed dynamical behaviours ranging from symmetric, apparently gravitationally bound gas to asymmetric bulk motions not easily interpreted except by postulating non-equilibrium settling of the gas. Asymmetric morphology is also common although the work described here emphasizes the occurrence of rather symmetric arcs and rings of gas. Clearly there is modelling work to be done.

To try to force all of these phenomena into a single system for the origin and the excitation of the gas is probably unrealistic. We surely see at least two different excitation mechanisms - photoionization from a nuclear source, and in situ ionization energized by the radio jet. The origin of the observed gas may also require different mechanisms or, at least, different scenarios for different objects. The location of PKS 0812+02 at the centre of a rich cluster of galaxies provides a completely different setting from that for PKS 0634-20 which seems to have very few neighbours. It is also true that we are lacking good statistical samples of radio galaxies for which deep X-ray observations are available. It should be noted here that two obvious characteristics of the extended gas, viz. reasonably normal chemical abundances and high specific angular momentum content, do not provide a satisfactory discriminant among various ideas for the origins of the gas. See, for example, Cowie et al. (1980).

The link between radio galaxies with extended gas and quasars with extended gas is being sought but still remains elusive. This link may, if elucidated, in the future provide a better insight into how significant is the role of cooling flows in the radio galaxy phenomenon.

We gratefully acknowledge the continuing collaboration with R. Fosbury which has generated many of the ideas for work in this area.

REFERENCES

Bertola, F., Bettoni, D., Danziger, J., Sadler, E., de Zeeuw, T.: 1987, Mon. Not. R. astr. Soc., in press.
Cowie, L.L., Fabian, A.C., Nulsen, P.E.J.: 1980, Mon. Not. R. astr. Soc. **191**, 399.
Danziger, I.J., Fosbury, R.A.E., Goss, W.M., Bland, J., Boksenberg, A.: 1984, Mon. Not. R. astr. Soc. **208**, 589.
Danziger, I.J., Ekers, R.D., Fosbury, R.A.E., Goss, W.M., Shaver, P.A.: 1983, in "Astrophysical Jets", A. Ferrari & A.G. Pacholczyk, eds. (Reidel, Dordrecht), 131.
Danziger, I.J., Goss, W.M., Frater, R.H.: 1978, Mon. Not. R. astr. Soc. **184**, 341.
Fosbury, R.A.E., Tadhunter, C.N., Bland, J., Danziger, I.J.: 1984, Mon. Not. R. astr. Soc. **208**, 955.
Fosbury, R.A.E., Boksenberg, A., Snijders, M.A.J., Danziger, I.J., Disney, M.J., Goss, W.M., Penston, M.V., Wamsteker, W., Wellington, K.J., Wilson A.S.: 1982, Mon. Not. R. astr. Soc. **201**, 991.
Heckman, T.M., van Breugel, W.J.M., Miley, G.K.: 1984, Ap.J. **286**, 509.

Heckman, T.M., Smith, E.P., Baum, S.A., van Breugel, W.J.M., Miley,
     G.K., Illingworth, G.D., Bothun, G.D., Balick, B.: 1986, Ap.J.
     311, 526.
Hintzen, P., Romanishin, W.: 1986, Ap.J. 311, L1.
Phillips, M.M.: 1981, Mon. Not. R. astr. Soc. 197, 659.
Robinson, A., Binette, L., Fosbury, R.A.W., Tadhunter, C.N.: 1987,
     Mon. Not. R. astr. Soc. 227, 97.
Tadhunter, C.N., Fosbury, R.A.E., Binette, L., Danziger, I.J.,
     Robinson, A.: 1987, Nature 325, 504.
van Breugel, W., Miley, G., Heckman, T., Butcher, H., Bridle, A.:
     1985, Ap.J. 290, 496.
van Breugel, W., Heckman, T.: 1982, in "Extragalactic Radio Sources",
     D.S. Heeschen & C.M. Wade, eds. (IAU), 61.

HI ABSORPTION DETECTION IN THE PERSEUS COOLING FLOW

W. J. Jaffe[1], A. G. de Bruyn[2], D. Sijbreng[3]

[1] Sterrewacht Leiden, P.O. Box 9513, 2300 RA Leiden (NL)
[2] Radiosterrewacht te Dwingeloo, Oudehoogeveensedijk 4, 7991 PD Dwingeloo (NL)
[3] Kapteijn Laboratorium, P.O. Box 800, 9747 AE Groningen (NL)

ABSTRACT. With the Westerbork Synthesis Radio Telescope (WSRT) we detect HI absorption at 5250 km/s in front of the nucleus of NGC 1275, with an optical depth of .003 and a dispersion of 205 km/s. We do not detect HI associated with the 30" halo in absorption or emission.

The current evidence for cooling flows consists primarily of emission from ionized components in X-ray and optical bands. There should also be substantial amounts of neutral gas, either atomic or molecular, present. For a residence time in the neutral phase of 10^7 years, perhaps typical of our Galaxy, we would expect $10^9 M_\odot$ or so of neutral material in a cluster flow. So far this material has not been seen in molecular emission or absorption, (Jaffe, 1987, O'Dea and Baum, 1987) and only one, somewhat questionable, detection in HI absorption has been claimed in the 5200 km/s system in Perseus (Crane *et al.*, 1982).

We have confirmed and extended the Perseus result using the superior dynamic range of the Westerbork Synthesis Radio Telescope at HI. The continuum radio source near NGC 1275 consists of a strong nuclear point source (about 20 Jy at 21 cm, flat spectrum and variable), a 30" halo (about 2 Jy, normal spectrum $\alpha \simeq -0.7$), and various more extended components of no interest here. In this preliminary study we looked for absorption in front of the nuclear component, and absorption or emission from the 30" region.

The quality of the nuclear absorption spectrum was limited by receiver drift. To minimize this we used a relatively short observation (2 hr) followed immediately by a similar observation of an absorption free calibrator, 3C 48. We divided the nuclear fluxes by those of the calibrator on a channel by channel basis, and, assuming the wings of the spectrum to be absorption free, corrected for the relative continuum flux of NGC 1275/3C 48 (0.861) and for the relative spectral index of the two sources ($\alpha_{1275} \simeq \alpha_{48} + 2.4$). We observed 16 independent channels, each with a velocity width of 132 km/s, for a total spectral width of 2100 km/s. This procedure was repeated 6 weeks later after the movable telescopes at Westerbork had been shifted. The results were consistent with each other (after correction for the 1.6% variation of the flux of NGC 1275 Perseus in the meantime), and their average is shown in Fig. 1.

With the exception of the end channels, the achieved dynamic range is about one part in 3×10^3. The observed absorption profile is fit reasonably well with a gaussian centered at 5250 ± 10 km/s, with a peak optical depth of $.00288 \pm .00006$, and a dispersion of 205 ± 10 km/s. The integral under the profiles is 1.71 K km/s. For optically thin clouds this corresponds to a surface density of $3.1 \times 10^{18} T_s$ atoms cm^{-2} where T_s is the spin temperature. Typical spin temperatures of Hydrogen in spiral galaxies are 100-1000 K. If the gas is very close to the nucleus, and not too dense, the nonthermal microwave radiation will drive up the spin temperature, for example by about

145

A. C. Fabian (ed.), Cooling Flows in Clusters and Galaxies, 145–147.

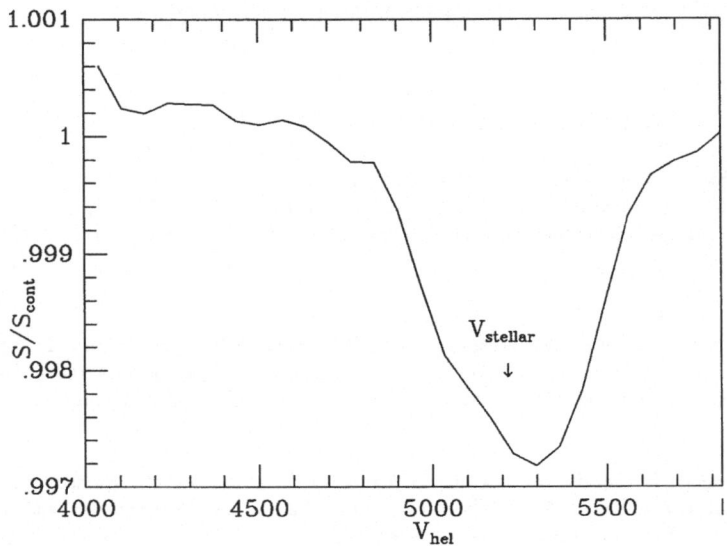

Figure 1. HI absorption spectrum in front of the nuclear source in NGC 1275.

$350K/n_H$ at a distance of 3 kiloparsec. The implied total mass of HI depends critically on the spatial distribution, which cannot be derived from these data. If distributed evenly over a 10 kpc sphere, the total mass would be $3 \times 10^9 (T_s/300K)M_\odot$

The most interesting features of the absorption profile are its smoothness and its symmetry with respect to the stellar velocity profile of NGC 1275. The HI must consist of many clouds, since a single cloud with a velocity spread of 200 km/s would soon evaporate. The smoothness of the profile indicates that the number of clouds along the line of sight must be much larger than the number of independent measured channels, so of order 100. The proximity of the velocity centroid to the 5200 km/s stellar velocity of NGC 1275 is also surprising since the gas is seen only in absorption, on the near side of the galaxy. Hence we are not seeing gas smoothly dropping out of a centripetal flow, which would show a positive velocity offset. Apparently we are seeing a large number of clouds in highly turbulent motion. The clouds must be relatively short lived. 100 or so clouds in turbulent motion along the line of sight would collide every 10^6 years or so if the total path length through the system is 10 kpc.

In an attempt to locate the gas, we looked for it in emission or in absorption in the 30" halo using an entire 12 hr observation. We constructed velocity profiles at each point in the halo, averaged them together, in an annulus about the nucleus and removed a linear baseline. We detected no emission or absorption at a level of 7 mJy (3σ). The interpretation of mixed emission and absorption can be tricky and is very sensitive to the exact geometry. If the HI coincides with the continuum emission and is further homogeneous, then these data place rough upper limits of 0.014 on the peak optical depth through the galaxy, and of 300 K on the spin temperature. Thus the data do not as yet exclude the nuclear absorption gas from being in the 30" (~10 kpc) halo. It is also possible that the absorption gas is closer to the nucleus. The velocity dispersion is similar to that found in the narrow line regions of Seyfert galaxies, and, if the material is confined to a region of, say, 300 pc radius about the nucleus, the total mass ($\sim 10^6 M_\odot$) is similar to that of the ionized material. In this case the neutral component of the cooling flow gas is still missing. Combining all our data on this source (4 × 12 hours) should clarify this situation.

Our further work in the area will include combining all the available data to improve the sensitivity in the halo, higher spectral resolution measurements of the nuclear absorption, searching for HI emission from the spiral galaxies hear NGC 1275, and reduction of similar observations made near M 87.

References

Crane, P.C., van de Hulst, J.M., Haschick A.D., 1982 in *Extragalactic Radio Sources, IAU Symposium 97*, Reidel, Dordrecht, p. 307

Jaffe, W. 1987, *Astron & Astrophys 171*, 378

O'Dea, C.P., Baum, S.A. 1987, *Astron. J.* in press

RADIO OBSERVATIONS OF NGC1275 AT 90 AND 20cm

A. Pedlar, H. Ghataure, R.D.Davies, B. Harrison
Nuffield Radio Astronomy Laboratories
Jodrell Bank, Macclesfield
Cheshire, SK11 9DL, U.K.

R. Perley & P.C. Crane
National Radio Astronomy Observatory,
VLA project, Socorro, NM
USA 87801

Summary We present high dynamic range VLA maps of NGC1275 at 20cm and 90cm , with resolutions ranging from 1" to 40". Over the central 30" there is evidence for collimated ejection in PA 160 degrees. Outside this the radio structure bends rapidly by approximately 90 degrees before merging into the 10' radio halo. It is suggested that the relativistic gas, which is responsible for the radio emission, may have a considerable effect on any cooling inflow , and should be taken into account in models of such phenomena.

1 INTRODUCTION

NGC1275 is the principal galaxy of the Perseus cluster of galaxies and appears to be a nearby example of a core dominated radio source (Browne *et al.*, 1982). It has extensive radio and Xray haloes, and is surrounded by a system of Hα filaments with two distinct velocity fields at 5000km/s and 8000km/s. The most plausible explanation for the high velocity filaments is that they are an infalling spiral galaxy (Rubin *et al.* 1977). The low velocity filaments have been interpreted as evidence for a pressure driven accretion flow onto NGC1275 (Fabian & Nulsen 1977).

The radio data on the halo of NGC1275 has been severely limited by dynamic range problems caused by the presence of the intense VLBI core. Thus initially, the structure in the halo was mapped only at low frequencies with angular resolutions of order 1' (e.g. Gisler & Miley 1979). Subsequently, using new techniques, Noordam & deBruyn (1982) produced a high dynamic range maps of this source at 20cm and 6cm, and Pedlar *et*

A. C. Fabian (ed.), Cooling Flows in Clusters and Galaxies, 149–154.

al. (1983) produced a 1" resolution MERLIN map at 73cm which showed evidence for collimated ejection in PA 160 degrees.

2 THE OBSERVATIONS.

In order to further investigate the radio structure we have used the full VLA at 20cm in A,C and D configurations, resulting in angular resolutions ranging from 1"-40". Maps with dynamic ranges exceeding 10000:1 were obtained for all 3 configurations. We also used 13 antennas of the VLA to measure the structure at 90cm using A and B configurations. This presented particular problems caused by the presence of a large number of confusing sources in the 3 degree primary beam. This was partially overcome by using narrow bandwidths and short integration times, which enabled us to map most of the confusing sources simultaneously with NGC1275.

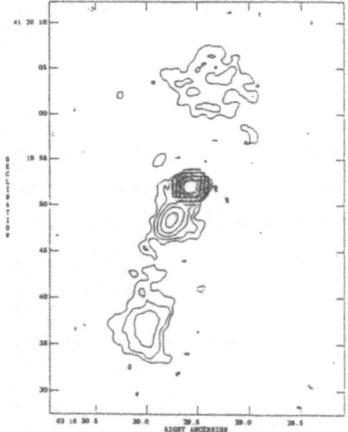

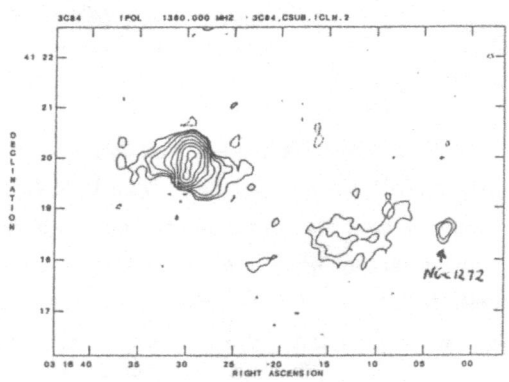

Figure 2 "C" array 1380MHz VLA map. The contour levels are -2, 2, 4, 8, 16 etc mJy per beam. The restoring beam is 10" HPFW. The compact source at the nucleus has been removed from the data.

Figure 1 "A" array 1380MHz VLA map. The contour levels are -2.5, 2.5, 5, 10, 20 etc mJy per beam. The restoring beam is 1.2" HPFW.

In Figures 1-3 we show maps obtained from A,C and D array observations at 1380 MHz. The A array map agrees well with the MERLIN 73cm map, and by comparing the flux of the two southern components we estimate a spectral index of -0.5 for the inner knot, and -0.7 for the outer. Both the MERLIN and VLA maps show an asymmetrical structure, with only weak emission to the north. This initially appeared to differ from the 6cm map of Noordam & deBruyn (1982) which showed a more symmetrical structure. This discrepancy is somewhat resolved if the higher resolution VLA data is tapered to lower resolution, which results in the total flux of the northern component being approximately the same as the southern. The C array map, with a resolution of 10" shows an almost symmetrical N-S structure with two features extending to the east and west of the 30" central source. Both extend at least 1' and the western component may continue as part of a diffuse feature which extends in the direction of NGC1272. The D array map has a dynamic range of 80000:1 and shows two new weak features extending 5' east of the nucleus.

The 90cm map shown in Figure 4 has been produced by combining the A and B array data which gives reasonable coverage of the Fourier plane and enables reliable maps to be made out to several arcminutes. The resolution of order 5" and confirms the S type symmetry of the 30" source suggested in the 20cm C array map. The sudden change in PA from 160 to 235 degrees in the south is particularly clear in these data.

3 DISCUSSION

i) The Hα filament in PA 172 degrees and the radio structure.

The MERLIN 73cm data (Pedlar *et al.* 1983) suggested that collimated ejection was occurring in PA 160 degrees. Although NGC1275 is surrounded by a complex series of Hα filaments (Rubin *et al.* 1978), only three show large velocity gradients (i.e. greater than 100km/s) and of these, the filament in PA 172 degrees shows by far the largest gradient (600km/s) and extends over 1' (20kpc) from the nucleus. The precise alignment of this filament with the VLBI structure (Romney *et al.* 1981), together with its unique dynamics, led Pedlar *et al.* (1983) to speculate that the filament was tracing out the position of the beam along which energy was transferred from the active nucleus into the radio halo. Marginal evidence for this hypothesis could be seen in the 20cm map of Miley & Perola (1975) in which the extended radio structure appeared to enclose the above filament.

The new VLA maps, however, show little evidence for radio structure associated with the filament beyond the MERLIN structure (knot 2). This is particularly clear in the 90cm map (Figure 4) and the C array 20cm map (Figure 2) in which neither map shows radio emission associated with the outer part of the filament. Rather than continuing along PA 160 degrees, either as knots or continuous structure, the radio emission beyond 20" from the core shows a sudden bend to the west along PA 235 degrees. This could be interpreted as evidence that the radio jet bends, although it is not clear that beyond this point we have a collimated flow, rather than clouds of relativistic particles drifting under the influence of external forces. It is possible that the structure extending towards NGC1272 (see Figure 2) is an extension of the above feature.

ii) Problems with pressure driven inflows

There is considerable evidence (e.g. Fabian & Nulsen 1977) that pressure driven inflows are occurring in the hot gas associated with clusters and elliptical galaxies. NGC1275 is often cited as an archetypal case for such cooling flows (e.g. Fabian *et al.* 1981, Branduardi-Raymont *et al.* 1981, Hu *et al.* 1983), largely on the evidence from X-ray observations alone, with the Hα filaments cited as further evidence. The effect of the relativistic particles and magnetic fields, which are responsible for the radio emission, upon the cooling inflow has not been fully investigated. Elsewhere in this volume Bohringer & Morfill have reported a preliminary investigation of this problem. As can be seen in Figure 3, the radio halo has an extent of at least 10', which includes the region for which there is good evidence for a decrease in temperature characteristic of a cooling flow (Fabian *et al.* 1981). Using the density and temperature profiles derived by Fabian *et al.* the thermal pressure ranges from 2×10^{-10} to 10^{-11} dynes cm^{-2} over this region. We can estimate the pressure of the relativistic gas using the minimum energy assumption (e.g. Moffat 1975). The radio halo has a flux density of approximately 5Jy at 1380MHz and hence an average pressure of 3×10^{-12} dynes cm^{-2}. Although this is an order of magnitude lower than the typical thermal pressures in the X-ray emitting gas, the radio derived pressures are by no means negligible. Using the same method, we can estimate the total energy in relativistic particles in the radio halo to be 10^{59} ergs (for k=100). If we assume the source to have an age of order 10^{8} years, then the average rate of supplying energy to the relativistic particles in the halo is at least 10^{43} ergs s^{-1}. The most likely source for this energy is via collimated ejection from the active nucleus. The detailed mechanism for transferring energy generated in the nucleus ,via the jet or whatever, to relativistic particles in the halo is not understood, but it is difficult to envisage a total process which is more than a few percent efficient. Thus of order 10^{44} ergs s^{-1} could easily be released into the cluster environment as a consequence of the production of the relativistic particles.

Within the 5Jy halo a large fraction (2Jy) of the extended radio flux originates within a 1' component which surrounds the nucleus, and thus has an average pressure of 3×10^{-11} dynes cm^{-2}. The pressures within individual knots associated with the radio 'jet' rise to 10^{-10} to 10^{-9} dynes cm^{-2} (Pedlar *et al.* 1983). The radio structure over this region suggests a series of components which are leaving the nucleus and expanding into a lower pressure medium. Thus it seems likely that the radio emitting plasma over the central 20kpc has a pressure at least comparable or greater than the thermal pressure of the X-ray emitting gas, and clearly any models of the pressure driven inflow must take this into account. Scenarios in which the cooling flow fuels the nuclear activity (Sarazin and O'Connell 1983) are particularly difficult to reconcile with these observations. Also the radial Hα filaments, which are often cited as strong evidence of the cooling flow, are largely concentrated in this same region. However, in view of the high pressure of the radio emitting plasma, it seems likely that any simple radial inflow will be, at the very least, highly perturbed. In fact, in view of the total amount of energy being injected into this region by the active nucleus (see previous paragraph), it seems more likely that the radial filaments are indicative of localised *outflows*, similar, but on a larger scale, to those seen in M82 (Axon & Taylor 1978). Alternatively they may also be formed as a consequence of Rayleigh Taylor instabilities forming between the relativistic and thermal gas as suggested by Bohringer and Morfill (this volume).

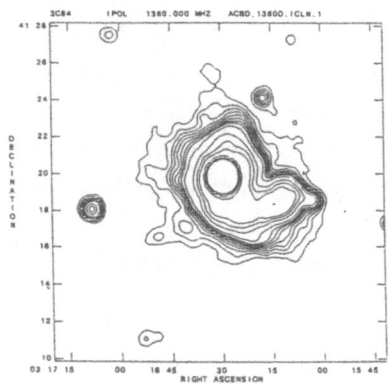

Figure 3 "D" array 1380MHz VLA map. The contour levels are -2, 2, 4, 6, 8, 10, 12, 14, 16, 20, 30, 40, 50, 100, 150, 200, 250, 300, 350 etc mJy per beam. The restoring beam is 40" HPFW.

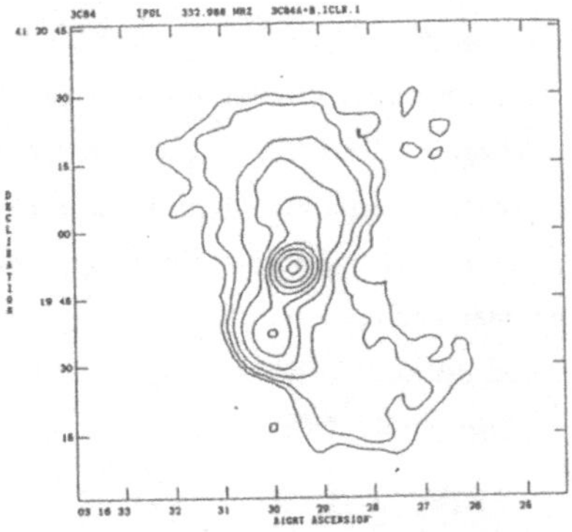

Figure 4 Combined "A" and "B" array VLA map at 323MHz. The contour levels are 25, 50, 100, 200, 400, 800 etc mJy per beam. The restoring beam is 5" HPFW.

At 20" from the core, the radio structure changes rapidly from a well collimated Fanaroff-Riley type I 'jet' in PA 160 degrees, to an ill defined 'tail' bending to the west (see Figure 4) and possibly continuing for at least 80kpc in the direction of NGC1272 (Figure 2). Although this phenomena could be explained by a variety of ad-hoc models, it seems to us that the change in character of the radio structure marks the boundary where the pressure of the relativistic gas becomes comparable to the pressure in the thermal gas. Hence, at this point, the motion of the relativistic gas ceases to be dominated by collimated ejection from the nucleus, and may become influenced by the mass motions in the cluster medium. The pressure of the radio emitting plasma, in the eastern section of the tail where it bends away from the 'jet', can be estimated by the usual minimum energy arguments. At 327MHz the extension visible in Figure 4 has a flux density of 1.5Jy and an approximate size of 20", from which we estimate its pressure to be of order 10^{-11} dynes/ cm^{-2}, in reasonable agreement with the pressure of the thermal gas implied by the densities and temperatures at this radius derived by Fabian *et al.* 1981). If the energy density in the relativistic gas were comparable, or less than, the energy density

154

in the thermal gas, then we could expect mass motions in the thermal gas to strongly influence the distribution of radio emitting gas. The asymmetric structure of the radio halo, particularly the tail stretching towards NGC1272, suggests either that the effect of radial flows is small compared with non-radial flows in the cluster gas, or that NGC1275 has a significant linear motion with respect to the cluster gas. If we assume that the 'tail' is bent away from the jet by the ram pressure of an east-west flow of thermal gas with a density of 0.03 cm^{-3}, then we can estimate the flow velocity to be at least 200km s^{-1}. Such a flow is subsonic and could be a consequence of the passage of NGC1272, or may simply represent cluster weather possibly generated by the massive activity in NGC1275. Whatever the cause, the flow would have a velocity more than an order of magnitude greater than the velocities expected from radial cooling flows at this radius, and brings into question whether simple cooling radial inflows can persist in the presence of strong nuclear activity.

A detailed account of this project will be submitted shortly to Monthly Notices of the Royal astronomical Society.

REFERENCES

Axon D.J. & Taylor, K.E. 1978 , *Nature*,**274**, 37.

Branduardi-Raymont *et al.*,1981,*Astrophys. J.*,**248**, 55.

Browne, I.W.A. *et al.*,1982,*Mon. Not. R. astr. Soc.* ,**198**, 673.

Fabian, A.C. & Nulsen P.E.,1977,*Mon. Not. R. astr. Soc.* ,**180**, 479.

Fabian, A.C. *et al.*,1981,*Astrophys. J.*,**248**, 47.

Gisler,G.R. & Miley,G.K.,1979,*Astron. & Astrophys.* ,**76**, 109.

Hu, E. *et al.*,1983,*Astrophys. J.*.,**275**, L27.

Miley, G.K. & Perola,G.C.,1975,*Astron. & Astrophys.*,**45** , 223.

Noordam, J. & deBruyn, A.G., 1982, *Nature*,**299**, 598.

Pedlar,A. *et al.*,1983,*Mon. Not. R. astr. Soc.* ,**203**, 667.

Romney *et al.*, 1981, *IAU Symposium no.* 97, p291.

Rubin, V. *et al.*,1977,*Astrophys. J.*,**221**, 693.

Rubin,V. *et al.*,1978,*Astrophys. J. Suppl.* ,**37**, 235.

Sarazin,C.L. & O'Connell,R.W.,1983,*Astrophys. J.* .,**268**, 552.

Photoionization versus shocks in the filaments of NGC 1275

C. Boisson and D. Péquignot
DAEC
Observatoire de Paris, Section de Meudon
92195 Meudon Principal Cédex
France

Abstract. We show that the excitation of the filaments observed around NGC 1275 is more probably due to photoionization by the active nucleus radiation than to local radiative shocks. The photoionized gas is in pressure equilibrium with the hot ambient gas of the cooling flow.

1. Introduction

The nucleus of the giant elliptical galaxy NGC 1275 shows numerous signs of activity which led to its classification as a Seyfert although it looks atypical (e.g., Véron 1978). Located at the centre of the Perseus cluster it is a strong radio and X-ray emitter (e.g., Mushotzky et al., 1981) and displays a remarkable giant filamentary gaseous envelope (Lynds, 1970), which is most often interpreted as infalling intergalactic gas (e.g., Fabian and Nulsen, 1977).

The spectrum of the filaments obtained by Kent and Sargent (1979) (KS79) at 10 kpc from the nucleus ($\sim$ 20") is of very low excitation ($[OIII]/H_\beta \sim 1$, $[OII]/H_\beta \sim 6$).

The question we address here is whether the emission from the filaments is due to the cooling and recombination of shocked gas or to photoionization by the central active nucleus. Stars are unlikely in view of the large strength of e.g.[NI]5200Å and [OI]6300Å.

The theoretical analysis by KS79 is obsolete. Under the assumption of photoionization by the Seyfert nucleus, they conclude to optically thin clouds at density over 10^3 cm^{-3}. But their calculations do not include charge exchanges, which, in their conditions, would modify some of their line ratios by orders of magnitudes.

A. C. Fabian (ed.), Cooling Flows in Clusters and Galaxies, 155–158.

2. A possible cooling scenario

X-ray observations from hot gas in the Perseus cluster (Fabian et al., 1981) around NGC 1275 give electronic temperature and densities consistent with approximately constant pressure $n_e \, T_e \sim 6 \; 10^5$ K cm^{-3}.

The following scenario is partly inspired by the review of Fabian et al., 1984.

The radiative energy loss due to the emission of the observed X-rays is sufficient to cool the 10^8 K intra-cluster gas in Hubble time scale. With the above quoted numbers this leads to :

Phase 1 : Cooling flow in quasi hydrostatic equilibrium down to 10^6 K ($n_H \sim n_e \sim 0.6$ cm^{-3}).

Phase 2 : Fast isochoric cooling ($t_{cooling} \ll t_{hydr}$) down to T $\sim 10^4$ K ($n_H \sim 0.6$ cm^{-3}).

Phase 3 : The surpression of the environment by a factor of ~ 100 triggers violent compression at the sound speed of the hot medium, i.e. ~ 10 times that of the cool gas. The compression is isothermal rather than adiabatic as t_{cool} always $\ll t_{hydr}$.

Phase 4 : Repressurization lasts until the density of the cold zone gives an $n_e \, T_e \sim 6 \; 10^5$, that is $n_H \sim 60$ cm^{-3}.

Phase 5 : The compression stops suddenly leading to a radiative shock and so to heating of the cool gas. For shock velocity ~ 100 km s^{-1} (sound speed at 10^6 K), the temperature after the shock is $\sim 10^5$ K and the density jumps by a factor 4, $n_H \sim 240$ cm^{-3}.

Phase 6 : Again the gas cools down rapidly to T $\sim 10^4$ K, at which temperature hydrogen begins to recombine, the density increasing by an extra factor ~ 20 (Cox, 1972) : $n_H \sim 5 \; 10^3$ cm^{-3}. At such a density the photoionization by emission from the nucleus is certainly negligible. One can conjecture that the filaments are thick enough that their inner part is shielded from the radiation, so that the temperature there, drops to T ~ 100 K very rapidly.

Phase 7 : But the "skin" of the filaments is anyhow illuminated by the UV radiation and is not cold. Photoionization by the nucleus continuum (ionization parameter U = (n° ionizing photons) / (4 π r^2 c n_H) $\sim 10^{-6}$ at 5 10^3 cm^{-3}) stabilizes the temperature at T $\sim 3 \; 10^3$ K (gas essentially neutral). The skin of the filaments, in overpressure, starts expanding in the hot medium: U increases and gas ionization starts again when U $\geq 3 \; 10^{-5}$.

The initial question is then: do we observe phase 6 or phase 7 (other phases produce much less optical emission). The main difference is in the density of the gas emitting the [OII] and [SII] lines (temperature effects are expected for [OIII] but [OIII]4363Å is too faint to be observed).

3. Discussion

Kent and Sargent (1979), combining optical and X-ray measurements, inferred a spectrum for the nucleus of $L_\nu = 2.8 \ 10^{28} \ (h_\nu \ / \ 13.6)^{-1.1}$ erg s^{-1} Hz^{-1}. Using IUE archive data, we find a fairly flat UV continuum and $L_{(1300\text{\AA})} \sim 1.4 \ 10^{29}$ erg s^{-1} Hz^{-1} (adopting E(B-V) = 0.43 based on $H_\alpha \ / \ H_\beta$ ratio). This is 3.4 times the KS79 luminosity. But this is not too surprising if one considers that NGC1275 is a Seyfert galaxy, where so called "blue-bumps" are common. The filaments far from the centre, are probably seeing a radiation field with E(B-V) intermediate between 0 and 0.43, probably closer to 0.43. So in the discussion, we adopt the original KS79 ionizing spectrum.

Our spectroscopic observations with the INT on La Palma, Canaries of a filament at PA = 162 (Boisson, Péquignot, in preparation) gives at $\sim$ 20 kpc from the nucleus :

$$[SII] \ 6716 \ / \ [SII] \ 6731 \sim 1.35 \pm 0.30 \ (1)$$
$$[SII] \ 4070 \ / \ [SII] \ 6767 + 6731 \prec 5 \ 10^{-2} \ (2)$$

Ratio (1) implies $n_e \prec 500$ cm^{-3} and ratio (2) $n_e \prec 900$ cm^{-3} if T_e ([SII]) $\sim 7 \ 10^3$ K, and $n_e \prec 100$ cm^{-3} if T_e ([SII]) $\sim 9 \ 10^3$ K. The latter temperature being more likely in the shock assumption. These constraints seem to reject phase 6 as the main contribution to the filament emission.

Concerning photoionization by a central source (phase 7), the [OIII] / [OII] and [OIII] / H_β ratios imply an ionization parameter U $\sim 5 \ 10^{-4}$. For such a low value, U is almost insensitive to abundances, ionizing spectrum shape, and even the exact value of the above line ratios. The observed intensities of [OIII], [OII], [OI], [NII], [NI] and a few others are in reasonable agreement with photoionization calculations assuming ionization bounded filaments.

The ionizing continuum of KS79 and U gives 100 cm^{-3} at 20 kpc from the centre. (This can be $\sim$ 3 times greater if one assumes a bump as indicated by IUE, after reddening correction. This low density is in agreement with the constraints found observationally (from [SII]).

Since the photoionized gas is at T $\sim 10^4$ K, the density $n_H \sim 100$ cm^{-3} derived from the ionization parameter corresponds to a pressure n T $\sim 10^6$ in approximate agreement with the pressure of the cooling flow, suggesting that the photoionized filaments observed at that distance reached maximum expansion compatible with ambient pressure.

References

Cox, D. P., 1972, *Astrophys. J.*, **178**, 143.
Fabian, A. C., Nulsen, P. E. J., 1977, *Mon. Not. R. astr. Soc.*, **180**, 479.
Fabian, A. C. et al., 1981, *Astrophys. J.*, **248**, 47.
Fabian, A. C., Nulsen, P.E.J., Canizares, C. R., 1984, *Nature*, **310**, 733.

Kent, S. M., Sargent, W. L. W., 1979, *Astrophys. J.*, **230**, 667.
Lynds, R., 1970, *Astrophys. J.*, **159**, L151.
Mushotzky, R. F. et al., 1981, *Astrophys. J.*, **244**, L47.
Véron, P., 1978, *Nature*, **272**, 430.

COOLING FLOW FILAMENTS AND THE EMISSION-LINE GAS IN NGC1275

Roderick Johnstone
Institute of Astronomy
Madingley Road
Cambridge CB3 0HA
U.K.

ABSTRACT. A brief review of possible mechanisms for ionizing cooling flow filaments is given. Emission line ratios from spatially resolved spectra of NGC 1275, covering [OII]λ3727, Hβ, [OIII] λ5007 and [SII]$\lambda\lambda$6717, 6730 are presented. The line ratios are shown to be incompatible with the simplest model in which the gas is photoionized by a central active nucleus and lies at its projected distance from the nucleus.

1 INTRODUCTION

Many central cluster galaxies have been known for many years to show optical line emission (e.g. Heckman 1981). When cooling flows where discovered around these objects it seemed clear that the optical line emission arose as large blobs cooled isochorically from X-ray emitting temperatures and shocks formed as the gas repressurized (Cowie, Fabian & Nulsen 1980).

Recently, Johnstone, Fabian & Nulsen (1987), amongst others have pointed out a severe energy problem with this interpretation since the luminosity in the Hβ line is given by:

$$L(\text{H}\beta) = 9.3 \times 10^{36} \dot{M} H_{rec} \text{ erg s}^{-1}$$

where $\dot{M}$ is the mass deposition rate in solar masses per year and H_{rec} is the number of times each hydrogen atom recombines. Using the $\dot{M}$ appropriate to their spectrograph aperture H_{rec} is required to be between 500 and 5000 in the centres of many cooling flows. These figures must be compared with values of 2-3 given from simple models of cooling gas in which repressurizing shocks reproduce the observed emission line spectrum quite well. An additional source of heat is therefore required.

2 CANDIDATE SOURCES OF EXTRA HEAT

A more sophisticated treatment of the cooling of the X-ray emitting gas on to an already cooled blob, in which the temperature / radius profile is used to weight the addition of hot gas emission spectra from a range of temperatures can give H_{rec} up to $\sim$ 130 in an extreme case. The enhancement of a factor of 40 or more comes essentially from the fact that the cooling gas can reionize already cooled gas. Hu *et al.* (1985, and this workshop) however, favoured multiple shocks as the source of extra heat: they require all of the mass deposition to flow into the inner few kiloparsecs and all of the gas to shock several hundred times. Heckman *et al.* (this workshop) suggest that

159

A. C. Fabian (ed.), Cooling Flows in Clusters and Galaxies, 159–163.

the emission-line gas is shock heated and the shocks are driven by the interaction of the cooled gas with the relativistic plasma seen in radio maps.

Hot stars are a copious source of ionizing photons and Kennicutt (1983) has shown that H_{rec} for spiral galaxies is typically 4000. Johnstone et al., (1987) in their study of the emission lines and continuum colours of southern central cluster galaxies, found that the star formation rate inferred from the Hβ luminosity was very similar to that inferred from the weakening of the 4000Å break if the excess blue light is produced by a stellar population with a disk-galaxy initial-mass-function (IMF). Figure 1 shows a plot of the star formation rate inferred from the Hβ luminosity against the star formation rate inferred from the continuum colours in their sample of southern cooling flows. Only a few percent (much less than ten percent in most objects) of the total mass deposition rate is required to go into stars with this form of IMF.

Figure 1

Comparison of Star Formation Rates Deduced from Hβ Luminosity and 4000Å Breaks

Southern Cooling flows

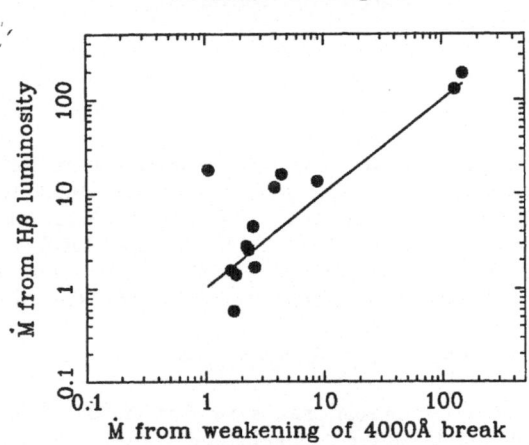

Robinson et al. (1987) have shown that the line ratios seen in cooling flow filaments form the low ionization parameter end of a sequence containing more highly excited objects such as the Extended Emission Line Regions commonly observed in radio galaxies. They therefore suggest photoionization by an active nucleus.

The diagnostics which are available for distinguishing between these possibilities are the emission-line ratios, the line luminosities and their spatial variation. Almost all of the models can give the correct line ratios although the hot stars model may require an additional source of heat to produce sufficient luminosity in the neutral oxygen lines. Johnstone et al. (1987) suggested that a contribution to the ionizing radiation from the cooling gas was required.

3 NUCLEAR OR DISTRIBUTED IONIZATION IN NGC 1275

Line ratios are presented from two long-slit spectra taken by Fabian and Arnaud in March 1985 at the Isaac Newton Telescope. In both observations the slit was aligned at position angle 104° and passed through the nucleus. Table I lists the details of these observations.

The spectra were flux calibrated by reference to standard stars. Reddening within our Galaxy was corrected for using $A_v = 0.92$. Emission line fluxes were measured with a profile

TABLE I
Log of Observations

Detector	Exposure (s)	Slit width (arcsec)	Pixel length (arcsec)	Coverage (Angstroms)
IPCS	1500	2.00	0.75	3650-5600
CCD	2000	2.00	0.31	6660-6858

fitting technique which allowed the flux in the extended narrow components to be evaluated even in projection across the nucleus which produces broader components to many of the same lines. The [OIII]λ5007 line was corrected for contamination by the [OIII]λ4959 line from the high velocity emission-line system (Kent & Sargent 1979). The filled circles in Figures 2-4 show the line ratios [OIII]/Hβ, [OII]/Hβ and [SII]λ6717/[SII]λ6730 as a function of distance from the nucleus, along the slit.

Figure 2   **Figure 3**

[OIII]/Hβ vs. Distance from Nucleus   [OII]/Hβ vs. Distance from Nucleus

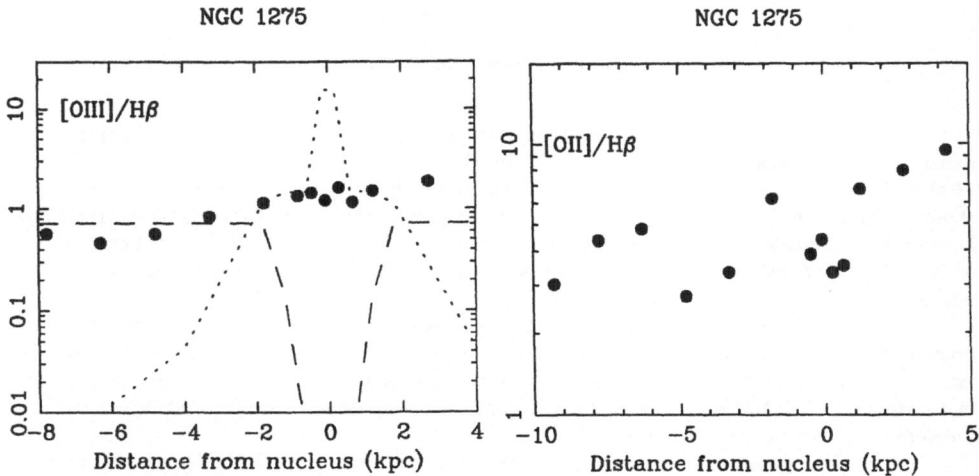

It is clear that over a wide range in projected distance from the nucleus that the [OIII]/Hβ ratio (which is particularly sensitive to the excitation state of the gas in cooling flows) remains remarkably constant. Under the assumption of a centrally located photoionizing source this would require either that the gas density follows closely an r^{-2} form, or that all the gas is confined to lie close to the surface of a sphere where the density is essentially constant. Figure 4 shows that, at least within the central 2 kpc that there is a significant change in the [SII] ratio. This ratio is predominantly sensitive to the electron density (although there is a small temperature dependence as well). The indicated change is about an order of magnitude, from $\sim 50 - 100\,\mathrm{cm}^{-3}$ at distances more than 2 kpc from the nucleus (a value which yields a total gas pressure within a factor of 2 of the value measured by Fabian *et al.* (1981) from the X-ray data at 20 kpc from the nucleus) to $\sim 500 - 1000\,\mathrm{cm}^{-3}$ in the centre.

Ferland's CLOUDY (Ferland & Truran 1981) program has been used to check two simple models. In both we make the naive assumption that the gas seen at a given projected distance from the nucleus really is at that distance. In the first model we use an $F(r) \propto r^{-2}$ law for the flux of ionizing photons. In the second we use $F(r) = const$.

Figure 4
[SII]λ6717/[SII]λ6730 vs. Distance from nucleus

NGC 1275

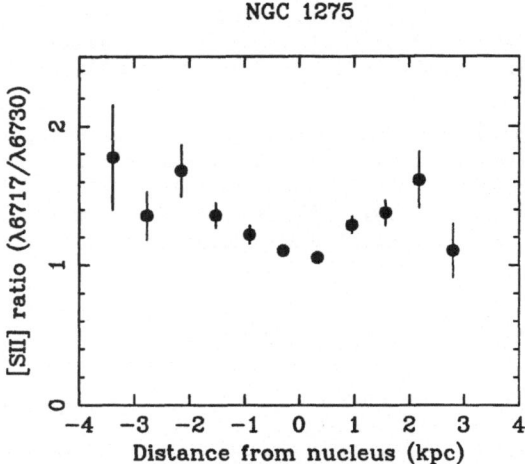

Plane parallel geometry and constant pressure clouds were assumed. The form of the ionizing spectrum was chosen to be a blackbody with $T = 1.3 \times 10^5$ K as suggested by Robinson *et al.* (1987). In the first model, the normalization given by $L(> 1\mathrm{Ryd}) = 6.2 \times 10^{42}\,\mathrm{erg\,s}^{-1}$ was chosen to match the line ratios at –0.6 kpc from the nucleus as well as possible. This spectrum does however over-produce the [OII] lines by about a factor of 2. In the second model, the normalization of $1.1 \times 10^{-2}\,\mathrm{erg\,cm}^{-2}\,\mathrm{s}^{-1}$ was chosen to match the line ratios between –8 and –2 kpc from the nucleus.

We have computed the [OIII]/Hβ line ratio at a number of points along the slit for both models. The pressure was set by adjusting the hydrogen density to match the observed [SII] ratio within 2 kpc of the nucleus. At this distance the pressure is within a factor of 2 of that measured at 20 kpc by Fabian *et al.* (1981) from the X-ray data. Therefore, at larger radii the pressure at 20 kpc ($1.2 \times 10^6\,\mathrm{cm}^{-3}\,\mathrm{K}$) is used.

In Figure 2 the dotted line joins the points generated from the r^{-2} law and the dashed line joins points generated by the constant ionizing flux model. It is quite clear that neither of these naive models can explain the line ratios over the full region observed. The r^{-2} model seems to work quite well between 0.5 and 2 kpc while the constant flux model is quite successful between 2 and 6 kpc. The emission line region is not consistent with being linear, in the plane of the sky and photoionized by a centrally located source. It is however quite possible that if the emitting region has a peculiar geometry as viewed by us that it can be reconciled with the emission line ratios. The total Hα luminosity produced in this model is only $1.5 \times 10^{41}\,\mathrm{erg\,s}^{-1}$ for a unit covering fraction. The total Hα luminosity observed by Cowie *et al.* (1983) is $9.6 \times 10^{42}\,\mathrm{erg\,s}^{-1}\,(H_0 = 50)$ when a factor of 1.9 increase in the observed flux is taken into account for line-of-sight absorption in the Galaxy. The deficit of over a factor of 64 means that for the hot blackbody model to work, the gas seen in projection at a distance of 0.6 kpc must be at at least 8 times this distance: ~ 5 kpc. An alternative way around this problem would be to use a photoionizing spectrum which is much less efficient at producing [OIII] flux. A higher ionization parameter is then required to give the same [OIII]/Hβ ratio and the Hβ luminosity is thereby increased. For example, a power law with an energy slope of –1.9 admits a 50% increase in Hα luminosity.

CONCLUSION

The emission line ratios in NGC 1275, presented in this paper, have been shown to be incompatible with the simplest model in which gas is photoionized by a central active nucleus and lies at its projected distance from the nucleus. The model fails to reproduce the correct spatial distribution of emission line ratios and sufficient luminosity. An alternative model in which the ionizing flux is constant with distance from the nucleus also fails. A more sophisticated analysis in which other geometries and many more aspects of the ionization of cooling flow filaments are considered is currently in preparation.

ACKNOWLEDGEMENTS

Andy Fabian is thanked for many helpful discussions. SERC is thanked for financial support.

REFERENCES

Cowie, L.L., Fabian, A.C. & Nulsen, P.E.J., 1980. *Mon. Not. R. astr. Soc.*, **191**, 399.

Cowie, L.L., Hu, E.M., Jenkins, E.B. & York, D.G., 1983. *Astrophys. J.*, **272**, 29.

Fabian, A.C., Hu, E.M., Cowie, L.L. & Grindlay, J., 1981. *Astrophys. J.*, **248**, 47.

Ferland, G.R. & Truran J.W., 1981. *Astrophys. J.*, **244**, 1022.

Heckman, T.M., 1981. *Astrophys. J.*, **250**, L59.

Hu, E.M., Cowie, L.L. & Wang, Z., 1985. *Astrophys. J. Suppl. Ser.*, **59**, 447.

Johnstone, R.M., Fabian, A.C. & Nulsen, P.E.J., 1987. *Mon. Not. R. astr. Soc.*, **224**, 75.

Kennicutt, R.C. Jr., 1983. *Astrophys. J.*, **272**, 54.

Kent, S.M. & Sargent, W.L.W., 1979. *Astrophys. J.*, **230**, 667.

Robinson, A., Binette, L., Fosbury, R.A.E. & Tadhunter, C.N., 1987. *Mon. Not. R. astr. Soc.*, **227**, 97.

EXOSAT DATA ON CLUSTERS

Niels J. Westergaard
Danish Space Research Institute
Lundtoftevej 7
DK-2800 Lyngby
Denmark

ABSTRACT. Published papers and preprints on EXOSAT observations of cluster of galaxies such as Coma, Perseus, Virgo, Fornax, A1060, A1367, Ophiuchus, PKS 0745-191, Cygnus-A, and 2A0335+096 are reviewed. The observational results comprise X-ray images and spectra for surface brightness and temperature determinations.

1. INTRODUCTION

The European X-ray satellite EXOSAT had a highly eccentric orbit in contrast to earlier X-ray satellites. Such an orbit allows ~ 70 h long uninterrupted exposure times which is ideal for the study of temporal variations of X-ray sources, which consequently constituted a fair fraction of the observation programme. The sensitivity was rather limited which favoured observations of galactic sources. Still, a number of extragalactic objects such as AGN's and clusters of galaxies were observed during EXOSAT's three years of lifetime.

EXOSAT was equipped with two low energy telescopes (LE) (0.02-2 keV), 8 medium energy (1-20 keV + 5-80 keV) proportional counters (ME) with collimators (FWHM 45'), and a gas scintillation proportional counter (GSPC), also with collimator (FWHM 45'). The telescopes also had transmission gratings. Unfortunately, only one LE detector, a charge multiplying array (CMA), survived the entire lifetime.

Among the main questions addressed in the cluster observations are the intracluster gas temperature and its distribution and the surface brightness distribution, which is almost a direct measurement of the total emission measure in the direction in question. The answers could come from good X-ray spectra and a good image.

This review of EXOSAT cluster observations also includes some results on individual galaxies or other interesting phenomena discovered in a cluster observation.

A. C. Fabian (ed.), Cooling Flows in Clusters and Galaxies, 165–173.
© 1988 by Kluwer Academic Publishers.

2. COMA

The observation is described in Branduardi-Raymont et al. (1985a+b).
A contour map is shown in Fig. 1. The diffuse emission of the Coma
cluster is clearly visible in the SW 'corner'. This observation,
however, does not add much to our previous knowledge of the cluster
itself through Einstein observations.

The attention is directed to five point sources in the field,
clearly marked in Fig. 1. The optical identifications of these X-ray
emitters prove four of them to be background AGN not associated with
the cluster, because they have much higher redshifts. The fifth source
(# 5) is a starburst galaxy of redshift z = 0.023. The result of this
discovery together with comparison with Einstein observations is

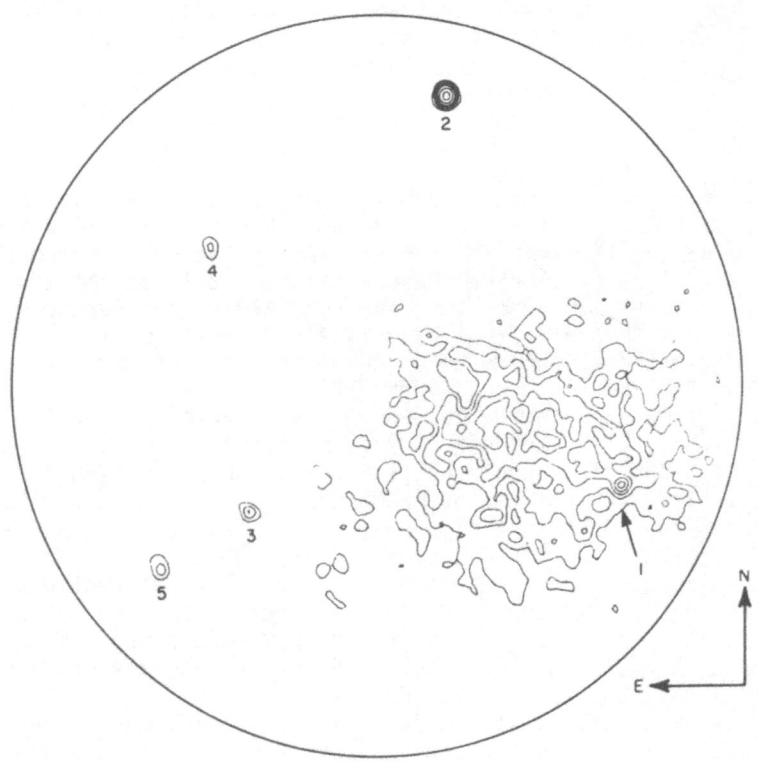

Fig. 1.   Contour map of the Coma cluster
           (Branduardi-Raymont et al. 1985b).
           The five point sources are numbered
           (1-5). The data has been smoothed
           with a Gaussian of 1 arcmin FWHM.

twofold: There is a 'hole' in the galactic absorption in the direction of Coma and the AGN spectra very likely consist of two components. The CMA on EXOSAT is sensitive down to 0.02 keV, so EXOSAT favours low absorption 'soft excess' sources.

3. PERSEUS

This 'nearby' cluster was observed with pointings in five different directions: on NGC1275, the central galaxy, and roughly 0.5° to the north, east, south and west of this galaxy as reported by Branduardi-Raymont et al. (1985c and these proceedings).

The LE brightness data were deconvolved and compared to earlier Einstein data showing good agreement. The emphasis, however, was put on the GSPC data to find the high energy intensity and the strength of the Fe line emission at ~ 7 keV. Fig. 2 shows the GSPC spectra obtained. The emission from the northern point was too weak to give a good spectrum. The main results are that the asymmetry in the surface brightness profile found at low energy extends to higher energy (6-10 keV) and that the measured temperatures are consistent with a single gas temperature. The latter result is supported by the ratio FeK_α to FeK_β, which also varies very little across the cluster surface.

Fig. 2.   The EXOSAT GSPC spectra (data and fit) from the
           Perseus cluster (Branduardi-Raymont et al. 1985c)
           Northern spectrum was too weak.

4. VIRGO

Due to the apparent large size of the Virgo cluster it was 'imaged' by the ME experiment as presented by Smith and Stewart (1985) and Edge, Stewart, and Smith (1987). Fig. 3 shows the ME pointings. One problem with such an extended source is that the off set angle for the detectors monitoring the background is not sufficiently large to avoid 'contamination' from the source.

The temperature gradient derived is very small indeed. The small but significant temperature peak is situated at the optical center and close to M86. The binding mass derived on the assumption of hydrostatic equilibrium increases proportionally to the radius. At 440 kpc

(H_0 = 50 km/s/Mpc) value is 8×10^{13} $M_\odot$.

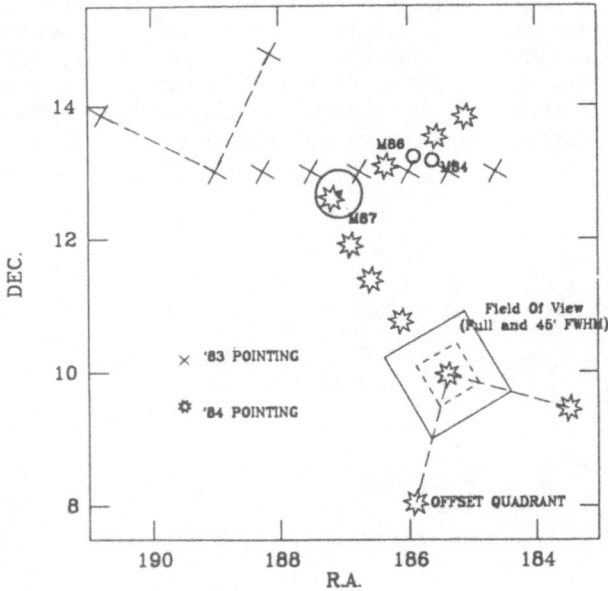

Fig. 3.   Schematic diagram of the pointings at the Virgo
           cluster (Edge, Stewart, and Smith 1987). The
           highest temperature is found near M86.

5. FORNAX

The observation of this cluster reported by Mason and Rosen (1985) revealed three sources: the galaxies NGC1399 and NGC1404 plus an unidentified source precisely midway between the two galaxies. The latter perhaps represents the gravitational neutral point between the galaxies.

Diffuse emission from the region seems to be below the detection limit.

The X-ray emission of the galaxies cannot be resolved, but from the upper limit to their extent the mean gas density is larger than 0.8 cm^{-3}. Also the X-ray luminosity is unusually high when compared to the optical luminosity. That has not been explained but the high gas density might be a result of a cooling flow.

6. OPHIUCHUS

This cluster, studied by Arnaud et al. (1987), as well as the following two, is situated close to the plane of our galaxy, so the absorption is quite strong. The temperature from ME is 9.4 keV and the Fe abundance 26% of the solar value. The brightness distribution is rather smooth so a deprojection has been carried out which gave a central electron density of ~ 0.02 cm^{-3}, a rather common value for a cooling flow cluster. The central cooling time is indeed much less than a Hubble time and the accretion rate is determined to be 150-220 $M_\odot/yr$.

7. PKS0745-191

Here the EXOSAT results are particularly important for the resolution of an earlier discrepancy between HEAO-1 and Einstein MPC data concerning the intracluster gas temperature. Arnaud et al. (1987) find a temperature of 8.6 keV, 33% Fe abundance and a central electron density of 0.018 cm^{-3} where the Einstein HRI gave 0.037 cm^{-3} (Fabian et al. 1985) with better angular resolution.

The radio source is found to be rather close to the X-ray centroid of the cluster; in fact, they might coincide.

This cluster presumably also contains a massive cooling flow with a rate of 440-930 $M_\odot/yr$.

8. CYGNUS-A

The LE does not contribute new information on this cluster (Arnaud et al. 1987). However, the presence of the double radio source makes it particularly important to separate the thermal and non-thermal components. A power law component in the model spectrum does indeed give a better fit to the ME spectrum than a single bremsstrahlung model. The absence of a point source in the LE image (as well as in the HRI image) is explained by strong intrinsic absorption. The derived values for the temperature (4.1 keV) and relative Fe abundance (0.6) are rather poorly determined due to the extended number of free parameters by the inclusion of the power law model.

9. A1367

Nørgaard-Nielsen et al. (1985a) found two sources in this irregular cluster in the LE image. The brightest source, NGC3862, is a radio source. The countrates in the 3LX and ALP filters are consistent with a temperature of 1 keV.

The other source is a blue object which is either variable or has a very soft spectrum because the 3LX countrate is twice the Einstein HRI rate.

10. A1060

The surface brightness of this cluster is rather low in the LE as reported by Nørgaard-Nielsen et al. (1985a). A more extensive study (Singh et al. 1986 and 1987a) gave a temperature of 2.9 keV from the ME and by deprojection a central electron density of 0.01 cm^{-3}. Fig. 4 gives the density and temperature profile. The derived mass accretion rate at 17 kpc is ~ 4 $M_\odot$/yr. Unfortunately, the background in the ME detector was unusually high at the period of observation so that the Fe abundance could not be determined.

11. 2A0335+096

This is a poor cluster of galaxies with a quite small angular extent. Optically, the dominant galaxy NGC3311 shows interesting features. Close to the galaxy there is a dwarf galaxy and the rotation curve obtained with a slit through the centers of the two galaxies is quite irregular, perhaps showing signs of a merging process (Nørgaard-Nielsen, these proceedings). A preliminary study of the X-ray observations is presented by Nørgaard-Nielsen et al. (1985).

Singh et al. (1987b) have performed a deprojection of both the LE and IPC data and have obtained a good match. The central electron density was found to be 0.02 cm^{-3} and the Fe abundance to be roughly half solar. The cooling mass flow rate drops from ~ 150 $M_\odot$/yr at 60 kpc to ~ 14 $M_\odot$/yr at 25 kpc. Temperature and density profiles are shown in Fig. 5.

12. FUTURE RESULTS

More clusters have been observed by EXOSAT, but they are quite faint objects and require a careful analysis. A. Edge (Leicester) is engaged in an analysis of 24 clusters and also A. Morini (Palermo) is reducing EXOSAT cluster observations.

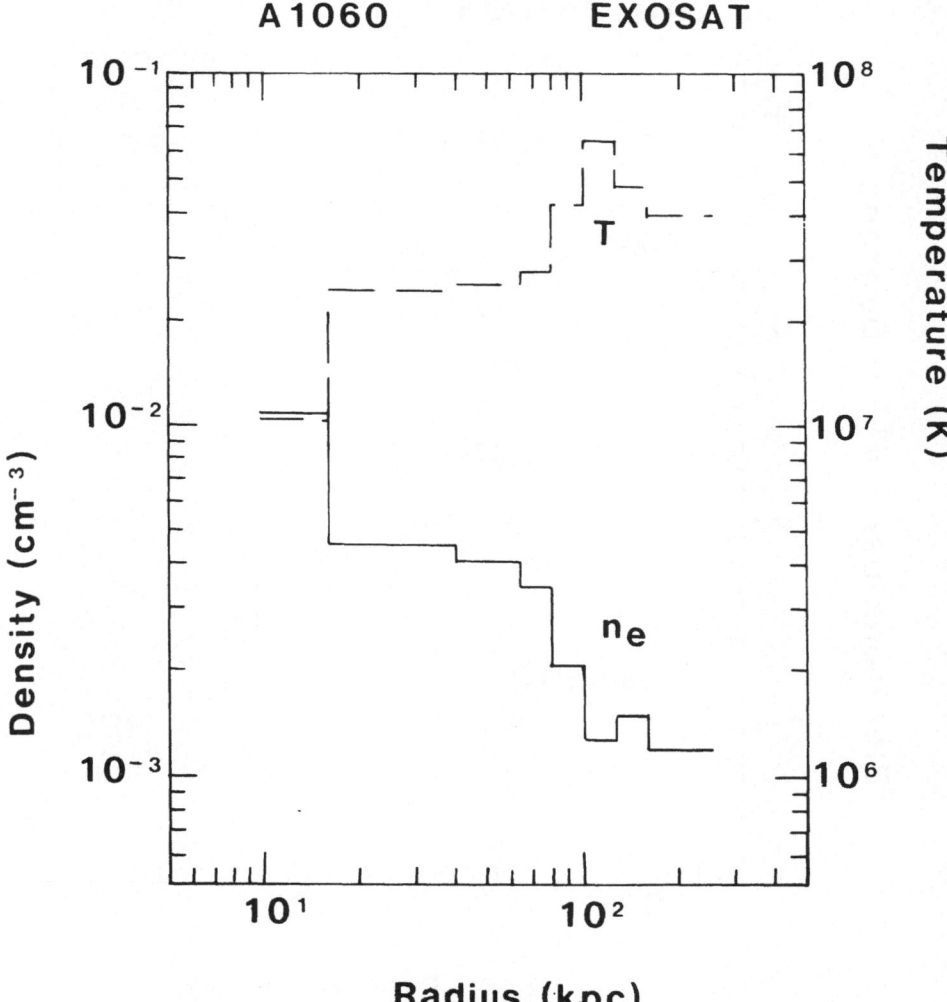

Fig. 4.   Temperature and electron density as function of
radius for A1060 (Singh, Westergaard, and Schnopper
1987a). The electron density is rather independent
of the model parameters in the deconvolution in
contrast to the temperature.

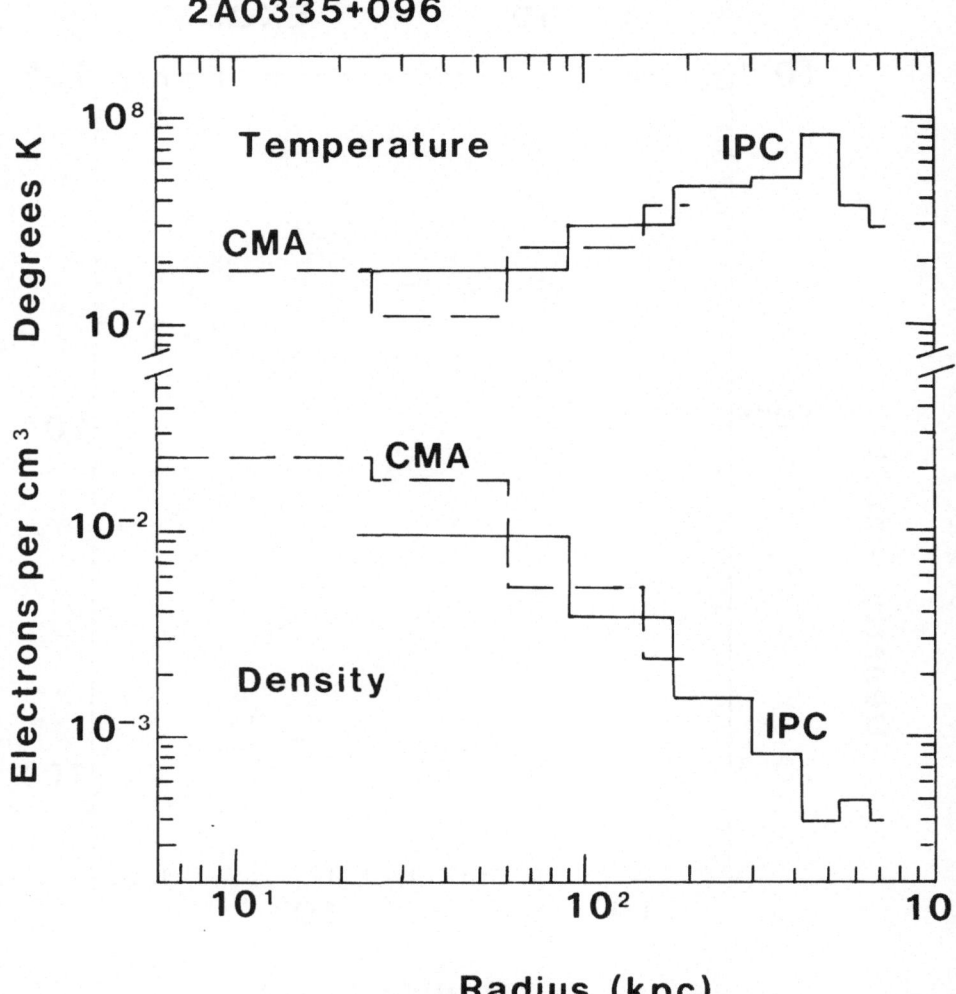

Fig. 5.   Temperature and electron density as function of radius for
2A0335+096 (Singh, Westergaard, and Schnopper 1987b). The
EXOSAT results match the Einstein results rather well.

REFERENCES

K.A. Arnaud, R.M. Johnstone, A.C. Fabian, C.S. Crawford, P.E.J. Nulsen, R.A. Shafer, and R. Mushotzky 1987, MNRAS **227**, 241.

G. Branduardi-Raymont, K.O. Mason, P.G. Murdin, and C. Martin 1985a, MNRAS **216**, 1043.

G. Branduardi-Raymont, K.O. Mason, P.G. Murdin, C. Martin, and S.P. McKechnie 1985b, Space Sci. Rev. **40**, 647.

G. Branduardi-Raymont, B. Kellett, A.C. Fabian, T.McGlynn, G. Manzo, and A. Peacock 1985c, Adv. Space Res., **5**, 133.

A.C. Edge, G.C. Stewart, and A. Smith 1987, preprint.

A.C. Fabian, K.A. Arnaud, P.E.J. Nulsen, M.G. Watson, G.C. Stewart, I. McHardy, A. Smith, B. Cooke, M. Elvis, R.F. Mushotzky 1985, MNRAS **216**, 923.

K.O. Mason and S.R. Rosen 1985, Space Sci. Rev. **40**, 675.

H.U. Nørgaard-Nielsen, N.J. Westergaard, L. Hansen, H.E. Jørgensen, and H.W. Schnopper 1985, Proc. ESA Workshop: 'Cosmic X-Ray Spectroscopy Mission', Lyngby, Denmark, June 1985 (ESA SP-239, Sep. 1985).

H.U. Nørgaard-Nielsen, N.J. Westergaard, and L. Hansen 1985, Space Sci. Rev. **40**, 669.

K.P. Singh, N.J. Westergaard, and H.W. Schnopper 1986, Astrophys. J. **308**, L51.

K.P. Singh, N.J. Westergaard, and H.W. Schnopper 1987a, submitted to Astrophys. J.

K.P. Singh, N.J. Westergaard, and H.W. Schnopper 1987b, submitted to Astrophys. J.

A. Smith and G. Stewart 1985, Space Sci. Rev. **40**, 661.

THERMAL INSTABILITY IN COOLING FLOWS

P.E.J. Nulsen,
Mount Stromlo and Siding Spring Observatories,
Private Bag, Woden P.O.
ACT 2606
Australia

ABSTRACT. In order to explain the widespread deposition of cooled gas in cooling flows the gas must be significantly inhomogeneous. Equations governing the development of an inhomogeneous flow are set out. It is argued that gas at different densities in the flow must be largely comoving, and with this assumption an integral set of flow equations are derived and solved approximately. The resulting mass deposition profile is sensitively dependent on only the dimensionless ratio of the mass deposition rate to the total radiation rate per unit volume, $\xi = 5\beta p/2\bar{\rho}R$. This result is little affected by the small amount of residual slip in the flow. Plausible values for ξ are confined to a fairly narrow range, including the value required by observations, but still allow considerable latitude in the resulting mass deposition profile. ξ is determined by the initial density distribution of the gas. Low temperature line emission from cooling flows directly maps the distribution of the gas deposition rate, providing a stringent test of cooling flow models.

1. INTRODUCTION

There are two observational indications that the gas in cooling flows is inhomogeneous. First, is the existence of the diffuse systems of line emitting filaments in many cooling flows (*e.g.* Lynds 1970; Heckman 1981; Cowie *et al.* 1983; Hu, this proceedings; Heckman, this proceedings). It has long been suggested that the filaments are the product of thermal instability in the flows (Fabian & Nulsen 1977; Mathews & Bregman 1978; Cowie, Fabian & Nulsen 1980) and this requires some inhomogeneity in the gas. The stationary filaments in the cooling flow around NGC 1275 extend to at least 100 arcsec from the centre of that galaxy (see photo by Lynds 1970), suggesting that thermally unstable gas must be cooling over a region comparable in size to the whole of the steady cooling flow.

In the absence of a detailed understanding of how the filaments are formed and powered we cannot determine the distribution of density fluctuations in the gas required to produce them. This is better addressed through the second line of evidence for inhomogeneity of the gas, the morphology of the X-ray source.

175

A C. Fabian (ed.), Cooling Flows in Clusters and Galaxies, 175–187
© 1988 by Kluwer Academic Publishers.

The basic physics of steady cooling flows is very simple and in a cluster cooling flow the X-ray luminosity from within a sphere of radius r can always be written

$$L_X = \dot{M} \left(\frac{5kT}{2\mu m_H} + \Delta\phi \right), \tag{1}$$

where $\dot{M}$ is the mass flow rate into the spherical region, T is the temperature of the inflowing gas and $\Delta\phi$ is the mean change in the gravitational potential of the gas before it cools. This last term is always found to be comparable to or smaller than the thermal contribution to L_X, so that using (1) we can get a robust estimate for the mass flow rate even when the potential is not well known.

In a steady cooling flow with no gas deposition the mass flow rate must be independent of the radius, however, analysis of the best X-ray data shows that it is not. A simple analysis (*e.g.* Fabian, Nulsen & Canizares 1984) and a more detailed analysis allowing for inhomogeneity in the gas (Thomas, Fabian & Nulsen 1987) both result in estimates for the mass flow rate which are strong functions of the radius,

$$\dot{M} \sim r^{\eta}, \tag{2}$$

with η about 1 (*e.g.* Arnaud, this proceedings).

Thermal instability is, perhaps, a misnomer in the context of cooling flows and has led to misunderstanding of the nature of the process involved. If the gas comoves then there is indeed a thermal instability as described by Field (1965), however, as Cowie, Fabian & Nulsen (1980) pointed out, perturbed regions tend to move relative to the mean flow, suppressing the instability in a lot of cases. Further to this Nulsen (1986) argued that convection of perturbed gas "blobs" leads rapidly to disruption and, ultimately, to their becoming effectively pinned in the mean flow (see section 3). This reestablishes amplification due to cooling of density contrasts in the gas. The misunderstanding arises because it is not widely appreciated that this description is of the *non-linear* development of the thermal instability. A linearised analysis cannot reproduce this description of the growth of the instability, and there do not generally appear to be any linearly unstable modes.

Malagoli, Rosner & Bodo (1986) analysed the linear development of the instability including effects of buoyancy and radiative heat loss. They found overstable growing modes, but the growth times were comparable to the flow time in the mean flow making their approximations inconsistent. A more thorough analysis casts considerable doubt on the existence of any linearly unstable modes in a steady, highly subsonic cooling flow (Nulsen 1987, in preparation).

The strong radial variation of the mass flow rate in cooling flows cannot be understood in terms of the unstable growth of small perturbations in the cooling gas. Instead we must consider the development of an initially inhomogeneous gas. Bearing in mind the caveats of this introduction, I will continue to use the phrase "thermal instability" to describe the amplification of density inhomogeneities caused by cooling of the gas. The title of this talk might more properly have been "Inhomogeneous Cooling Flows".

2. GENERAL THEORY OF INHOMOGENEOUS FLOWS

2.1. Governing equations

In order to describe the behaviour of an inhomogeneous gas which is cooling radiatively we introduce a volume distribution function, $f(\mathbf{r}, \rho, t)$, defined so that the mass of gas in a small volume δV, in the density range ρ to $\rho + d\rho$ is

$$\delta M = \rho f(\mathbf{r}, \rho, t) \, d\rho \, \delta V. \tag{3}$$

With this distribution function the equation of mass conservation is

$$\frac{\partial \rho f}{\partial t} + \nabla.\rho \mathbf{v} f + \frac{\partial}{\partial \rho} \rho f \dot{\rho} = \text{source term}, \tag{4}$$

where $\dot{\rho}$ is the convective rate of change of the density and $\mathbf{v}$ is the gas velocity. The source term is to allow for mass injection. It is generally negligible in cluster cooling flows, but must be incorporated for a full treatment of the cooling flows in isolated elliptical galaxies.

To complete the set of equations describing an inhomogeneous cooling flow we need to determine $\dot{\rho}$ and $\mathbf{v}$ from suitable energy and momentum equations.

In all known cooling flows the cooling time in the bulk of the flow is much longer than the sound crossing time. Pressure equilibrium is therefore maintained locally to a good approximation. Large blobs can cool fast enough to drop out of pressure equilibrium at temperatures less than about 10^6 K, but this has very little effect on gas close to the mean temperature and we ignore it. Local pressure equilibrium means that the pressure is independent of the density, $p = p(\mathbf{r}, t)$, and does not fluctuate rapidly. The energy equation may be written

$$\frac{3}{2} p \frac{d}{dt} \ln \frac{p}{\rho^{5/3}} = -n^2 \Lambda, \tag{5}$$

where n is the electron density corresponding to ρ and Λ is the cooling function. Note that this equation is applied for all ρ to solve for the flow.

In general the distribution function will depend on a number of quantities besides those explicitly shown. For example, thermal conduction must be suppressed in order for cooling flows to exist (e.g. Bregman, this proceedings) and magnetic fields are generally invoked as the cause of the suppression. A detailed description of the effect of thermal conduction in the presence of a tangled magnetic field would require keeping a great deal of detail in the distribution function. This would have to include the dependence of the distribution on the the size and shape of density fluctuations, magnetic field orientation and more. To avoid this problem here I assume that thermal conduction is completely negligible.

It is always possible to obtain a distribution function of the form given in (3) by integrating out the other dependencies, but the information discarded will make simplifying assumptions necessary in the construction of a closed set of equations governing the flow. The momentum equation provides a particular example of this problem, since the velocity behaviour of the blobs depends on a great many factors besides the magnitude of the density fluctuations. I do not attempt to write a momentum equation here, but note that in general the velocity of the gas is a function of its density as well as position and time. The gas does not generally comove.

There is one more technical matter to consider before leaving this section. The distribution function describing the exact state of the gas is a delta function in the density,

$$f_{\text{exact}}(\mathbf{r}, \rho, t) = \delta \left[\rho - \rho(\mathbf{r}, t) \right], \tag{6}$$

since the gas has a well defined density at any given point and time. f_{exact} may be regarded as the "fine grained" distribution function by analogy with stellar dynamics, but as in stellar dynamics it is of little value. We are interested in the "coarse grained" distribution function, obtained from f_{exact} by taking small scale or short term averages, denoted by

$$f = \langle f_{\text{exact}} \rangle. \tag{7}$$

Using the coarse grained distribution function eliminates the large and rapid fluctuations that can occur in the fine grained distribution. Under the average the form of the mass conservation equation (4) will change unless

$$\mathbf{v}(\mathbf{r}, \rho, t) = \langle \mathbf{v}_{\text{exact}} f_{\text{exact}} \rangle / f \tag{8}$$

and

$$\dot{\rho}(\mathbf{r}, \rho, t) = \langle \dot{\rho}_{\text{exact}} f_{\text{exact}} \rangle / f. \tag{9}$$

These conditions define locally averaged values for the velocity and $\dot{\rho}$ respectively, however, the result for $\dot{\rho}$ is very simple when local pressure equilibrium is maintained. In that case the averaged energy equation is identical in form to (5), so that we may ignore the averaging when determining $\dot{\rho}$. The assumption of local pressure equilibrium greatly simplifies the determination of the flow structure.

2.2. The distribution function at high densities

The rate of cooling increases rapidly with density, so that at high densities it is always the fastest process. This requires

$$\rho f \dot{\rho} \to \text{constant}, \tag{10}$$

at high densities in order to satisfy the mass conservation equation (4). Physically, $\rho f \dot{\rho}$ is the rate per unit volume at which gas cools through a given density, so that

$$\beta = \lim_{\rho \to \infty} \rho f \dot{\rho} \tag{11}$$

is the rate at which cooled gas is deposited per unit volume.

$\dot{\rho}$ is determined from the energy equation (5). At high densities the rate of change of the pressure is negligible, so that we get a closed expression for $\dot{\rho}$ and inverting (11) then gives

$$f \to \frac{5\beta p}{2\rho^2 n^2 \Lambda}. \tag{12}$$

We see that the density distribution at high densities only depends on the rate of deposition of cooled gas. This is not very important for the structure of inhomogeneous cooling flows, but does have significant observational consequences (see below). Note that the argument only depends on the assumptions of local pressure equilibrium and continuity of f. It does not require a steady flow. At constant pressure the cooling rate increases rapidly as the temperature falls, so that the asymptotic form will apply quite well even for gas which is only 2 – 3 times the mean density.

2.3. Steady, Spherically Symmetric Flow

By analogy with homogeneous cooling flows we expect that the flow will be approximately steady inside the region where the mean cooling time is less than the age of the flow (see section 4). It is therefore of some interest to determine the structure of steady inhomogeneous flows. Under the additional assumption of spherical symmetry we can formally solve the mass conservation equation (4). We introduce the mass flux function

$$\dot{m}(r,\rho) = - \int_0^\rho 4\pi r^2 v_r(r,\rho')\, \rho' f(r,\rho')\, d\rho', \qquad (13)$$

which is the mass flow rate of gas with density less than ρ into a sphere of radius r. The total mass flow rate is $\dot{M}(r) = \dot{m}(r,\infty)$.

Integrating the mass conservation equation (4) with respect to ρ gives

$$v_r \frac{\partial \dot{m}}{\partial r} + \dot{\rho} \frac{\partial \dot{m}}{\partial \rho} = 0. \qquad (14)$$

In words, the convective derivative of $\dot{m}$ in the (r,ρ) plane is zero and hence $\dot{m}$ is conserved on streamlines in the (r,ρ) plane. Following a fluid element in the flow, if its density at r_1 is ρ_1 and at r_2 it is ρ_2, then $\dot{m}(r_1,\rho_1) = \dot{m}(r_2,\rho_2)$. Alternatively, we may say that streamlines do not cross in the (r,ρ) plane. This is not generally correct for the fine grained gas distribution and applies only because we have assumed is that the flow is steady in the mean.

3. WHY THE GAS COMOVES

This problem is discussed in more detail in Nulsen (1986) and I only summarize the arguments here.

As pointed out above, the gas does not generally comove. A blob of gas that is more dense than the mean will be negatively buoyant and will tend to fall in faster than the mean flow, while one that is underdense tends to move outward relative to the mean. Because the inflow velocity is small compared to the local Keplerian velocity most blobs move relative to the mean flow at close to their terminal velocities. The relative motion creates internal stresses in the blobs which induce internal motions. There is no force holding a blob together, so that the internal velocities lead rapidly to breakup. Ultimately, the fragmentation results in the blobs becoming pinned in the flow, either effectively when their terminal velocities fall below the mean flow velocity, or actually by magnetic stresses.

This process is highly nonlinear. The linearised stability analysis can only be applied in describing displacements relative to the mean flow which are small compared to the size of a blob (or the wavelength of a disturbance).

Note that the fragmentation requires some residual drift between gas at different densities. Thus, although the bulk of the gas is very nearly comoving, the locally averaged velocity will depend weakly on the density. The drift has its greatest effect on the low density wing of the distribution, but its overall effect on the flow will generally be small (see below).

4. INTEGRAL FLOW BEHAVIOUR FOR COMOVING FLOWS

4.1. The Integral Flow Equations.

It is useful to construct a set of integral flow equations in order to investigate the general behaviour of inhomogeneous cooling flows. We derive them here for a comoving flow. First, integrating the mass conservation equation (4) with respect to the density gives

$$\frac{\partial \bar{\rho}}{\partial t} + \nabla . \bar{\rho}\mathbf{v} = - \lim_{\rho \to \infty} \rho f \dot{\rho} = -\beta, \tag{15}$$

where the mean density

$$\bar{\rho} = \int_0^\infty \rho f d\rho, \tag{16}$$

the mean velocity is defined by

$$\bar{\rho}\bar{\mathbf{v}} = \int_0^\infty \rho \mathbf{v} f d\rho, \tag{17}$$

and β is the rate of deposition of cooled gas from the flow. Apart from the mass deposition term this is just the usual equation of mass conservation for a fluid.

Once a significant fraction of an initially inhomogeneous gas has been deposited, the mass deposition rate is given by

$$\beta = \xi \frac{\bar{\rho}}{t_{\rm cp}}, \tag{18}$$

for some ξ which is generally of order unity. The constant pressure cooling time

$$t_{\rm cp} = \frac{5p}{2R}, \tag{19}$$

where the total radiation per unit volume from the gas is

$$R = \int_0^\infty f n^2 \Lambda d\rho. \tag{20}$$

We therefore parameterise the mass deposition rate in terms of the dimensionless parameter ξ. Note that ξ is usually a function of position in the flow.

The global energy equation is obtained by multiplying the mass conservation equation by $1/\rho$ and integrating with respect to ρ. Under the assumption that the gas comoves this gives

$$\nabla.\bar{\mathbf{v}} + \int_0^\infty f\frac{\dot{\rho}}{\rho}d\rho = 0, \tag{21}$$

where the first term has vanished since f is normalised. We use the energy equation (5) to get an expression for $\dot{\rho}$, and $\nabla.\bar{\mathbf{v}}$ is eliminated using the integral mass conservation equation (15), giving finally

$$\frac{3}{2}p\frac{d}{dt}\ln\frac{p}{\bar{\rho}^{5/3}} = \frac{5\beta p}{2\bar{\rho}} - R = -(1-\xi)R. \tag{22}$$

This result depends critically on the assumptions of local pressure equilibrium, and that the flow comoves. If the flow does not comove then the mean velocity $\bar{\mathbf{v}}$ is not equal to

$$\tilde{\mathbf{v}} = \int_0^\infty \mathbf{v} f d\rho, \tag{23}$$

which occurs several times in the energy equation just derived. In spherically symmetric flows we can allow for this by introducing another parameter, say ν, defined by

$$\tilde{\mathbf{v}} = \nu\bar{\mathbf{v}}. \tag{24}$$

For a steady flow the inflow velocity will generally increase with the density, so that $\nu \leq 1$. The effect of drift on the flow may be included in the power law solutions below by replacing the parameter ξ with $\nu\xi$, provided that ν varies slowly throughout the flow.

Equation (22) can be compared with the energy equation (5) which is identical to that for a homogeneous gas. Identifying the total radiation per unit volume, R, with $n^2\Lambda$ for the homogeneous gas, we see that the only difference is in the extra term $+\xi R$ on the right hand side. The quantity $\ln(p/\bar{\rho}^{5/3})$ is a measure of the mean entropy of the inhomogeneous gas, so that the additional term reduces the effect of radiative heat loss on the mean entropy. It can cause the entropy to increase as the gas radiates if $\xi > 1$. This is a consequence of the deposition of cooled gas. The effect is analogous to that of dropping ballast from a hot air balloon. Cooling reduces the mean entropy of the uncooled gas, but the deposition of cooled gas counters this and may be sufficiently rapid to cause the mean entropy to rise. It is easy to construct distribution functions giving any value for ξ, so that this is a real possibility (see below).

The momentum equation for a locally comoving gas will just be the usual one for a fluid at the mean density, i.e.

$$\bar{\rho}\frac{d\bar{\mathbf{v}}}{dt} = -\nabla p - \bar{\rho}\nabla\phi, \tag{25}$$

where ϕ is the gravitational potential. There are two assumptions which have gone into this equation. First, the gas which is deposited, i.e. has cooled to very low temperatures, is assumed to rapidly decouple from the mean flow. If this were not the case, the cooled gas may significantly increase the mean density. There is currently little evidence for the large quantities of cool gas that this would require, but we cannot dismiss it completely. The other assumption is that the cooled gas is comoving when it decouples from the flow and requires little justification.

4.2. Power Law Solutions

We now look for approximate power law solutions for a steady, spherically symmetric flow. We begin by assuming that the total mass flow rate

$$\dot{M} \sim r^{\eta}, \tag{26}$$

for some η which we will treat as constant. Substituting this into the equation of mass conservation (15) gives

$$\frac{r}{\bar{v}} = \frac{\eta}{\xi} t_{\text{cp}}, \tag{27}$$

so that the flow time is roughly equal to the cooling time for inhomogeneous flows provided that $\eta/\xi \sim 1$. Note that $\xi = 0$ corresponds to a homogeneous flow, so that η/ξ should asymptote to a finite value of order unity as $\xi \to 0$. This is confirmed below.

The total radiation rate may be written

$$R = K \bar{n}^2 \Lambda(\overline{T}), \tag{28}$$

where K is another dimensionless parameter which is always of order unity. Provided that both the distribution function and the cooling function are reasonably smooth K will vary slowly as the gas cools. On the assumption that the cooling function can be represented by a power law,

$$\Lambda(T) \sim T^{\alpha}, \tag{29}$$

and that ξ, K and α all vary slowly with radius, equation (27) gives

$$\bar{\rho} \sim \overline{T}^{\frac{1-\alpha}{2}} r^{\frac{\eta-3}{2}}. \tag{30}$$

Using (27) to eliminate $\bar{v}$ from the time independent form of (22), the integral energy equation, gives

$$\frac{d}{dr} \ln \frac{\overline{T}}{\bar{\rho}^{2/3}} = \frac{5\eta(1-\xi)}{3\xi r}, \tag{31}$$

and this equation is consistent with the expression (30) for $\bar{\rho}$ only if

$$\eta = \frac{\xi[3 + (2+\alpha)\theta]}{5 - 4\xi}, \tag{32}$$

where θ is is the logarithmic derivative of the temperature with respect to the radius,

$$\theta = \frac{r}{\bar{T}} \frac{d\bar{T}}{dr}. \tag{33}$$

This result is exact if we interpret η as a logarithmic derivative rather than as an exponent (see 26), and if the quantity $\eta/\xi K$ is independent of the radius. Indeed, as commented above, ξ/η asymptotes to a constant as $\xi \to 0$, so that the approximation is even reasonable at the onset of cooling.

As discussed above, the effect of slip in the flow is to reduce ξ by a factor ν (equation 24), which is a measure of the amount of slip. For fixed ξ this reduces the value of η, making the gas deposition more centrally concentrated. Although η is quite sensitive to ξ (and hence ν) the amount of any slip is not likely to be sufficient to substantially alter the mass deposition profile in the bulk of the flow.

Notice that the result for η only depends on the parameter ξ, and not on K. K plays a role in determining the absolute values of the density and mass flow rate, but is not important to the structure of the flow. In contrast, the result is sensitive to ξ, which we recall is defined as

$$\xi = \frac{5\beta p}{2\bar{\rho}R}, \tag{34}$$

where R is the total radiation rate and β is the mass deposition rate. The value of ξ is therefore the parameter of chief physical interest in pursuing solutions for inhomogeneous flows.

If $\xi > 1$ then the mean entropy increases as the gas cools (equation 22). Were we simply to solve for steady inflows with $\xi > 1$ this would lead to convectively unstable atmospheres, so that such flows are not physically realizable. If $\xi > 1$ in the initial density distribution, an inhomogeneous cooling flow will undergo large-scale convection which will mix the gas and push the effective value of ξ closer to 1. (This is possible because the mean flow velocity is small compared to the Keplerian velocity, see Nulsen 1986.) As in convectively unstable stellar atmospheres, the mean temperature profile is nearly adiabatic and the mass flow rate will be determined from the expression (32) for η with $\xi = 1$. This would give rise to a universal structure for cooling flows which is insensitive to the initial density distribution in the gas, but it does not appear to be consistent with available X-ray data (Thomas et al. 1987). We conclude that the actual density distributions must have $\xi < 1$.

5. EXACT FLOW SOLUTIONS

What are reasonable values for the parameter ξ? Nulsen (1986) discusses a range of exact solutions for comoving, steady, inhomogeneous cooling flows. The distribution functions considered there are self-similar and give constant values for ξ in the range $0 - 1$, so that the result of the previous section applies exactly. We begin with a summary of these solutions.

The key to the exact solutions is the assumption that the cooling function may be modelled as a power law,

$$\Lambda(T) = DT^{\alpha}, \tag{35}$$

for constant D and α. Assuming local pressure equilibrium, the energy equation (5) may then be integrated exactly to give the density of a fluid element as a function of the radius and its value ρ_i at the initial radius r_1,

$$\rho(r, \rho_i) = \left[\frac{p(r)}{p_1}\right]^{.6} \left(\rho_i^{-\overline{2-\alpha}} - \rho_c^{-\overline{2-\alpha}}\right)^{\frac{-1}{2-\alpha}}, \tag{36}$$

where $p(r)$ is the pressure and p_1 is $p(r_1)$. The critical density ρ_c is a function of the radius alone, defined so that $\rho_c(r)$ is the density of the gas at r_1 which just cools to infinite density at r. Thus any gas that was initially more dense than ρ_c cools to infinite density before reaching r, while gas which was less dense still has finite density at r.

Equation (36) determines the streamlines for the flow in the (r, ρ) plane. We see that the effects of cooling and compression separate, with all of the effect of cooling being contained in the single function $\rho_c(r)$, and the pressure change simply causing an adiabatic compression. Knowing ρ_c and p, the distribution function, and hence any other property of the gas, is fully determined in terms of the initial distribution function $f(\rho, r_1)$. From section (2.3), the mass flux function

$$\dot{m}(r, \rho) = \dot{m}(r_1, \rho_i), \tag{37}$$

in the present notation. Differentiating this with respect to the density gives

$$f(r, \rho) \propto f(r_1, \rho_i) \left(\frac{\rho_i}{\rho}\right)^{4-\alpha}, \tag{38}$$

with normalisation determining the constant of proportionality.

Substitutions which make the notation more compact are

$$w = \rho^{-\overline{2-\alpha}}, \tag{39}$$

and

$$g(r, w) = \frac{1}{2-\alpha} \rho^{4-\alpha} f(r, \rho). \tag{40}$$

w is proportional to the cooling time at constant pressure and g is the mass distribution function in terms of w instead of ρ. From its definition (34) we get

$$\xi = \frac{(2-\alpha)g_0(w_c)\int_{w_c}^{\infty} g_0(w)(w - w_c)^{\frac{1}{2-\alpha}} dw}{\int_{w_c}^{\infty} g_0(w) dw \int_{w_c}^{\infty} g_0(w)(w - w_c)^{-\frac{1-\alpha}{2-\alpha}} dw}, \tag{41}$$

where $g_0(w)$ is $g(r_1, w)$ and $w_c(r)$ is the value of w corresponding to $\rho_c(r)$.

The self-similar distribution functions discussed in Nulsen (1986) may be written in terms of g and w as

$$g(w) \propto \begin{cases} (w_0 - w)^{k-1}, & w < w_0, \text{ and } k > 0; \\ e^{-w/w_0}, & (k = \infty); \\ (w_0 + w)^{k-1}, & k < \frac{-1}{2-\alpha}. \end{cases} \tag{42}$$

(The last of these is not discussed in Nulsen 1986, since it gives $\xi > 1$.) Being self similar, all these distributions give a fixed value for ξ, independent of w_c. The result may be written collectively in terms of k as

$$\xi = \frac{\overline{2 - \alpha k}}{2 - \alpha k + 1}. \tag{43}$$

We see that a distribution function giving any positive value for ξ is possible.

Those distributions with $k > 0$ vanish for densities less than the cutoff ρ_0, corresponding to w_0. Close to ρ_0 they behave as $f \sim (\rho - \rho_0)^{k-1}$. Those for $k < 0$ are finite for all positive densities, and behave at low densities as $f \sim \rho^{\overline{-2-\alpha k}-2}$. The exponential distribution ($k = \infty$) is intermediate, with f being finite for all positive densities, but vanishing faster than any power law as $\rho \to 0$ ($w \to \infty$).

More generally, ξ is a function of ρ_c. It seems likely that there is a finite maximum temperature and hence minimum density in any real gas mixture, so that after a substantial fraction of the gas has cooled we can expect the density distribution to be well modelled by one of the self-similar distributions with positive k. The density distribution is probably continuous at the cutoff, requiring $k > 1$. In other words, for a realistic density distribution ξ will asymptote to a finite value less than 1 and most likely $\gtrsim (2 - \alpha)/(3 - \alpha)$ as the gas cools. Even with these constraints there still remains a great deal of latitude in the form of the mass deposition profile (i.e. in η, see equation 32).

Given that ξ varies with position, and assuming that it does not fluctuate wildly, we identify the value of ξ applying when half of the gas initially present has cooled as a characteristic value for the flow. Restricting attention to distributions which, in terms of $g_0(w)$, are singly peaked, smooth and symmetric about the peak gives a narrower range still for ξ at the half mass point. For example, if $\alpha = 0$ and g_0 is Gaussian then ξ varies from .52 to .88 while the middle 70 percent of the gas is deposited, and is .76 at the mid-mass point.

As noted previously, the assumption that the flow comoves is an idealization. A simple model for the slip is obtained if we assume that the inflow velocity

$$v(r, \rho) \propto (\rho/\bar{\rho})^{\overline{2-\alpha}\epsilon} \tag{44}$$

for a constant ϵ between 0 and 1. Small values for ϵ give small amounts of slip. Much of the detail of the exact solutions is preserved and we again find a family of solutions which are self-similar and give constant ξ and ν. As discussed in the previous section, the mass deposition profile depends only on the product $\xi\nu$, with $\nu = 1$ for the case of no slip. I give no details here, except to note that there are distributions giving any value for $\xi\nu$ in the range $0 - 1$, and that convective stability limits this quantity to a max imum of 1. Thus any mass deposition profile from the flows with no slip can be reproduced under this model for the slip. As a rule, however, for a given initial density distribution the mass deposition is more centrally concentrated when slip is introduced. To get a given mass deposition profile, that is a given value of $\xi\nu$, the self-similar distribution function must be broader when there is slip in the flow.

6. OBSERVING INHOMOGENEOUS FLOWS

An obvious prediction of the inhomogeneous flow models is that substantial quantities of gas are cooling to low temperatures throughout cooling flows, not just close to the centres. We are already looking forward to space missions capable of doing medium to high resolution X-ray spectroscopy with a spatial resolution of 1 arcmin or better and I want to emphasize their significance for the study of inhomogeneous cooling gas.

As shown above, at high densities the distribution function

$$f \to \frac{5\beta p}{2\rho^2 n^2 \Lambda} \tag{45}$$

because cooling is the fastest process. This result only depends on the assumption of local pressure equilibrium. It does not require a comoving or steady flow. At constant pressure the cooling time decreases rapidly with increasing density so that the asymptotic form of f will apply reasonably well, even at only twice the mean density. For a line emitted by gas at less than about half the mean temperature the emission per unit volume is therefore

$$\ell_{\text{line}} = \int n^2 \Lambda_{\text{line}} f d\rho = \frac{5\beta k}{2\mu m_H} \int_0^\infty \frac{\Lambda_{\text{line}}(T)}{\Lambda(T)} dT, \tag{46}$$

where Λ_{line} is the cooling function due to the line, and the upper limit of the integration is taken as ∞ on the assumption that the integrand in negligible for T greater than half the mean temperature. With these assumptions ℓ_{line} only depends on β and the element abundances. For given abundances we can use spectroscopic observations to map the distribution of the gas deposition rate, β.

This result has been applied elsewhere to determine total cooling rates (*e.g.* Canizares, this proceedings). For a homogeneous cooling flow it must be applied with the assumption that the cooling is isobaric, whereas for a spatially resolved inhomogeneous gas we have a criterion for determining when it is applicable.

The importance of this result is that it gives a method for determining the gas deposition rate which is independent of the morphological determinations and requires only one reasonable assumption concerning the state of the gas. It will provide a stringent test of the assertions that the gas flow is steady and comoving, and will allow determination of gas deposition rates in places where the flow is not steady. It will also be interesting to look for low temperature line emission from non-cooling flow clusters to determine whether the gas in them is inhomogeneous, as seems likely.

7. DISCUSSION AND CONCLUSIONS

Thermally unstable growth of initially small density fluctuations cannot account for the widespread deposition of cooled gas that is observed in cooling flows. The gas must already be significantly inhomogeneous before it starts to cool. The rms value of $\delta\rho/\bar{\rho}$ required to give the observed profile for $\dot{M}$ is about 60 percent for the self-similar density distributions. Thomas *et al.* (1987) find that the initial density fluctuations are about 30 percent. Although the density fluctuations are substantial, the associated temperature spread will not be easy to detect unless individual gas blobs can be resolved.

Buoyant motions tend to suppress the thermal instability, but mostly they cause inhomogeneities to fragment and be pinned to the mean flow. As a result there is little relative motion of gas at different densities in the flow, which may be treated as comoving.

We can construct a set of integral flow equations for comoving flows similar to those governing homogeneous flows (equations 15, 25 and 22). Like homogeneous flows, inhomogeneous cooling flows are approximately steady inside the region where the mean cooling time of the gas (19) is less than than the age of the flow. The integral flow equations are not closed, since the mass deposition rate β and total radiation rate R depend on the details of the density distribution. However, we may parameterize these quantities in terms of the mean flow properties and the structure of steady flows is found to depend principally on the dimensionless ratio

$$\xi = \frac{5\beta p}{2\bar{\rho}R}. \tag{47}$$

Representative values for ξ for the region of steady flow fall in a fairly narrow range, although this leaves considerable latitude in the mass deposition profile. Observed deposition profiles do require physically reasonable values for ξ, however, it is not currently obvious what factors determine the initial density distribution of the gas and hence also ξ. For the present we can use ξ as a model parameter in the integral flow equations in order to investigate inhomogeneous cooling flows. These equations can easily be modified to include the effects of mass injection, so that we may use them to model cooling flows in isolated galaxies as well.

Slip between gas at different densities reduces the effective value of ξ, making the mass deposition more centrally concentrated. It is possible that the increased density spread due to cooling may increase the slip enough to ultimately determine the mass deposition profile (see section 5).

Low temperature X-ray line emission from galaxies and clusters of galaxies will provide a very powerful tool for the study of cooling flows. Under very general conditions such line emission traces the distribution of the rate of gas deposition. Amongst other things, it may be used to look for gas deposition in clusters that do not have cooling flows.

REFERENCES

Cowie, L.L., Fabian, A.C. & Nulsen, P.E.J., 1980. *Mon. Not. R. astr. Soc.*, **191**, 399.

Cowie, L.L., Hu, E.M., Jenkins, E.B. & York, D.G., 1983. *Astrophys. J.*, **272**, 29.

Fabian, A.C. & Nulsen P.E.J. 1977. *Mon. Not. R. astr. Soc.*, **180**, 479.

Fabian, A.C., Nulsen, P.E.J. & Canizares, C.R., 1984. *Nature*, **310**, 733.

Field, G.B. 1965. *Astrophys. J.*, **142**, 531.

Heckman, T.M. 1981. *Astrophys. J.*, **250**, L59.

Lynds, R., 1970. *Astrophys. J.*, **159**, L151.

Malagoli, A., Rosner, R. & Bodo, G. 1986. Preprint.

Mathews, W.G. & Bregman, J.N. 1978. *Astrophys. J.*, **244**, 308.

Nulsen, P.E.J., 1986. *Mon. Not. R. astr. Soc.*, **221**, 377.

Thomas, P.A., Fabian, A.C. & Nulsen, P.E.J. 1987. *Mon. Not. R. astr. Soc.*, in press.

RADIO EMISSION IN COOLING FLOWS

Edwin A. Valentijn
Kapteyn Astronomical Institute
P.O. Box 800
NL-9700 AV Groningen
The Netherlands

ABSTRACT. It is reviewed whether cooling flows could fuel central radio sources in spheroidal galaxies as a general, universal mechanism. The radio luminosity functions of S0, E, gE and cD galaxies show a similar scaled dependence on L_{opt}, demanding a non-exclusive radio powering mechanism for cD's. If cooling flows generate the central radio sources in these objects, then a non-exclusive X-ray halo formation process should occur. Gravitational accretion of gas from the inter galactic medium is identified as such a mechanism for both E+S0's and cD's. This requires a large mass for the cD's, to generate a potential well deep enough to bind the gas. Evidence for a large cD mass is discussed. In such a context it is justified to combine the data of several major samples of cD, E and S0 galaxies in order to study the relevance of the input scenario. It is found, that for a sample of 111 objects, L_{opt} is correlated with L_x, $\log L_{opt} \sim 0.13 \log L_x$, in a *continuous* way for the different optical types. The slope of the correlation is consistent with a gravitational accretion scenario. The L_r is strongly correlated with L_x, also in a continuous way over five decades in both luminosities. In the time variable radio source output interpretation, the *one-to-one* correlation between the integral fraction of radio detections $(F(> P))$ and L_x implies a one-to-one correspondence between the peak amplitude of a radio source and the X-ray luminosity of its host galaxy. This is strong evidence for gravitationally driven cooling flows fuelling central radio sources in spheroidal galaxies in general and the results are confirming the input scenario. The cosmological evolution of the scenario is discussed and found to predict $L_r \sim (1 + z)^5$.

INTRODUCTION

Some evidence that cooling flows centered on first ranked cluster galaxies, mostly of the cD type, provide the material for non-thermal central radio sources in these galaxies has been presented by Valentijn and Bijleveld (1983, hereafter VB83) and by Jones and Forman (1984, hereafter JF84). However, it has been known for a long time that the probability of an elliptical galaxy containing a central radio source of a certain power is related to the optical luminosity (L_{opt}) of that object (Auriemma et al.,1977). This probability function, normally called the Radio Luminosity Function (RLF), was found to scale continuously as function of L_{opt} over a range of four optical magnitudes in M_v and over a range of 5 decades

189

A. C. Fabian (ed.), Cooling Flows in Clusters and Galaxies, 189–198.
© 1988 by Kluwer Academic Publishers.

in radio luminosity L_r (see figure 1). The brightest optical objects for which these functions have been determined correspond to the cD galaxies and indeed their RLF represents a *smooth upscaling* of the RLF's of elliptical galaxies of lower optical luminosity. This should be a crucial point when discussing whether cooling flows in cD galaxies can provide the material for the mass to energy conversion in the centers of these galaxies. If we were to accept that at least a major fraction of the radio sources in cD galaxies are powered by a cooling flows, then the continuity of the RLF's would require this process to occur in scaled down versions in less bright elliptical galaxies as well.

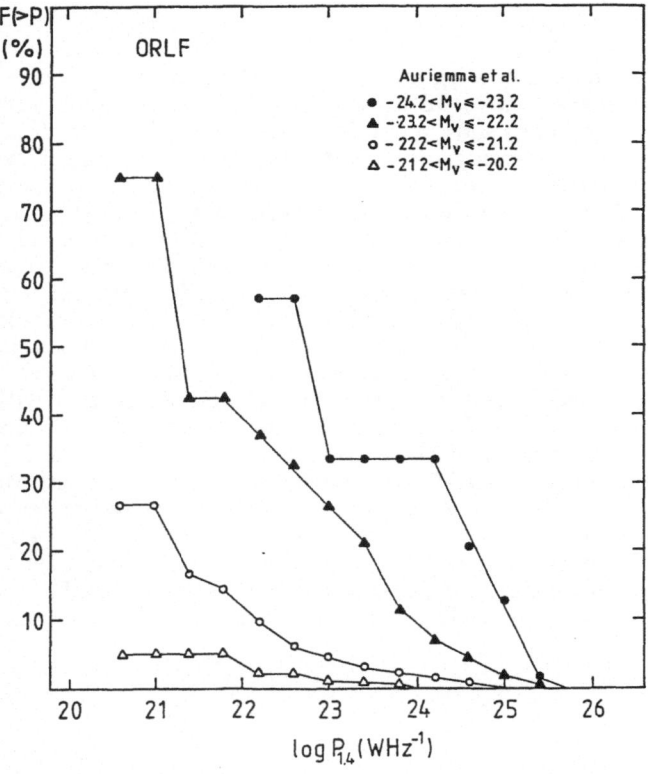

figure 1. *The low radio luminosity part of the Auriemma et al. (1977) RLF of E, gE and cD galaxies, is presented in different absolute visual magnitude intervals. The integral fraction of radio detections at 1.4 GHz is plotted as function of the monochromatic radio power at that frequency.*

In this context, I will address the question of whether there exists evidence, beyond that presented in the VB83 and JF84 papers, for cooling flows to power central radio sources in E, gE and cD galaxies. I will combine the data obtained for the cD sample of the previous work with the more recent very interesting radio and X-ray studies of samples of normal E and SO galaxies published by Canizares, Fabbiano and Trinchieri (1987, hereafter CFT87) and Fabiano, Klein, Trinchieri and Wielebinsky (1987, hereafter FKTW87).

It should be emphasized that the samples discussed here were *optically* selected samples or are truncated to that. The samples contain only those 3C radio sources which received their 3C status due to their proximity to us. The radio luminosities are derived from the *central* emission and generally do not include emission from the few extended radio lobes found. Thus we are dealing with samples of at least optically representative rather *normal* galaxies. Before presenting the merging of the different samples I will summarize the main results of older RLF studies. A value for H_o of $50 \, \text{km} \, \text{sec}^{-1} \, \text{Mpc}^{-1}$ will be used throughout this paper.

THE OLD RLF RESULTS

The Auriemma *et al.* RLF is reproduced in figure 1. The prime conclusion from this and many related projects is that an elliptical galaxy has a certain probability of containing a radio source of power L_r, which is a function of L_{opt}. There is no *a priori* way given L_{opt}, to determine what L_r would be expected for a given object, but for *samples* of objects this probability function is well defined and was found not to vary within at most a factor of two between all sorts of different environments, *i.e.* rich clusters and poor groups of galaxies (the Westerbork Survey of Rich Clusters, see Valentijn,1980 for a review). This implies that correlative studies using L_r as one parameter should always include the upper limits of non-detections in order to derive fractions of detections for samples of objects. Hence, power-power plots are of limited significance.

It is believed that the reason for this particular radio source property is an intrinsic variation of L_r on time scales of $10^6 - 10^7$ yr. Radio sources go on and off, as illustrated by the distinct separated blobs in head-tail radio sources. If radio sources in general are fuelled by cooling flow material, then it is possible that a flow switches on a central radio source, which then by its radiation pressure, temporarily stops the flow until its fuel is exhausted, giving the flow the opportunity to start the cycle again, etc.

THE GALACTIC MASS / CLUSTER POTENTIAL PROBLEM

As mentioned in the Introduction, the continuous scaling of the RLF's, when computed as function of L_{opt}, suggests a non-exclusive radio powering mechanism for massive cD galaxies. Many cD's are found to be exactly centered on large scale X-ray haloes (JF84) and recently, E and gE galaxies were also found to have X-ray components (Forman *et al.*,1985 and CFT87). Comparing the properties of the X-ray sources centered on E galaxies with those on cD galaxies will only be meaningful if the basic formation mechanism of the X-ray emitting gaseous components are similar for both galactic types. Thus the major task, to evaluate the question of whether cooling flows fuel a central radio source, is to study whether a uniform X-ray halo formation process can be identified and whether that process can give rise to the eruption of the central radio sources with the properties as dictated by the RLF's. The continuity requirement can be satisfied when the X-ray haloes are formed by the gravitational pull of the individual central galaxy, acquiring the gas from the cluster and supercluster prevading medium. In the VB83 work on cD's we were able to interpret the L_x, L_{opt}, L_r relations in terms of a gravitationally driven cooling flow scenario. The consequence of all this is that the large scale (at least 50 kpc radius) cD X-ray haloes should be bound to the host cD galaxy by its own gravitational force, rather than a by a central cluster potential well resulting from the contribution of many smaller galaxies. Do we really have evidence for large massive haloes around cD galaxies? The mass required to initiate binding of an X-ray halo of temperature T out to a radius R with an extended mass distribution parametrized with a softness parameter ϵ ($\simeq 5$) is:

$$M(< R) = \epsilon \, 17 \cdot 10^{10} \, R_{kpc} \, T_{3 \cdot 10^7 K} (M_\odot).$$

In the worst case scenario, $T = 10^8 K, R = 250 kpc$, the required mass is $\simeq 7 \cdot 10^{14} M_\odot$, while a more typical temperature of $3 \cdot 10^7 K$ requires a cD mass of $\simeq 2 \cdot 10^{14} M_\odot$. In the most optimistic case, when the dark matter has a typical scale size of 50 kpc and acts almost as a point mass at a 250 kpc radius , ϵ is close to unity and the required mass is $\simeq 4 \cdot 10^{13} M_\odot$. At present several investigations hint towards the existence of dark massive haloes on cD galaxies, but none with a significance of better than a 2σ level. Some examples are:
- The A496 cD, which has an X-ray halo radius of 250 kpc (Nulsen et al., 1982), is not coincident with the centre of the cluster (Mazure et al., 1986). This cD galaxy has only a few other galaxies of at least two optical magnitudes fainter brightness within that radius (at least projected on the sky), which appear to hardly contribute to the overall potential well.
- In the very few cases where the stellar velocity dispersion of cD galaxies has been measured out to large radii (50 kpc), they were found to rise significantly (Dressler, 1979, Carter et al., 1985).
- The Cowie et al. (1987) analysis of the cluster virial masses (as deduced from X-ray data) favoured a lower cluster virial mass but also indicated that a central high mass concentration is required.
- A recent study of the relative velocities of a sample of $\simeq 40$ dumbbell galaxies (double cD's !) showed that these objects perform rather different dynamics than normal cluster galaxies. If the dumbbells form bound systems indeed, then a statistical analysis rejects their orbits being in pure Keplerian motion on a 2σ level. A logarithmic type of potential is favoured and consequently points to large massive haloes around these systems (Valentijn and Casertano in preparation).

The evidence for massive haloes around cD galaxies is accumulating but still not compelling. Here, I assume the X-ray haloes of cD's to be linked to their gravitational well and in the following paragraphs will address whether such an approach leads to results.

L_{opt} VERSUS L_x

In figure 2 the L_{opt}, L_x data of the 55 objects of the CFT87 sample of E and SO galaxies, including the sample of Virgo cluster objects, is reproduced (open circles). The L_B represents the total blue luminosity. The optical data of the cD sample of VB83 has been converted to L_B by assuming a growth curve of 0.4^m between a 38 kpc and an infinite radius (total magnitude), and using an intrinsic $B - V = 1.05$. Thus the data of 43 Abell cluster cD's (filled circles) and 13 poor group cD's (crosses) could be added to figure 2. The straight line represents the regression analysis presented in VB83 for the cD sample alone: $log L_{opt} \sim 0.08 \, log L_x$. The extrapolation of this 'best fit' also describes the E/SO galaxies quite well. In VB83 it was suggested that such a steep L_{opt} dependence of L_x would be expected in a simple model, where of the X-ray halo is initially formed by gravitational accretion of the (super)cluster medium of density ρ_∞. For a nearly iso-thermal X-ray source with core radius r_c the $L_x \sim \rho_o^2 \, r_c^3$, with ρ_o the central gas density. Assuming a linear relation between r_c, accretion radius R_{acc} and galactic mass M_{gal}, and adopting a spherical initial gravitational accretion model ($\rho_o \sim M_{gal}^2 \rho_\infty$) the expected relation between galactic optical luminosity and L_x is

$$logL_{opt} \sim \frac{1}{7+7\alpha} logL_x,$$

where the mass to light ratio has been parameterized as $(M/L) \sim L_{opt}^\alpha$. For α equals 0, 0.5 or 0.75 the expected slope is 0.14, 0.10 or 0.08 respectively. The observed slope for the cD's alone was $0.08 \pm .02$, while the regression analysis of the total sample has a slope 0.13.

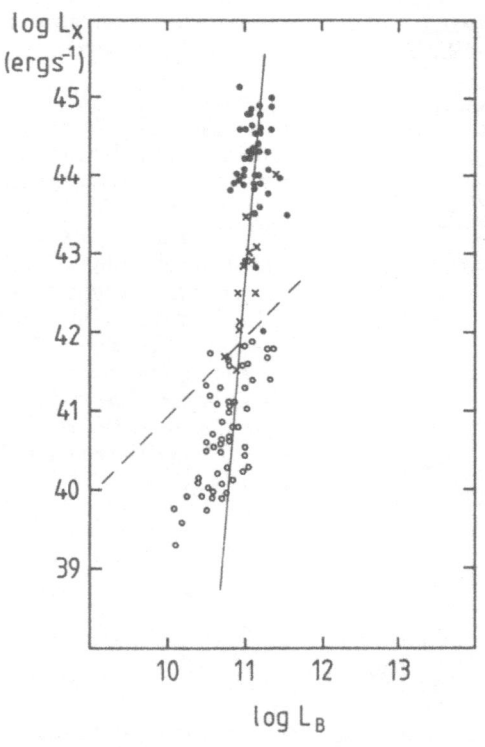

figure 2. *Blue total optical luminosity versus total X-ray luminosity for a sample of 111 objects, including rich cluster cD's (filled circles, VB83), poor group cD's (crosses, VB83) and E/S0 galaxies (open circles, CFT87). The straight line represents the old fit for the cD sample alone. The dotted line is taken from CFT87 and represents their estimate of the maximum L_x that could be caused by supernovae heating.*

The conclusion is, that in general the E/S0 galaxies are found to follow and extrapolate the L_{opt} - L_x regression of the cD's and that the overall slope appears consistent with a gravitational accretion model for the X-ray source formation for both E's and cD's with a ratio of $M/L \sim L_{opt}^{0.5}$ for the gravity inducing mass. The continuity between E's and cD's, which is a strong requirement set by the RLF's in order that cooling flows power central radio sources, seems fullfilled. However, there is a large scatter around the L_{opt}-L_x regression. During gravitational accretion the mass inflow rate is $\sim v^{-3}\rho_\infty$, at least in the subsonic regime, with v the velocity of the galaxy through the medium. Both a spread in environmental gas density and the peculiar motions of the accreting objects could cause the observed scatter.

L_r VERSUS L_x

The best way to assess correlations with radio luminosity is to calculate RLF's of samples of objects. In figure 3 I have reproduced from VB83 the integral fractions of radio detections ($F(>P)$) as function of the 1.4 GHz radio power ($P_{1.4}$) of Abell cluster and poor group cD's, separated in three intervals of their corresponding L_x. Using the new data for 27 E+SO's from FKTW87, I have computed the RLF's of this sample separating the objects in two sub-samples with their corresponding X-ray luminosities within $41 \leq logL_x < 42$ and $39.7 \leq logL_x < 41$ respectively.

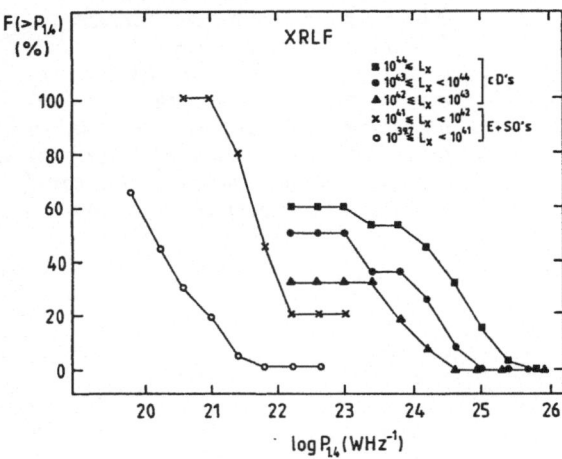

figure 3. *The RLF's of E,SO and cD galaxies presented at five different intervals of their corresponding L_x. Over five decades in both L_r and L_x the radio output of the central radio source is found to correlate with the total X-ray luminosity. The data of the cD's is reproduced from the VB83 paper. The radio and X-ray observations of the E+SO's were reported by FKTW87.*

Fornax A was removed from the original sample since it can not be considered as being optically selected. The only other object with extended radio lobes (also the only other 3C radio source) 3C296 was included in the $41 \leq logL_x < 42$ RLF. It is somewhat doubtful whether or not this galaxy satisfies the optical selection criterium. If omitted, the RLF in this L_x interval will be downshifted by $\sim 20\%$ and it will be somewhat better centered between the neighbouring RLF's. The 5 GHz flux densities reported by FKTW87 were converted to 1.4GHz assuming a spectral index of 0.7.

Figure 3 clearly indicates that the correlation between L_x of X-ray haloes on cD's and the power of *central* radio sources, as found before for the cD's alone, is extended continuously to E+SO galaxies at both lower L_r and L_x. This is illustrated in figure 4, which shows the regression of these five curves with respect to each other. For each of the five RLF's in figure 3 the mean L_x of an L_x interval was computed and this is plotted along the horizontal axis in figure 4. Then for three fixed values of $F(> P)$ the corresponding $P_{1.4}$ was read from figure 3 and is plotted along the vertical axis in figure 4. The $F(> P)$ - L_x plot shows, that at a fixed integral fraction of radio detections a linear correlation appears on log-log scale between the radio power, where that integral fraction of detections is reached, and L_x of the X-ray sources belonging to these objects.

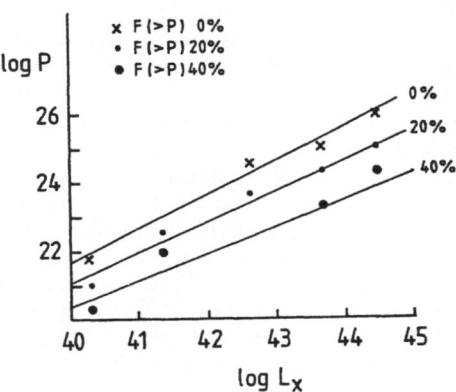

figure 4. *A condensation of data of hundreds of observations in the radio and X-ray bands of 104 cD galaxies and 27 E+SO's. The plot represents the regression of the five RLF's in figure 3 as a function of L_x at three choosen values of integral fractions of detections($F(> P)$).*

Most strikingly, the linear fits to the data have slopes between 0.90 and 1.0 over 5 decades both in L_x and L_r, implying a *one to one* correlation between L_x and the detection limit in the radio where certain fixed fractions of galaxies are detected. Note, that this statement is not identical to "L_x is one to one correlated with L_r".

The reason to undertake the trouble of computing RLF's was the intrinsic *on-off* nature of radio sources, which is not expected to occur with any similar periodicity in the thermal X-ray output. What we have found here is that there exists a one-to-one correlation between the amplitude of that time variable radio output and L_x. This is best illustrated by considering the $F(> P) = 0\%$ (read $0\% + \epsilon$) points in figure 4. In the time-variable radio source output interpretation, the 0% points in figure 4 represent the peak amplitude a radio source can produce when it has a a certain value of L_x. In fact, integrating the radio spectrum between 10MHz and 10GHz the $F(> P) = 0\%$ linear fit translates to

$$L_r^{peak} = 0.05\, L_x\, (ergs^{-1}).$$

The one-to-one correlation between L_x and the amplitude of L_r makes it much more difficult for confinement mechanism's to be the primary cause for the L_r - L_x correlation. For example, in the case of thermal confinement of central radio sources by the X-ray gas, one would expect $L_r \sim L_x^{0.5}$, while ram-pressure confinement on an expanding blob would result in a negative index.

In conclusion, a *continuous* correlation between L_x and L_r over five decades has been found and an important continuity criterium for the concept of cooling flows fuelling radio sources is satisfied. The one-to-one L_r^{peak}-L_x correlation supports a cooling flow/radio source scenario, since it requires a plausible linear conversion from L_x to M and subsequently from $\dot{M}$ to L_r.

Something should be said about the internal logics of the L_{opt} - L_x - L_r triangle. First of all, it should be mentioned that for both the cD's and the E+SO's, I have computed the RLF's as function of L_{opt} irrespective of their L_x or upper limits. These RLF's conform to the Auriemma *et al.* RLF in figure 1 and in this respect the samples studied are both optically and radio representive. We observed $L_{opt} \sim L_x^7$ and $L_x \sim L_r$. Naively, one would expect from these that $L_{opt} \sim L_r^7$, but the Auriemma *et al.* RLF indicates at most $L_{opt} \sim L_r^5$. What is happening is that the X-ray halo formation acts as a prime mechanism, and thus L_{opt}-L_x and L_x-L_r are prime correlations, naturally only settled for those galaxies that have been successful in building a X-ray source. The flatter slope of the supposed secondary L_{opt}-L_r relation, determined from a larger sample than used for the other two relations, indicates that not all galaxies have been that successful and other agents than the galactic mass are effecting the X-ray halo formation. In the previous Section I mentioned the effect of the two parameters v and ρ_∞ on the scatter of the L_{opt}-L_x correlation in a gravitational accretion scenario. The tentative estimate is that $\sim$ 30% of the objects, used for the calculation of the RLF as function of L_{opt} alone, are much less effective in building an X-ray halo and subsequently a radio source. This then explains the fact that the L_{opt}-L_x and the L_{opt}-L_r relations are not parallel. A sensitive test of this interpretation would be to observe to a low L_x a (RLF) representative sample of galaxies for which presently only X-ray upper limits are available. The present X-ray upper limits are to high for this exercise.

The observed relations and their interpretation as presented in this paper can be summarized in an overall grand radio source fuelling formula in terms of a gravitational accretion scenario:

$$L_x \sim \dot{M} \sim L_r^{peak} \sim G M_{gal}^7 v^{-3} \rho_\infty^2 \epsilon(a/b, \sigma_*),$$

where ϵ is an efficiency factor for the flow within the galactic body, which probably depends on both the axial ratio and the stellar velocity dispersion of the host object. At present the observational data appear consistent with this, and it is most provocative that the E+SO's are now found to be exactly located on both the L_x-L_{opt} and the L_r-L_x regressions as they have been predicted by downscaling the properties of cD galaxies.

COSMOLOGICAL EVOLUTION

The possible generality of gravitational binding of X-ray haloes to spheroidal galaxies giving rise to cooling flows and subsequently to the eruption of central radio sources, makes it interesting to analyse how that process is expected to function at higher redshifts z (Valentijn,1987). Gravitational accretion of hot X-ray gas will only occur if an object contains a component with a large M/L. Here, I assume this component to be *pre-existent* at $z > 6$ and that the visible galactic component forms after that and has a much lower mass. Typical mass inflow rates of $\sim$ 100 $M_\odot$/year imply that , if the major fraction of the matter is processed in star formation, then in a Hubble time a continuous acquisition of gas into the X-ray halo is needed. If this acquisition is done by means of gravitation from the intracluster medium and subsequently from the supercluster pervading medium, then we can examine how the input of gas into X-ray haloes is expected to evolve as the universe expands. Note, this approach is entirely complementary to the other cooling flow evolution models presented at this workshop, in which a static universe was assumed (Bertschinger, White , Loewenstein).

The only observations of the present density (ρ_{sc}) of the supercluster pervading hot gas comes from the radio observations of head-tails in the Coma and the Hercules superclusters (Valentijn,1979) and indicates $\rho_{sc} = 4\,10^{-6}\,cm^{-3}$. Since at the boundaries of the rich clusters the intra cluster medium is likely to be enriched by gas processed in star formation, I adopt a primordial component with $\rho_{sc} = 2\,10^{-6}\,cm^{-3}$. I assume a temperature for the gas of $T_{sc} = 5\,10^6\,K$, which is consistent both with the absence of *broad Lyα* absorption lines in quasar spectra (low z) and with the X-ray background spectrum (high z). The supercluster gas density is low enough to follow the Hubble expansion and thus

$$\rho_{sc} = 2 \cdot 10^{-6} (1+z)^3\,cm^{-3}$$

and

$$T_{sc} = 5 \cdot 10^6 (1+z)^2 K.$$

Now, let the supercluster gas accrete into a cluster with a simple linear density enhancement of a factor of 10 (Gunn and Gott, 1972). When that gas is subsequently accreted by a proto-type cD, another factor of 100 in density enhancement is actually observed between the outskirts and the centers of X-ray haloes. Thus a z dependence of the central gas density in a proto-type cD can be derived, which leads to a central cooling time of

$$\tau_c = 4.6 \cdot 10^9 (1+z)^{-2}\ year.$$

In all Friedmann cosmologies τ_c is short enough, compared to the age of the universe, to switch on cooling flows after gravitational binding has had a chance to build a halo. This can happen first for the heaviest dark potential wells at $z \sim 4$.

An estimate of the *central* $\dot{M}$ can then be made by dividing the total gas mass in a certain central volume by its expected average cooling time. For example, within 0.05 core radius:

$$\dot{M}_{cen} = \frac{2 \cdot 10^9 (1+z)^3}{\tau_c} = 2.6\,(1+z)^5\ M_\odot/year\ .$$

Kron *et al.* (1985) found, that radio source counts down to 10 mJy are dominated by gE galaxies and modelling the counts (Windhorst, 1984) indicated $P \sim (1+z)^{4.6}$ in the redshift range $z = 0.3$ to $z = 0.75$. It was also concluded, that if gE's are radio active at larger redshifts ($z > 1$) then they should have gone through a radio loud phase simular or equal to that of quasars. The density evolution of quasars is known to be proportional to $(1+z)^5$.

In conclusion it seems, if the central $\dot{M}$ can be linearly related to the radio power, that a gravitational cooling flow/radio source fuelling scenario predicts the strong cosmological evolution of radio sources and quasars, as suggested by the observations.

REFERENCES

Auriemma,C., Perola, G.C., Ekers, R., Fanti, R., Lari, C., Jaffe, W.J., Ulrich, M.-H.: 1977, Astron. Astrophys. **57**,41
Canizares, C.R., Fabbiano, G., Trinchieri, G.: 1987, Astrophys. J. **312**,503
Carter,D., Inglis, I., Ellis, R.S., Efstathiou, G., Godwin, J.G.: 1985: Mon. Not.

198

Roy. Astr. Soc. **212**,471
Cowie, L.L., Henriksen, M., Mushotzky, R.: 1987, Astrophys. J. **317**,593
Dressler, A.: 1979, Astrophys. J. **231**, 659
Fabbiano, G., Klein, U., Trinchieri, G., Wielebinsky, R.: 1987, Astrophys. J. **312**,111
Forman, W., Jones, C., Tucker, W.: 1985, Astrophys. J. **293**, 102
Gunn, J.E. and Gott III, J.R.: 1972, Astrophys. J. **176**, 1
Jones, C. and Forman, W.: 1984, Astrophys. J. **276**, 38
Kron, R.G., Koo, D.C., Windhorst, R.A.: 1985, Astron. Astrophys. **146**,38
Mazure, A., Gerbal, D., Proust, D., Capelato, H.V.: 1986, Astron. Astrophys. **157**,159
Nulsen, P.E.J., Stewart, G.C., Fabian, A.C., Mushotzky, R.F., Holt, S.S., Ku, W.H-M., Malin, D.F.: 1982, Monthly Notices Roy. Astron. Soc. **199**,1089
Valentijn, E.A.: 1979, Astron. Astrophys. **78**,362
Valentijn, E.A.: 1980, Highlights of Astronomy, Vol **5**,715, IAU, ed. P.A. Wayman
Valentijn, E.A. and Bijleveld, W.: 1983, Astron. Astrophys. **125**, 223
Valentijn, E.A.: 1987, Proc. of IAU Symposium No. 127, *Structure and Dynamics of Elliptical Galaxies*, ed. T. de Zeeuw, Dordrecht: Reidel Publ. Comp.
Windhorst, R.A.: 1984, Ph. D. Thesis, Sterrewacht Leiden

EMISSION FROM THERMAL INSTABILITIES

L. P. David
University of Virginia
P.O. Box 3818
Charlottesville, VA. 22901
U.S.A.

J.N. Bregman
NRAO
Edgemont Road
Charlottesville, VA. 22901
U.S.A.

ABSTRACT. We present the results of numerical simulations of the growth of thermal instabilities in cooling flows. These simulations begin with $10^7 - 10^8$K gas and follow the evolution of the instability as the gas cools to a terminal temperature of 10^4K. Only perturbations larger than 1 kpc develop shocks in the hot environment of cooling flows. These large perturbations evolve through three distinct phases: (1) the formation of a cold core, (2) a supersonic accretion phase onto the cold core, and (3) a subsonic accretion phase. We present the optical and ultraviolet line emission produced during the supersonic accretion phase. The emission lines have a lower luminosity and higher ionization level than the observed optical emission in cooling flows. These shocked condensations are a strong source of high ionization lines which may be detectable in the ultraviolet.

1. INTRODUCTION

All of the gas in a cooling flow cannot settle to the center of the accreting galaxy or the x-ray emission would be more centrally peaked than is observed. Thus, if cooling flows exist, most of the hot x-ray emitting gas must condense out of the flow via thermal instabilities before reaching the center of the accreting galaxy. The aim of this paper is to determine the dynamical behavior and the emission properties of thermal instabilities. Several papers have investigated the linear growth of thermal instabilities (Field 1965, Mathews and Bregman 1978, and Balbus 1986). However, the observational signature of thermal instability does not develop until after the perturbations have become nonlinear (i.e., during shock repressurization). Observations of the x-ray emission from cluster cooling flows imply that most of the hot gas condenses out of the flow at large radii, suggesting that the perturbations were

A. C. Fabian (ed.), Cooling Flows in Clusters and Galaxies, 199–204.

initially in the nonlinear regime. The nonlinear growth of thermal instabilities is thus an integral part of the cooling flow scenario.

2. METHOD

We have used a time-dependent, one-dimensional, hydrodynamics code to simulate the evolution of thermal instabilities in $10^7 - 10^8$K gas. Plane parallel symmetry is used in these calculations and the simulations are run in Lagrangian coordinates; gravity is not included. Initially, the gas is stationary and at constant pressure. An entropy perturbation is imposed on the gas with the form

$$n(x) = n_0(1 + A \ exp(-x/s)^2)$$

$$T(x) = T_0(1 + A \ exp(-x/s)^2)^{-1}$$

where n_0 and T_0 are the ambient density and temperature, A is the amplitude of the perturbation, and s is the length scale of the perturbation. Only half of the perturbation is modeled in the hydrodynamics code with the peak density occuring in the innermost cell. As gas cools below 10^6K, the radiative cooling time becomes less than the recombination time of some ions and the electron temperature is no longer equal to the ionization temperature. The actual cooling rate below 10^6K is less than the equilibrium value since the electrons cannot efficiently excite the highly overionized atoms. At these lower temperatures we have carefully followed the ionization and cooling of the gas using atomic data from Edgar (1987). Because the shocks encountered in these models are not very strong, the emission line strengths produced during the supersonic accretion phase are very sensitive to the preshock ionization state of the gas. Below 10^4K, the environment of the gas can significantly affect the thermal evolution of the instability and cooling is turned off below this temperature. A more detailed discussion of our method is given in David, Bregman, and Seab (1987).

3. RESULTS

3.1. Hydrodynamics

Perturbations in the intracluster of interstellar medium may have a particular size but are unlikely to be characterized by a single wavelength component. For this reason, we begin our simulations with a gaussian perturbation since the Fourier transform of a gaussian is also a gaussian in the wavenumber with most of the power at long wavelengths. The evolution of these thermal instabilities can be described as a superposition of all the Fourier components. The shortest wavelength modes cool isobarically while the longest wavelength modes cool isochorically. The first mode to cool is located near the origin. Larger wavelength modes also cool isobarically until the cooling time becomes less than the sound crossing time, after which they cool isochorically establishing a pressure gradient between the cold core and the ambient medium. As the gas cools to 10^4K, the gas velocities become supersonic. The gas is then decelerated onto the core through a shock. The shock front propagates outward as longer wavelength modes cool and accrete onto the core. The shock strength increases at first, then begins to decrease as hotter gas accretes onto the core, and eventually becomes a sound wave. The remaining gas accretes onto the core subsonically.

Only perturbations larger than 1 kpc develop shocks in conditions appropriate for cooling flows in elliptical galaxies, while perturbations must be greater than 5 kpc in order to develop shocks in conditions appropriate for cluster cooling flows. The temperature evolution of a model with $T_0 = 3 \times 10^7$K, $n_0 = 10^{-2}$cm^{-3}, A = 0.1, and s = 10 kpc is shown in figure 1. In this model, the cold core forms after one-half of the initial instantaneous cooling time has elapsed with an extent of 1 pc and n = 10 cm^{-3}. A shock then develops after a sound corssing time through the core. The accretion shock is easily identifiable as a sharp spike in figure 1. As gas continues to accrete onto the core, the shock front propagates outward, attaining a peak shock velocity of 130 km s^{-1} before being degraded to a sound wave after 10^5 years. The shock dissipates after the core has grown to 10 pc and 5×10^5K gas begins accreting subsonically.

3.2 SPECTRA

The origin of the optical emission in cluster cooling flows is still unknown. Two possible line excitation mechanisms are photoionization and shock excitation. In Table 1 we compare the time averaged line strength ratios produced during the supersonic accretion phase in our models with the observations of Hu, Cowie, and Wang (1985) and Johnstone, Fabian, and Nulsen (1987). Our line luminosities are based on the assumption that all of the gas in a cooling flow condenses out of the flow via thermal instabilities and is eventually shocked. The line luminosities given in Table 1 are normalized to an accretion rate of $\dot{M} = 100$ M$_\odot$ yr^{-1}. The spectra produced during the supersonic accretion phase is not very sensitive to the length scale of the perturbation as can be seen in Table 1. It is obvious from Table 1 that shocks do not produce enough emission in the lower ionization lines to match observations and overestimate the [OIII]λ5007/Hβ ratio. Hu et al. argue on the basis of shock models by Shull and McKee (1979), that the optical emission from cluster cooling flows can be understood in terms of shock excitation with shock velocities between 70-90 km s^{-1}. These models, however, assume preshock ionization equilibrium which is not valid in gas cooling below 10^6K. High ionization species like OIII and CIV are orders of magnitude overabundant before being shocked. Similar shock velocities are encountered in our models but the emission strongly weighted

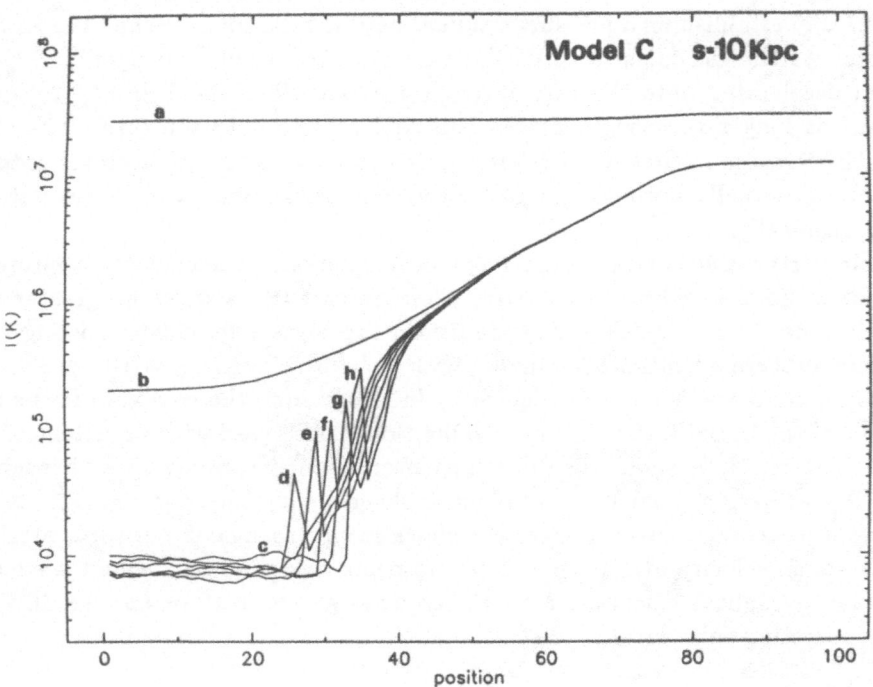

Figure 1: Temperature profiles for a model with $T_0 = 3 \times 10^7 K$, $n_0 = 10^{-2} cm^{-3}$, $s = 10$ kpc, A= 0.1 at the following times (10^9 yr): a, 0.0; b, 1.03543; c, 1.03560; d, 1.03562; 3, 1.03564; f, 1.03566; g, 1.03569; h, 1.03571. The horizontal axis correponds to the lagrangian cell number in the hydrodynamics code.

TABLE 1
LINE STRENGTH RATIOS

Model	Hα/Hβ	OIII/Hβ	NII/Hα	SII/Hα	OI/Hα
HCW	3.4	0.9	1.5	1.1	0.3
JFN		0.6	1.3	1.2	0.7
B	3.13	2.12	0.239	0.333	0.017
C	3.16	1.93	0.249	0.341	0.017
E	3.28	1.58	0.207	0.315	0.025

Note: The first row (HCW) contains averaged observed line ratios from Hu, Cowie, and Wang (1985). The second row (JFN) contains averaged line ratios from Johnstone, Fabian, and Nulsen (1987). The length scale of the perturbation S = 5, 10, 20 kpc in Models B, C, and E, respectively.

toward higher ionization lines. Another problem with emission from shocks producing the optical emission in cooling flows is that the observed Balmer line luminosities ($L_{H\beta} \sim 10^{40} - 10^{41}$ ergs s^{-1}) are roughly an order of magnitude greater than can be produced through shocks. Balmer line luminosities this large require each hydrogen atom to recombine approximately 10 times, which requires a steady source of ionization.

Even though the emission from shocked condensations cannot reproduce the optical emission observed in cluster cooling flows, these condensations are a source of high ionization lines and may be detectable in the ultraviolet. Only two observations of cluster cooling flows by the IUE have been reported in the literature. Fabian et al. (1984) and Norgaard-Nielsen et al. (1984) have detected Lyα emission in NGC1275 and A1795, respectively, and also place upper limits on CIVλλ1548,1551 emission. Table 2 contains a comparison of the Lyα and CIV emission from our thermal instability models with these observations. An accretion rate of 560 M$_\odot$ yr^{-1} is used for A1795 and 270 M$_\odot$ yr^{-1} for NGC1275 to determine the line luminosities given in Table 2. Our thermal instability models underestimate the Lyα emission by a factor of 3-20, while our CIV doublet emission is an order of magnitude below the upper limits attainable by the IUE. This suggest that most of the Lyα emission also arises from photoionized hydrogen. It is important to remember that our estimates on collisionally excited line luminosities assume that all of the gas in a cooling flow is eventually shocked and are thus upper limits. The actual fraction of gas that is shocked depends on the perturbation spectrum since small perturbations do not develop shocks.

TABLE 2
UV LINE STRENGTHS

Source	Lyα(obs) ergs/s	Lyα(Shocks)	CIV(obs)	CIV(shocks)
NGC1275	6.2×10^{42}	3.1×10^{41}	$< 5 \times 10^{41}$	4×10^{40}
A1795	2.7×10^{42}	6.4×10^{41}	$< 2 \times 10^{41}$	9×10^{40}

Note: A Comparison of observed (obs) line strengths with predictions from our shock models.

4. SUMMARY

The optical emission from cooling flows probably does not arise in shocked condensation for two reasons: (1) the observed Balmer line luminosities are roughly an order of magnitude greater than can be produced by shocks, and (2) the line emission produced in shocked condensations is of a higher ionization level than the observed optical emission. Since these shocks are strong sources of high ionization lines they may be observable in the ultraviolet.

The best candidates for the source of ionizing photons are post asymptotic giant branch stars and young O and B stars which have recently formed in the cooled filaments. Since shocks only survive for 10^5 years, star formation cannot proceed in a condensation until long after the shock has dissipated and significantly cooling has taken place. Cold condensations may only become visible if massive star formation occurs.

REFERENCES

Balbus, S.A. 1986, Ap.J. Lett., 303, L79.

David, L.P., Bregman, J.N., and Seab, C.G. 1987 (preprint).

Edgar, R.E. 1987 (private communication).

Field, G.B. 1965, Ap.J., 142, 531.

Hu, E.M., Cowie, L.L., and Wang, Z. 1985, Ap.J. Supp., 59, 447.

Johnstone, R.M., Fabian, A.C., and Nulsen, P.E.J. 1987, MNRAS, 224, 75.

Mathews, W.G., and Bregman, J.N. 1978, 224, 308.

Norgaard-Nielsen, H.U., Jorgensen, H.E., and Hansen, L. 1984, Astr. Ap., 135, L3.

Shull, J.M., and McKee, C.F. 1979, Ap.J., 227, 131.

RADIO SOURCES AND COOLING FLOWS

Lance Miller
Royal Observatory
Blackford Hill
Edinburgh EH9 3HJ

ABSTRACT. What effect do radio sources have on their surroundings, and in particular, what effect do radio sources in clusters of galaxies have on the cooling of the intracluster gas? It is shown here how a lower limit to the power input into the surrounding gas from extended radio sources can be calculated, and how for the most powerful radio sources this may have a significant effect. The consequences could be important both for cooling flows and for the discovery of extended emission-line regions around the most powerful radiogalaxies.

1. RECENT OBSERVATIONS

The motivation for asking these questions has been stimulated by the recent discoveries of extended continuum and line emission around the most powerful radiogalaxies (Djorgovski et al. 1987, McCarthy et al. 1987). It is likely that there will be suggestions made that these regions contain high-mass star formation in a cooling flow around the radio source, with that cooling flow also fuelling the active galaxy. But if the evidence of a correlation between the radio structure and the extended emission is confirmed (McCarthy et al. 1987, Chambers et al. 1987) then it seems far more likely that the radio sources are themselves directly responsible for high-mass star formation. To find out how this may arise, let's look at the power input from the extended radio lobes into the surrounding medium.

2. THE EFFECT OF RADIO SOURCES ON THEIR ENVIRONMENTS

Cygnus A is a typical powerful radio galaxy. We are fortunate that it has a sufficiently low redshift that we can see the cluster of galaxies and the hot intracluster gas which surround it. From the standard minimum energy argument for radio sources we can say that the extended lobes (excluding the hotspots) contain a minimum of 3×10^{52} joule in relativistic electrons and magnetic fields alone. The true total energy content could be much higher if there is energy stored in protons. In inflating the hot lobe material into the surrounding gas the lobes must have done work on the gas to the tune of at least 10^{52} joule (e.g. Longair, Ryle & Scheuer 1973). It is believed that the age of the lobes is about 6×10^6 years (Winter et al. 1980). So the implication is that the immediate surroundings of

205

A. C. Fabian (ed.), Cooling Flows in Clusters and Galaxies, 205–208.

Cygnus A have had an average heat input of more than 5×10^{37} Watts over that time period. Yet the X-ray luminosity from the region around the radio source is also about 5×10^{37} Watt (Fabbiano et al. 1979, Arnaud et al. 1984). So we can see that it is in principle possible for the heating by the radio source to have had a significant effect on the heat content of the gas. The next section will try to tie down these numbers a bit better.

3. THE MINIMUM POWER INPUT FROM EXTENDED RADIO LOBES

The minimum energy content of the lobes is obtained from the requirement that the energy stored in relativistic electrons and magnetic field be enough to produce the observed synchrotron luminosity. Thus the total stored energy requirement, U, comprises two terms, the first being the energy in electrons and the second being the energy in magnetic fields: each term comprises a factor incorporating the observed quantities and a dependence on the unknown magnetic field strength B

$$U = K_1 B^{-3/2} + K_2 B^2$$
$$= \frac{U_{eq}}{2}\left[\left(\frac{B_{eq}}{B}\right)^{3/2} + \left(\frac{B}{B_{eq}}\right)^2\right]$$

where the subscripted quantities refer to the values that provide equipartition of energy between these components, which results in a total energy requirement very close to the minimum value needed. If spectral aging of the synchrotron spectrum is observed, as in Cygnus A (Winter et al. 1980) then the age of the emitting region can be deduced. That age is dependent on several quantities, including the magnetic field strength:

$$t = t_{eq}\left(\frac{B_{eq}}{B}\right)^{3/2}.$$

Comparing these two quantities gives us the mean power supplied to the lobes, which since this is relativistic material is three times the PdV work done by the lobes on the external material. We can treat the lobe material as a fluid in this way because the magnetic gyro radius is very small, leading to a small effective mean path for ions and electrons inside the lobe. Thus the mean rate of work done by the lobe is

$$P = \frac{U}{3t} = \frac{U_{eq}}{6t_{eq}}\left[1 + \left(\frac{B}{B_{eq}}\right)^{1/2}\right].$$

Thus the mean rate of work has a **minimum** value of $U_{eq}/6t_{eq}$ independent of the assumed field strength. At equipartition the mean rate of work done is a factor two higher than the minimum value.

One key assumption which has been made in the above derivation is that the lobe is uniformly filled with emitting material. In fact the result is insensitive to how the lobe is filled. We can readily show that if the lobe comprises material with a filling factor f of uniform emissivity the minimum heat input is independent of f. At equipartition,

$$B_{eq} \propto \left(\frac{L_{radio}}{V_{lobe}}\right)^{2/7}$$

where L_{radio} is the synchrotron luminosity and V_{lobe} the emitting volume assumed. So

$$U_{eq} \propto B_{eq}^2 V_{lobe} \propto L_{radio}^{4/7} V_{lobe}^{3/7},$$

but

$$t_{eq} \propto B_{eq}^{-3/2} \propto L_{radio}^{-3/7} V_{lobe}^{3/7}$$

so the mean power input deduced is directly proportional to the observed radio luminosity and is independent of the assumed size of the emitting region. Thus, provided we apply this calculation to regions of constant energy density the minimum heat input can be securely derived.

4. THE CONSEQUENCES FOR COOLING FLOWS: DISRUPTION OR TRIGGERING OF COOLING FLOWS?

The consequences for cooling flows depend on the manner in which work is done on the intracluster gas. If the radio lobe is at a sufficiently high pressure with respect to its surroundings a shock will propogate into the surroundings at a speed of order $\sqrt{(P_{lobe}/\rho_{gas})}$. If the shock is strong this will cause a density jump of a factor four and an arbitrarily large temperature jump, so that the radio lobe will be surrounded by a cocoon of expanding hot gas whose extent will depend on the radio lobe pressure and intracluster gas density. Unless the cocoon temperature were very high ($>> 10^8$K) we would not expect to be able to observe the cocoon around Cygnus A. Such compression and heating in the intracluster gas will increase the gas cooling timescale, since the radiative losses $\Lambda(T)$ either fall with temperature (10^4K$< T < 10^6$K) or rise only slowly ($\Lambda \propto T^{1/2}$ at $T > 10^7$K). If $\Lambda \propto T^\alpha$ then the cooling time varies as $T^{1-\alpha}/\rho$ and the cooling time will be increased for $\alpha < 1$. Thus if the shock is sufficiently strong it could disrupt any cooling flow in the region around the radio lobes. Cygnus A has lobes which extend to a radius of 100 kpc, and the dominant cooling in the intracluster gas has been deduced as occurring within a radius of about 125 kpc (Arnaud et al. 1984). Thus the supposed cooling flow around Cygnus A, which has not been substantiated by direct X-ray spectroscopic observations of cooler gas, could have been disrupted in the last few million years. It is irrelevant that the X-ray emitting gas is observed over a much larger extent than this scale.

However, if the intracluster gas is simply compressed adiabatically as a radio lobe inflates subsonically into its surroundings there is the possibility that the gas cooling time will be shortened, the required condition being $\alpha > -1/2$ as at temperatures $T > 10^6$K. Thus the effect of a radio source expanding into a cluster could be to increase the value of the mass loss rate in an existing cooling flow, and possibly even to initiate a cooling flow in a cluster which previously was only close to cooling sufficiently quickly. One final point is that while the radio source is expanding the gas in the cluster will clearly not be in hydrostatic pressure equilibrium.

Whether any of these effects are significant will depend on whether radio sources can cause significant pressure variations in the intracluster gas. If they do, these pressure variations will exist for timescales of order 10^8 years, the sound crossing time of a cluster core. At present the equipartition pressures deduced in the radio lobes of radio sources in clusters are usually comparable to the intracluster gas pressure, but the true total pressure in those lobes could well be much larger. Even if the lobes are in approximate pressure balance with their surroundings coninuing pressure variations must be occurring in those clusters as the lobes inflate if the radio source is still active.

5. THE CONSEQUENCES FOR STAR FORMATION AROUND RADIO SOURCES

Returning to the observations which originated these thoughts, we can ask whether we might expect high-mass star formation to be enhanced around powerful radio sources. To achieve this we again require the expanding radio lobe to trigger cooling and fragmentation in the surrounding medium. This could be achieved if there are cool ($< 10^4$K) gas clouds surrounding the radio lobes, in which case a strong shock will propogate through the clouds. At low temperatures the cooling function is sufficiently steep that even a shock will promote cooling and may thus lead to star formation, as is thought to happen in our Galaxy.

6. THE PREVALENCE OF POWERFUL RADIO SOURCES IN CLUSTERS

Many clusters at low redshifts with deduced cooling flows contain moderately powerful radio sources. Two of the most well-known examples of powerful double radio galaxies occur in rich clusters (Cygnus A and 3C295) and there is now evidence that all extended radio sources which are as luminous occur in rich clusters (Yates *et al.* in preparation). The numbers of such powerful radio sources are low at the present day, but there were significantly more at earlier epochs. How many powerful radio source events have there been in the universe? By integrating the evolving radio luminosity function (Dunlop in preparation) for powerful, steep-spectrum radio sources ($P_{2.7GHz} > 10^{26}$Watt Hz^{-1} sr^{-1}) over epochs from $z = 3$ to the present day and assuming that each radio source lasts for a typical lifetime of 10^8 years, we find that there have been a total comoving number density of $2 \times 10^{-6} h_{50}^2$ Mpc^{-3} radio source events ($q_0 = 1/2$), ignoring the possibility of multiple events. This value compares with the present-day number density of clusters (Abell richness ≥ 1) of around $0.8 \times 10^{-6} h_{50}^3$ Mpc^{-3}: every present-day cluster could at one time have contained a powerful radio source.

References

Arnaud, K.A., Fabian, A.C., Eales, S.A., Jones, C. & Forman, W., 1984. *Mon. Not. R. astr. Soc.*, **211**, 981.
Chambers, K.C., Miley, G.K. & van Breugel, W., 1987. *Nature*, **329**, 604.
Djorgovski, S., Spinrad, H., Pedelty, J., Rudnick, L. & Stockton, A., 1987. *Astron. J.*, **93**, 1307.
Fabbiano, G., Doxsey, R.E., Johnston, M., Schwartz, D.A. & Schwarz, J., 1979. *Astrophys. J.*, **230**, L67.
Longair, M.S., Ryle, M. & Scheuer, P.A.G., 1973. *Mon. Not. R. astr. Soc.*, **164**, 243.
McCarthy, P.J., van Breugel, W., Spinrad, H. & Djorgovski, S., 1987. *Astrophys. J.*, **321**, L29.
Winter, A.J.B., *et al.*, 1980. *Mon. Not. R. astr. Soc.*, **192**, 931.

THE DEPOLARIZATION OF RADIO SOURCES BY CLUSTER GAS

S.T.Garrington
Nuffield Radio Astronomy Laboratories
Jodrell Bank
Macclesfield SK11 9DL, U.K.

ABSTRACT. Recent observations of powerful double radio sources with one-sided jets have shown that almost invariably the jet side depolarizes less with increasing wavelength than the opposite side. We discuss the possibility that this is due to a large scale X-ray halo surrounding the source: the asymmetry is then naturally explained as a geometric effect if the visible jet is relativistically beamed. Simple models suggest that if the gas distribution has a core of uniform density, the size of this core must be closely matched to the size of the radio source. This constraint may be relaxed if there is a cooling flow in the gas.

1. INTRODUCTION

In trying to understand why otherwise symmetric double radio sources (FR2 radio galaxies and quasars) only ever have one-side jets, Laing (1987) has discovered an intriguing new result: in 8 out of 9 sources with one-side jets mapped at high resolution the jet side depolarizes less rapidly with increasing wavelength than the opposite side.The same effect is present, to a lesser degree, in the component polarizations compiled by Strom and Conway (1985).

In order to establish this effect we have observed a selection of double sources with one-sided jets. The results for the 25 sources with small angular size have been presented by Garrington *et al.* (1987, hereafter paper 1): 23 sources clearly show the jet side depolarizing less. In paper 1 we show how, given the strength of this result, detailed polarization observations of just a few sources should determine conclusively whether jets are intrinsically one-sided or are relativistically beamed. In this paper I shall assume that the jets are relativistic and consider the implications for the existence of large scale distributions of hot or ionized gas around powerful extragalactic radio sources.

A. C. Fabian (ed.), Cooling Flows in Clusters and Galaxies, 209–213.
© *1988 by Kluwer Academic Publishers.*

2. OBSERVATIONS AND RESULTS

The sources were selected without reference to polarization characteristics, using high resolution maps available in the literature along with some unpublished data. We discuss here the 25 sources with largest angular size less than 30 arcseconds: they include 23 quasars and 2 radio galaxies with redshifts typically between 1 and 2. Snapshot observations were made with the N.R.A.O. Very Large Array using the A and B configurations at 20 and 6 cm respectively, giving the same resolution of about 1.3 arcsec. A full list of sources and further details of the observations and data reduction are given in paper 1.

In order to quantify the depolarization effect we calculate for each component the quantity m' which represents the arithmetic mean fractional polarization, weighted by the total intensity distribution. We define a depolarization ratio $DP = m'_{20}/m'_{6}$. Figure 1 shows that 22 out of the 25 sources have DP values significantly higher on the jet side. Detailed inspection shows that of the remaining 3 there are no clear counter examples.

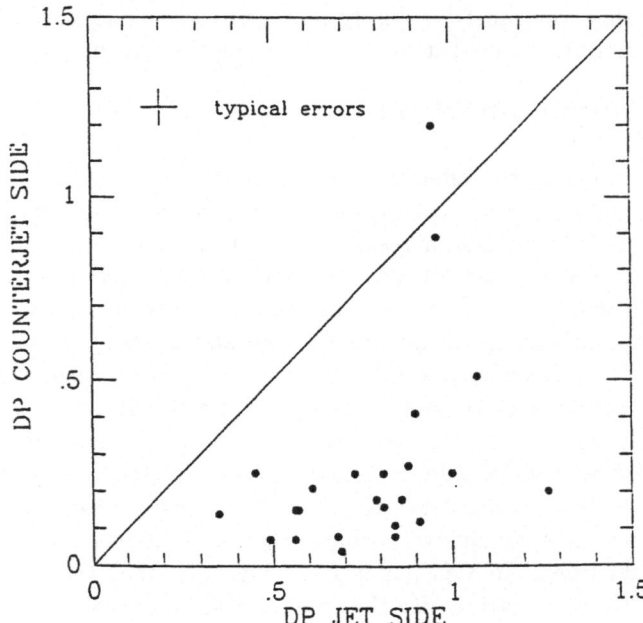

Figure 1. Plot of depolarization ratio DP for the jet side and counter-jet side components, showing that the values of DP are markedly lower on the counter-jet side for almost all sources.

3. DISCUSSION

Almost certainly, the depolarization we observe is due to differential Faraday rotation. We can define a Faraday Dispersion Δ as the standard deviation of the Faraday dispersion function $F(\Phi)$ which denotes the fraction of the polarized emission coming from a Faraday depth $\Phi \propto \int n\mathbf{B} \bullet \mathbf{dl}$ where n is the thermal electron density and $\mathbf{B}$ is the magnetic field strength (Burn 1966). Assuming a Gaussian form for $F(\Phi)$ we derive form the DP values the following approximate distributions of Δ

	lower quartile	median	upper quartile	
Δ_j jet side	35	65	105	$\mu G\,cm^{-3}\,pc$
Δ_{cj} counterjet side	130	180	250	$\mu G\,cm^{-3}\,pc$
$R_\Delta = \Delta_{cj}/\Delta_j$	1.9	2.6	3.2	

The thermal plasma responsible for the depolarization may occur within the radio components (internal depolarization) or in an irregular foreground screen (external depolarization). If jets are intrinsically one-sided, differences between the two lobes might naturally lead to different intrinsic Faraday depths and hence cause aymmetric depolarization. On the relativistic beaming hypothesis we expect the two lobes to be intrinsically similar. Asymmetric depolarization must then occur in a foreground screen. Since the visible jet always points towards us, the counterjet side is seen through more of this screen and hence depolarizes more. We now consider in more detail the nature of this screen.

Possible configurations for the screen include a disk of emission line gas perpendicular to the radio axis (*e.g.* Fosbury, 1986; see also Danziger, this volume), a halo of X-ray gas surrounding the central object or a cocoon of thermal plasma surrounding the emitting regions. Analysis of depolarization maps (Laing, 1987) which show a smooth variation of depolarization across the source suggest a smoothly distributed medium rather than a disc. We consider first an isothermal sphere of hot gas around the cental object. The radial density profile may be modeled as

$$n(r) = n_0(1 + \frac{r^2}{a^2})^{-3\beta/2}$$

where a is the core radius ($\sim$300 kpc for a cluster halo; 1-10 kpc for a galaxy halo) and β is found empirically to be $\sim$0.6 in clusters (Forman & Jones, 1984) and $\sim$0.5 in x-ray bright elliptical galaxies (Forman et al., 1985). The depolarization must be due to small scale (*i.e.* less than a typical beam size of 10 kpc) variations in nB. We assume that in a shell between r and $r + \Delta r$, $F(\Phi)$ is Gaussian with mean zero and variance $\sigma^2(r) = n^2(r)B^2(r)d(r)\Delta r$ where d is the scale size of the magnetic field reversals. For simplicity we assume that the properties of the magnetic field are either constant with radius or have the same form as the density variations so that $\sigma \propto n^m$. We then calculate the Faraday Dispersion Δ by integrating σ^2 along the line of sight to a radio component.

In order to produce sufficiently large Faraday Dispersions on the counter-jet side haloes with small cores ($a < 10$ kpc) would require unreasonably large central densities ($n_0 > 1\,cm^{-3}$) assuming $B_0 \sim 1\,\mu G$ and $d = 1$ kpc. This suggests that the depolarization is due to gas associated with the cluster environment rather then the individual host galaxy.

For a given source size D and combined exponent βm the observed distribution of the depolarization asymmetry ratio R_Δ may be used to place further independent constraints on the size of the halo. Figure 2 shows contours corresponding to the median and interquartile range of R_Δ in the (a, θ) plane, where θ is the angle between the radio source axis and the line of sight. Here we assume $\beta m = 1.0$ and $D = 200\,\mathrm{kpc}$. We see that the core radius cannot be more than 60% larger than the size of the radio source and the source axis must be within 70° to the line of sight. Steeper profiles (larger βm) would allow slightly larger core radii and angles to the line of sight. Thus the observed asymmetry would seem to exclude the possibility that the depolarization is caused by the large haloes ($a \sim 500\,\mathrm{kpc}$) observed by Abramopoulos and Ku (1983) in nearby clusters unless $\beta m \gg 2$ and the sources are at very small angles to the line of sight.

In order to model the effects of a cooling flow in such an x-ray halo we change the radial density profile to (Bertshinger and Meiksin, 1986)

$$n(r) = n_0 \frac{(r/a_1)^{-\alpha}}{(1 + r/a_1)}$$

whose size is characterized by a break radius a_1.

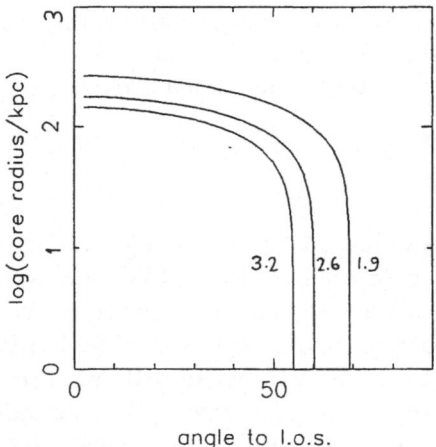 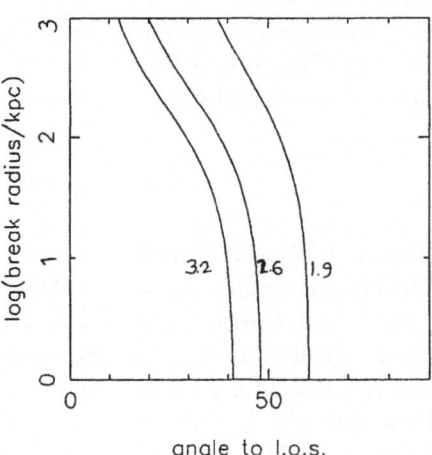

Figures 2 & 3. Contours corresponding to the median and interquartile range of R_Δ in the (a, θ) plane for the isothermal model *(left)* and the cooling flow model *(right)*.

Because this model does not have a completely flat core (for $\alpha > 0$) it produces sufficient depolarization asymmetry even for large radii a_1. Figure 3 shows contours of R_Δ for this model assuming $m = 1$ and $\alpha = 0.8$. For the M87 cluster Bertschinger and Meiksin (1986) estimate $n_0 = 1.46 \times 10^{-2}$ cm^{-3}, $a_1 = 30$ kpc and $\alpha = 0.81$ from the data of Stewart et al., (1984). Were all powerful double radio sources surrounded by such cooling flows (which would be presently undetected beyond redshifts of about 0.5), they would show the observed depolarization asymmetry provided the central magnetic field strength $B_0 > 3\mu$G and the source axes were mostly oriented within about 55° to the line of sight.

4. CONCLUSIONS

If the depolarization asymmetry in double radio sources recently discovered by Laing (1987) and confirmed by Garrington et al. (paper 1) is due to Faraday effects in large scale spherical gas clouds surrounding these sources, then simple models may be used to place the following constraints:

1. Galaxy-sized haloes are unable to produce sufficient depolarization.
2. Isothermal cluster-sized haloes are unable to produce sufficient depolarization asymmetry unless sources are at small angles to the line of sight and the core radius is not much larger than the size of the radio source.
3. If the cluster gas has a cooling flows the constraints on the size of the halo and the angle of the source to the line of sight may be relaxed.

ACKNOWLEDGEMENTS. I thank my collaborators R. G. Conway, J. P. Leahy and R. A. Laing for their help and S. E. R. C. for a research studentship.

REFERENCES

Abramopoulos, F. & Ku, W. 1983, *Astrophys. J.* **271**, 446

Bertschinger,E. & Meiksin, A. 1986, *Astrophys. J.* **306**, L1

Burn, B.J. 1966, *Mon.Not.R.astr.Soc*,**133** 67

Forman, W. & Jones, C. 1982, *Ann.Rev.Astr.Astrophys.* **20**, 547

Forman, W., Jones, C. & Tucker, W. 1985, *Astrophys. J.* **293**, 102

Fosbury, R.A.E. 1986 *Structure and Evolution of Active Galactic Nuclei*
        eds. G. Giuricin et al. p.297 (Reidel, Dordrecht)

Garrington, S.T., Leahy, J.P., Conway,R.G. & Laing, R.A. 1987 *Nature* submitted

Jones, C. & Forman, W. 1984 *Astrophys. J.* **276**, 38

Laing, R.A. 1987 *Nature* submitted

Stewart, G.C., Canizares,C.R., Fabian, A.C. & Nulsen, P.E.J. 1984 *Astrophys. J.* **278**, 536

Strom, R.G. & Conway, R.G. 1985 *Astr. Astrohys. Suppl.* **61**, 547

THE STRUCTURE OF ELLIPTICAL GALAXIES

Tod R. Lauer,
Princeton University Observatory

ABSTRACT. Recent imaging investigations highlight the structural diversity of elliptical galaxies. The distribution of light within the cores and envelopes shows a wide range in central and average system stellar densities at any galaxy luminosity. These variations appear to offer new information on galaxy formation mechanisms. Understanding the common structural properties of elliptical galaxies leads to interesting applications such as their use as metric distance indicators or as probes of tidal interactions in first-ranked galaxies.

The simplest picture of elliptical galaxies is that they are smooth, dynamically hot, stellar systems whose surface brightness distributions are well characterized by an $r^{-1/4}$ law. The characteristic size of any elliptical galaxy is then given by its half-light radius, R_e, which is related to the galaxy's total luminosity, L, by some sort of magnitude radius relationship. As we have studied elliptical galaxies in greater detail, however, we find that this picture is only a thumbnail sketch of systems whose true personalities are becoming visible with the richly detailed portraits provided by high signal to noise CCD imagery. An abundance of high-quality surface photometry of elliptical galaxies is now available; some of the larger surveys can be found in Kent (1984), Lauer (1985a), Djorgovski (1985), and Jedrzejewski (1987). I will not begin to summarize all that we have learned about the structure of elliptical galaxies from these and many other recent observational programs. Instead, I would like to concentrate on our improved understanding of the diversity of properties seen among these systems. Elliptical galaxies can no longer be regarded as simple systems, all alike, whose properties are dictated entirely by their total luminosities. Their cores show strong variations in their degree of concentration that may reflect a dissipational origin of ellipticals. New work on the envelopes of elliptical galaxies shows that luminosity is not the only parameter that sets the envelope scale (Lauer 1985b) — correct isolation of a "second" envelope structural parameter leads to the use of ellipticals as metric distance indicators (Dressler et al. 1987 and Djorgovski and Davis 1987). At large radii, well beyond the effective radii, ellipticals show additional variations in their brightness profiles that again reflect their formation or environment (Kormendy 1977 and Schombert 1987). Elliptical galaxies can also have dust or not, H I or not, and so on (see Schweizer 1987 for a review). One common structural element is that isolated ellipticals have concentric and nearly elliptical isophotes (Lauer 1985c), although the isophotes can

A. C. Fabian (ed.), Cooling Flows in Clusters and Galaxies, 215–223.
© 1988 by Kluwer Academic Publishers.

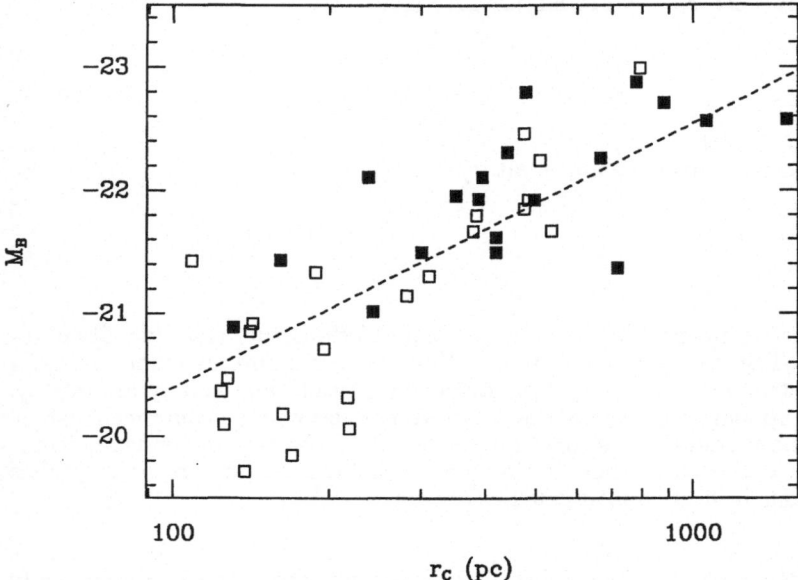

Figure 1: Core radii r_C versus total galaxy luminosity for the systems presented in Lauer (1985b). Well resolved galaxies are plotted as solid symbols — poorly resolved galaxies only provide upper limits to r_C. The mean $r_C - L$ relationship for the well resolved galaxies is also plotted.

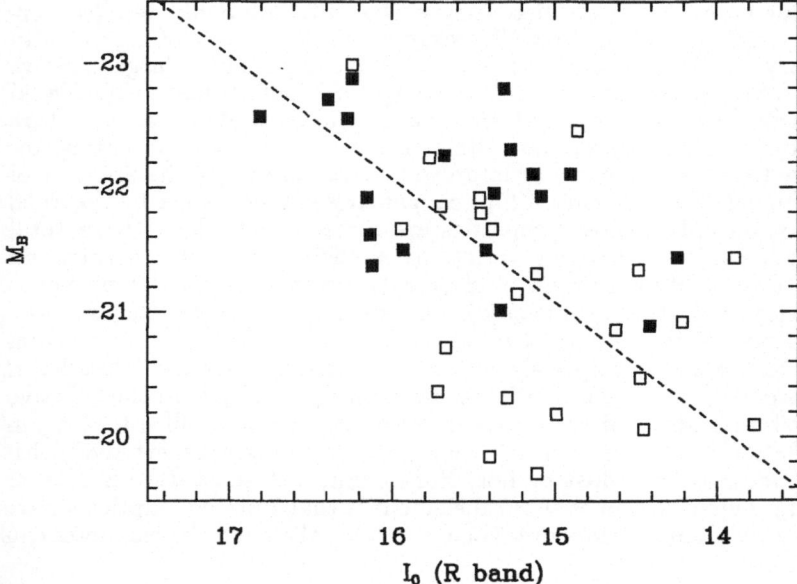

Figure 2: Central surface brightness I_0 versus total galaxy luminosity for the systems presented in Lauer (1985b). The mean $I_0 - L$ relationship for the well resolved galaxies is also plotted.

twist, flatten, or round out at different radii. Models of elliptical galaxies must reflect these variations in their properties, but once this is done, as shown below, it is possible to detect structural properties that are not normal and may reveal ongoing tidal disturbances or evolutionary changes within the systems.

Let me start with the cores of elliptical galaxies. Cores are the central portions of ellipticals where the brightness profiles flatten out and converge to a constant central surface brightness, I_0. The size of a core is characterized by its core radius, r_C, which is the half-width half-maximum of the brightness profile. We are interested in the cores because they give us a direct look at the bottom of the potential wells of ellipticals. The typical core mass for even giant ellipticals is only $10^8 - 10^9 M_\odot$, a small fraction of their total masses. Core structure therefore may be an excellent diagnostic for small amounts of central mass accretion or the effects of central massive black holes. Further, since cores are the densest portion of the galaxies, they may also provide clues on elliptical galaxy formation. What mechanism produced the high phase-space density cores in the first place? I should mention before I go further that study of the cores is tricky — Schweizer's (1979) observational cautions are still in effect since the apparent angular core radii of even the nearest ellipticals are only a few arcseconds at best, which means that the effects of seeing must be accounted for. The better resolved cores can be studied by Fourier deconvolution of the seeing PSF, although many ellipticals have cores that cannot be resolved even at Hawaii and will have to await the Space Telescope for study.

With the material we have available now, I believe we can already say some interesting things about elliptical galaxies. Kormendy (1985) and I (Lauer 1985b) have both shown that brighter ellipticals tend to have larger cores, but with dimmer central surfaces brightnesses. From my own work I get $r_C \propto L^{1.2}$ and $I_0 \propto L^{-1.0}$, as shown in Figures 1 and 2. There is significant large scatter in both relationships; however, but with tightly correlated residuals — a point that I will elaborate. At any galaxy luminosity excessively large cores are dim in the center while the more compact cores are brighter, which means that central core stellar spatial density, ρ_C varies widely at any total galaxy luminosity. This result is summarized graphically in Figure 3, which plots residuals from the mean $r_C - L$ relationship against those from the mean $I_0 - L$ relationship. Contours of constant central density excess or deficit with respect to the average expected from the core relationships are shown as dashed lines. As can be seen, variations in central stellar density readily explain the residuals in the core relationships. Luminosity independent variation in core structure is shown directly in Figure 4, which displays core profiles for five galaxies of the same luminosity (and at nearly the same distance). The cores of NGC 4621 and NGC 4552 are unresolved and thus even more compact than shown here.

Basically these results mean that while core structure in determined in part by the global properties of elliptical galaxies, there is an important "second parameter" independent of total luminosity that determines how densely stars are packed into cores. As I alluded to above, I think that dissipational formation of ellipticals provides a natural explanation for such variations in core structure. If even a small amount of gas is left over from the initial formation of the elliptical galaxy envelope, it may greatly change the structure of any primordial core as it settles down into the potential well and crystalizes into stars. Larson (1975), for example, has already suggested that cores form slowly following after envelope formation under a dissipational formation theory, and shows how the brightness profile becomes more concentrated as the proto-elliptical is allowed to hold onto its gas for increasingly long periods of time. Other models based on more modern kinematic pictures of

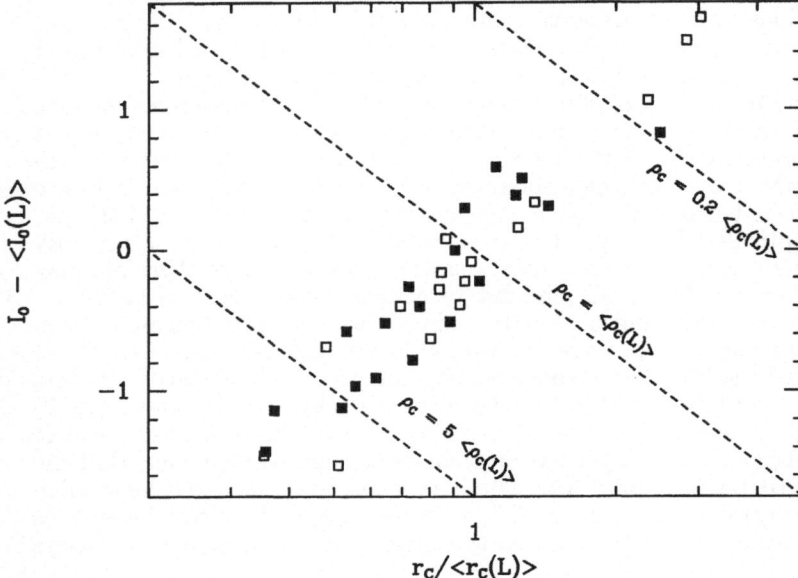

Figure 3: Residuals of the mean $r_C - L$ relationship (Figure 1) versus those of the mean $I_0 - L$ relationship (Figure 2). Dashed lines mark contours of constant core luminosity density ρ_C excess or deficit from the average ρ_C at any luminosity.

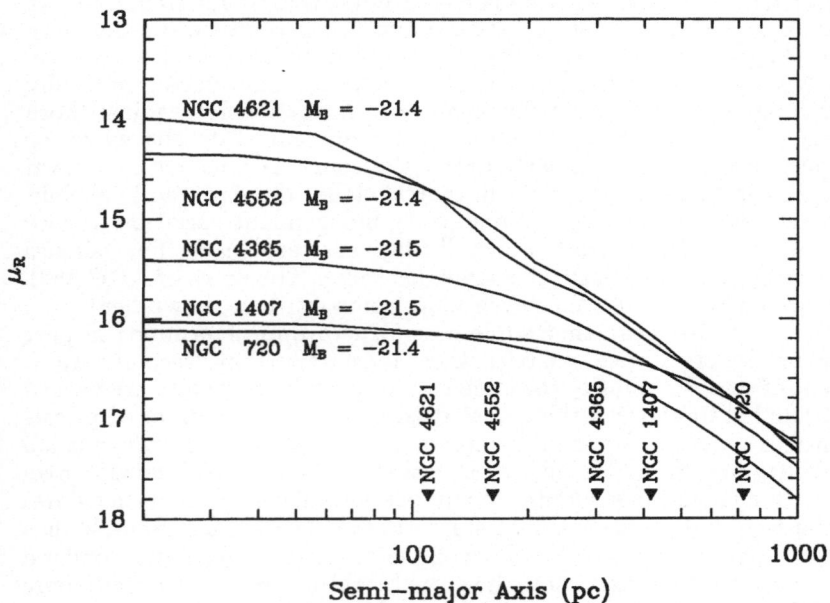

Figure 4: Core surface brightness profiles for five galaxies with $M_B \approx -21.5$ and at similar distances. Core radii of the galaxies are indicated on the horizontal axis.

elliptical galaxies also show that central structure may be influenced by variations in the importance of dissipation during formation. In the violent collapse models of van Albada (1982), for example, the central density is determined by how clumpy and tightly bound the initial star forming regions within the protogalaxy are prior to its collapse. It is also noteworthy that there appears to be no relationship between the ratio R_E/r_C and galaxy luminosity. The size of the core knows little about the scale of the surrounding envelope. This argues against the formation of elliptical galaxies by any sort of dissipationless hierarchy of merging, since simulations of such processes (e.g. Farouki, Shapiro, and Duncan 1983) predict R_E/r_C to increase as a strong function of L. Merging with dissipation may work provided if enough gas is left to form stars and mask the density distribution determined by preexisting stars.

Next I would like to turn to recent work done on the global structure of elliptical galaxies. Much work has gone into investigating the luminosity – effective radius relationship. Typically, one finds $R_e \propto L^{0.8}$ (see Davies et al. 1983 for a historical review). As with relationships between core structure and luminosity, the radius – luminosity relationship shows significant scatter. Figure 5 plots radius versus luminosity for 106 elliptical galaxies studied by Djorgovski and Davis (1987). This is the tightest radius-luminosity relationship I've seen so far, but yet a factor of three scatter in effective radius is still visible at any luminosity, well in excess observational errors. Similar scatter exists in the effective surface brightness – luminosity plot. In analogy to the cores, this scatter appears to be due to large variation in the average bulk stellar density $\rho_E \equiv 0.5 I_E/R_E$ of ellipticals at any luminosity. Figure 6 plots ρ_E against L; at any luminosity there is an order of magnitude scatter in the galaxies' bulk stellar densities. As with the cores, two parameters are needed to characterize the global structure of elliptical galaxies.

This improved understanding of elliptical galaxy structure leads to their use as metric distance indicators as introduced separately by Dressler et al. (1987) and Djorgovski and Davis (1987). In fact, the relationships between central velocity dispersion and metric radius are just as expected from the virial theorem under the assumption that a galaxy's luminosity and density are dependent on two independent parameters (Faber et al. 1987). Dressler et al. (1987), for example, define a metric radius D_N in which the average surface brightness is 20.75 (B-band), and find that $D_N \propto \sigma^{4/3}$. If surface brightness at the effective radius, I_E is used instead, then an equivalent relationship relating L, I_E, and σ is found:

$$L \propto \sigma^{8/3} I_E^{-3/5}. \tag{1}$$

Now with $L \propto \rho R_E^3$ and $I_E = 2\rho_E R_E$, then Equation (1) can be written as:

$$\sigma \propto \rho^{0.6} R_E^{1.38}. \tag{2}$$

The virial theorem gives $\sigma \propto \rho_E^{0.5} R_E (M/L)^{0.5}$, where (M/L) is the mass-to-light ratio. With Kormendy's (1987) result that $(M/L) \propto L^{0.2}$ based on his core observations, then Equation (2) can be reduced directly to the result expected from the virial theorem.

As with the cores, the problem remains as to what the two-dimensional nature of elliptical galaxy envelope structure says about their formation. Returning to Figure 6, I am intrigued by the the curvature in the relationship between ρ_E and L. For low luminosity elliptical galaxies ρ_E shows little if any dependence on L, while

220

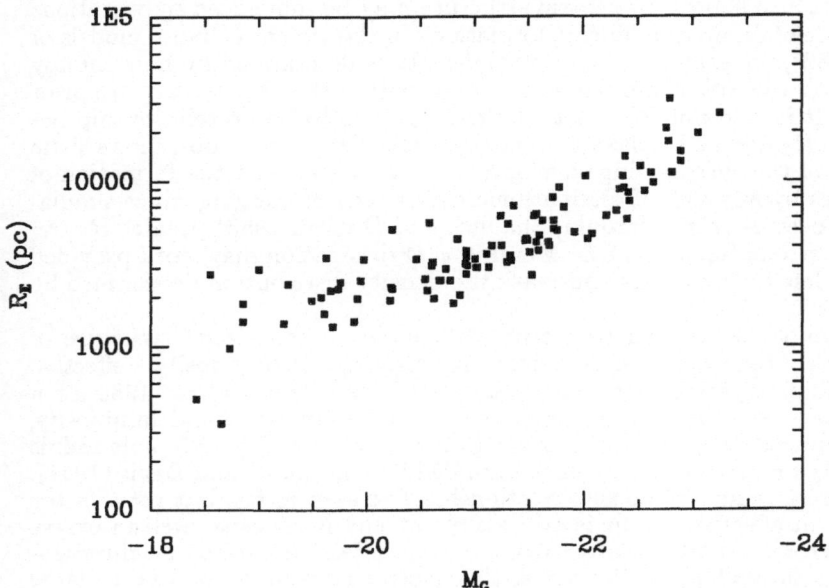

Figure 5: Effective radius versus total luminosity for the elliptical galaxies surveyed by Djorgovski and Davis (1987).

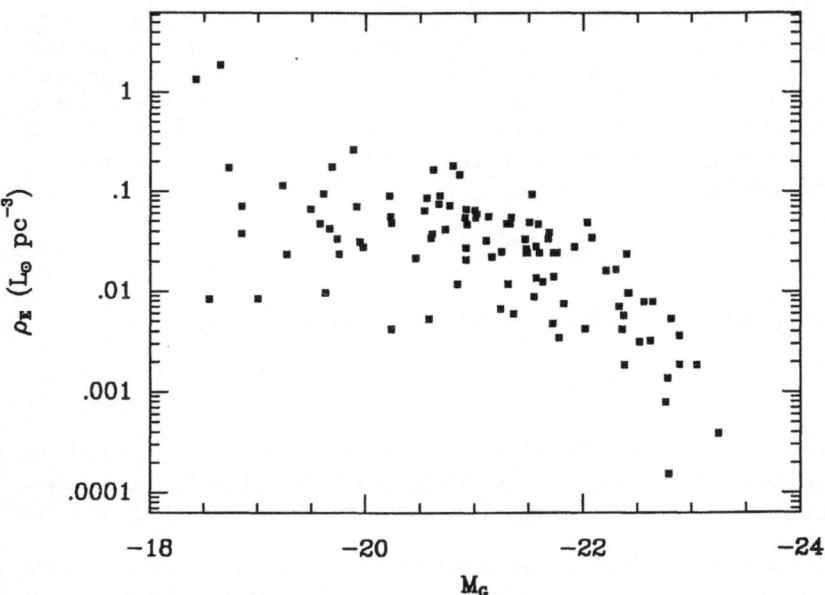

Figure 6: System-average stellar luminosity density versus total luminosity for the elliptical galaxies surveyed by Djorgovski and Davis (1987).

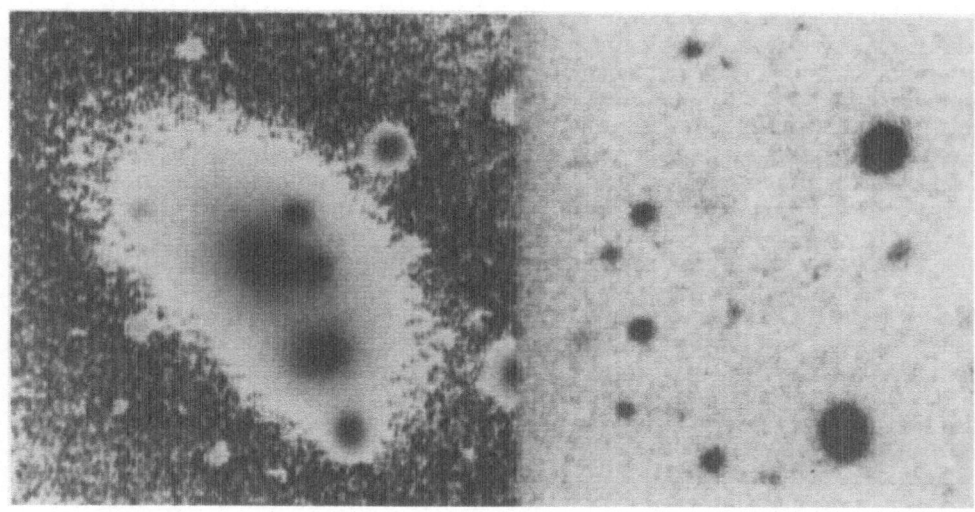

Figure 7: CCD image and residual map of the A1927 system. Each panel is 60"
on a side. The central cD and three nearby companion galaxies were fitted with a
superposition model Note the flatness of the residual image.

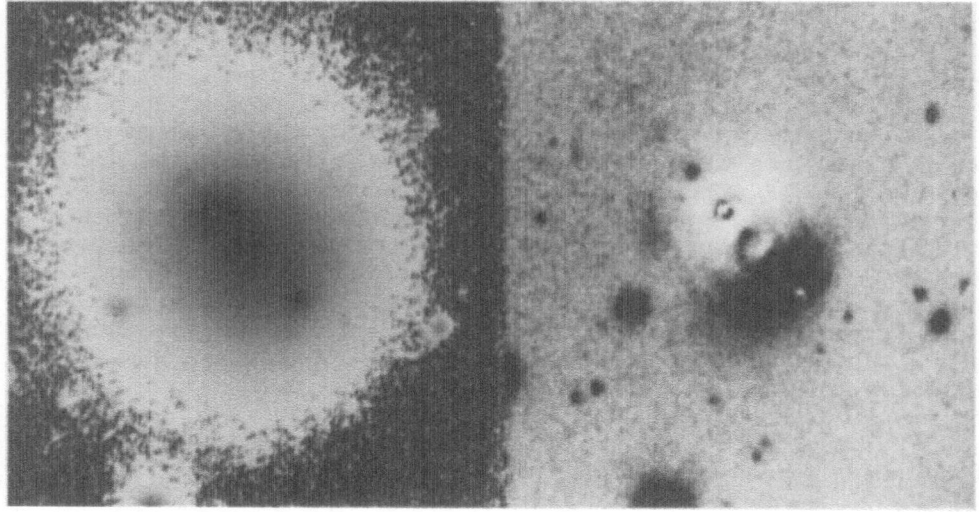

Figure 8: CCD image and residual map of the A1185 system. Each panel is 86"
on a side. The central giant elliptical and two nearby companion galaxies were
fitted with a superposition model. Both panels show the strongly non-concentric
isophotes of the central galaxy.

at high luminosities ρ_E decreases rapidly with L. This last relationship is similar to what is expected from homologous mergers of galaxies. One possible picture is that we are looking at a time sequence of galaxy formation — low luminosity elliptical began forming first and were allowed to complete their collapse and star formation, while the larger-scale perturbations that gave rise to more massive galaxies did not have time to collapse dissipationally before their internal smaller-scale perturbations completed their own collapse and star formation. Continuing collapse of the large-scale perturbation would then involve dissipationless merging of newly formed low luminosity galaxies.

Formation of the first-ranked members of clusters is of particular interest to students of cooling flows, given their frequent association with the X-ray emitting cores of rich clusters. A number of the galaxies plotted in Figure 6 are cDs, first-ranked galaxies and the like; however, the morphological properties of these systems are not particularly different from other bright giant ellipticals that have no special position within their clusters. Dressler (1984), and Merritt (1985) have both argue that first-ranked galaxies formed in regions of strong *local* density enhancements of galaxies that collapsed and merged over a relatively brief period of time to make a single system, consistent with the picture discussed above.

A related question is to what extent first-ranked galaxies are growing *now* by cannibalization of their fainter fellow cluster members. I have recently investigated this problem by direct morphological analysis of first-ranked galaxies (Lauer 1988). I will conclude with a brief discussion of this work both to illustrate the detailed analysis that can be done with modern material and to present what I hope are useful constraints on a least one evolutionary effect acting on first-ranked galaxies. My approach to determining the rate of cannibalism is to look for evidence of tidal interaction between the components of multiple-nuclei first-ranked galaxies, systems that have been suggested as sites of cannibalism in progress (Hausman and Ostriker 1978). Two multiple systems, A1927 and A1185, are shown in Figures 7 and 8. Tonry (1985) and Merritt (1984) have argued alternatively that many of these systems may be simple line-of-sight superpositions of normal cluster galaxies. Beers and Tonry (1986) further argue that the cores of clusters are cuspier than previously realized, which would greatly enhance the likelyhood of superposition. I test the superposition hypothesis directly by modelling the systems as a simple sum of light from normal isolated elliptical galaxies. In some cases the procedure works, as in A1927, while in other cases it fails as in A1185. The residuals from the fit to A1927 are shown on the right half of Figure 7. As can be seen they are flat, testifying to the success of the model; further, none of the components fitted have abnormal brightness profiles or any other detectable deviations from normality. On the other hand, the central galaxy in A1185 is strongly disturbed in such a way as to implicate ongoing interaction with at least one of its companion galaxies. In a total sample of 17 multiple systems studied so far, nine show at least some evidence of tidal interaction between their components, of which five are kinematically consistent with ongoing cannibalistic mergers. The cannibalism rate for a typical first-ranked galaxy can then be inferred based on the prevalence of multiple systems in all Abell clusters (see Hoessel and Schneider 1985). My estimate of this rate is that the typical first-ranked galaxy eats $0.2L^*/10^9 yr$ at this epoch. For the Beers and Tonry (1986) model of cluster cores, this implies a total cannibalized luminosity $L_C \approx 2L^*$ over the age of the universe, in excellent agreement with the theoretical calculations of Merritt (1985). Since the typical total luminosity of a first-ranked galaxy is $L \approx 12L^*$. cannibalism will significant growth of a first-ranked galaxy, but is insufficient to produce its total luminosity.

REFERENCES

Beers, T. C. and Tonry, J. L. 1986, *Ap. J.*, **300**, 557.

Davies, R. L., Efstathiou, G., Fall, S. M., Illingworth, G., and Schechter, P. L. 1983, *Ap. J.*, **266**, 41.

Djorgovski, S. 1985, PhD. thesis, University of California, Berkeley.

Djorgovski, S. and Davis, M. 1987, *Ap. J.*, **313**, 59.

Dressler, A. 1984, *Ann. Rev. Astr. Ap.*, **22**, 55.

Dressler, A., Lynden-Bell, D., Burstein, D., Davies, R. L., Faber, S. M., Terlevich, R. J., Wegner, G. 1987, *Ap. J.*, **313**, 42.

Faber, S. M., Dressler, A., Davies, R. L., Burstein, D., Lynden-Bell, D., Terlevich, R. J., Wegner, G. 1987, in *Nearly Normal Galaxies,* ed. S. M. Faber (New York: Springer-Verlag), p. 175.

Farouki, R. T., Shapiro, S. L., and Duncan, M. J. 1983, *Ap. J.*, **265**, 597.

Hausman, M. A. and Ostriker, J. P. 1978, *Ap. J.*, **224**, 300.

Hoessel, J. G. and Schneider, D. P. 1985, *A. J.*, **90**, 1648.

Jedrzejewski, R. I. 1987, *M. N. R. A. S.*, **226**, 747.

Kent, S. M. 1984, *Ap. J. Suppl.*, **56**, 105.

Kormendy, J. 1977, *Ap. J.*, **218**, 333.

Kormendy, J. 1985, *Ap. J.*, **295**, 73.

Kormendy, J. 1987, in *IAU Symposium 127, Structure and Dynamics of Elliptical Galaxies,* ed. T. P. De Zeeuw (Dordrecht: Reidel), p. 17.

Larson, R. B. 1975, *M. N. R. A. S.*, **173**, 671.

Lauer, T. R. 1985a, *Ap. J. Suppl.*, **57**, 473.

Lauer, T. R. 1985b, *Ap. J.*, **292**, 104.

Lauer, T. R. 1985c, *M. N. R. A. S.*, **216**, 429.

Lauer, T. R. 1988, *Ap. J.*, in press.

Merritt, D. 1984, *Ap. J. (Letters)*, **280**, L5.

Merritt, D. 1985, *Ap. J.*, **289**, 18.

Schombert, J. M. 1987, *Ap. J. Suppl.*, **64**, 643.

Schweizer, F. 1979, *Ap. J.*, **233**, 33.

Schweizer, F. 1987, in *IAU Symposium 127, Structure and Dynamics of Elliptical Galaxies,* ed. T. P. De Zeeuw (Dordrecht: Reidel), p. 109.

Tonry, J. T. 1985, *Ap. J.*, **291**, 45.

van Albada, T. S. 1982, *M. N. R. A. S.*, **201**, 939.

DYNAMICS OF E GALAXIES AND CLUSTER SOURCES

James Binney
Department of Theoretical Physics
Keble Road
Oxford OX1 3NP
U.K.

ABSTRACT. Spherical stellar systems with large numbers of stars on nearly radial orbits are prone to be bar unstable. It seems likely that stars formed from
a cooling flow tend to populate such orbits, thus ensuring that the underlying
galaxy is barred; a simple estimate suggests that a mass deposition rate in excess
of $5\,\mathrm{M_\odot\,yr^{-1}}$ should suffice for this purpose. We expect the figures of such bars to
rotate slowly. The rate of radiation of accoustic energy by a rotating bar into the
intergalactic medium proves to be small, but it is suggested that the radiation of
energy into g waves in the cluster gas is likely to be an important process.

1. INTRODUCTION

The last few years has seen the settling of the dust raised a decade ago in the field
of elliptical galaxies by the discovery of Bertola & Capaccioli (1975) and Illingworth
(1977) that the rotation speeds of elliptical galaxies are at best poorly correlated
with their shapes. In the new dogma (which in my view is about as securely based
as was the old) giant ellipticals are triaxial, slowly rotating systems whose dynamics
are dominated by non-classical integrals. No entirely satisfactory models of such
systems are available, but Schwarzschild (1979, 1982) has revealed how such things
might work, several groups have constructed n-body models of this type (Aarseth &
Binney 1977; Wilkinson & James 1982,; McGlynn 1984), and Statler (1986, 1987)
has studied a special but highly interesting subclass in some detail. A full account of
the current state of the modern picture of ellipticals will be found in the proceedings
of IAU symposium 127 (de Zeeuw 1987).

Since de Zeeuw's book is still so new, I shall not give a general review of
elliptical galaxy dynamics, but concentrate on two aspects of the modern theory of
ellipticals that may be particularly significant for what, notwithstanding the peril
of confusing previously established terminology, x-ray astronomers call "central
dominant" galaxies; that is, galaxies such as NGC 1275 that are the focus of dense
cooling flows. Both topics relate to the structure of galaxies at radii $r > r_e$, the
effective radius, since it is at such radii that the mass density of the hot gas is
largest relative to the mass of the luminous stars.

A. C. Fabian (ed.), Cooling Flows in Clusters and Galaxies, 225–233.

226

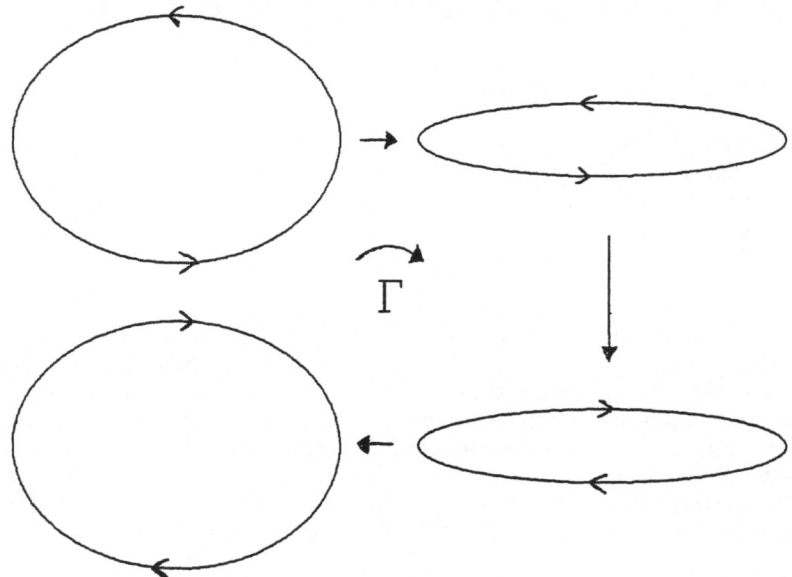

Figure 1. Orbits respond to a retrograde torque by reversing their sense of circulation and precession. A near Kepler orbit would start by precessing with the torque and end by precessing in opposition to the torque, while the behaviour of a near harmonic orbit would be quite the reverse.

2. BAR INSTABILITY

In a rather enigmatic paper Antonov (1973) argued analytically that a spherical stellar system in which all stars were on purely radial orbits would be unstable to the formation of a bar. This conjecture was subsequently confirmed and extended to systems in which stars have non-zero angular momentum by numerical integration of the collisionless Boltzmann equation (Polyachenko & Shukhman 1981), n-body integrations (Barnes *et al.* 1986; Merritt & Aguilar 1985) and a more complete analytic calculation (Palmer & Papaloizou 1987). The basic physics of the process is cute and well worth a brief digression.

2.1 Basic Physics of the Bar Instability

Imagine the effect of applying a small torque Γ to a star on an elliptical orbit in a Kepler potential. Suppose the vector Γ is oppositely directed to the star's angular momentum vector L. Then Γ will eventually reverse L. Figure 1 illustrates this evolution: the shape and orientation of the orbit are unchanged by Γ, but the star's sense of rotation has been reversed.

A bound Kepler orbit is always a closed ellipse because its radial frequency κ is equal to its azimuthal frequency Ω. Galactic potentials, by contrast, always have $\kappa > \Omega$, though at large enough radii we must have $\kappa \simeq \Omega$. Thus far out in a galactic potential the star returns to its apocentric radius a little before it has completed a complete revolution in azimuth, and the orbital ellipse appears to precess retrogradely. Applying to this galactic orbit a torque in the same sense as this precession, that is in opposition to the star's rotation, reverses both the orbit's sense of rotation and of precession, and thus produces an orbit which precesses in the opposition to the applied torque. In Donald Lynden-Bell's (1979) memorable phrase, near Kepler orbits behave like donkeys—pull them forwards and they edge backwards.

Now consider the effect of applying a small torque to a star orbiting in a nearly simple-harmonic axisymmetric potential. This orbit is almost an ellipse centred on (rather than focused on) the centre of attraction; κ is less than but comparable to 2Ω and the ellipse precesses in the same sense as the star rotates. When we apply a torque in opposition to the star's rotation, we again reverse both rotation and precession, but now an opposed precession is transformed into a sympathetic precession. Thus near harmonic orbits, behaving like sheep rather than donkeys, move in the direction they are driven.

One may show that in the limit $L \to 0$ of highly eccentric orbits in any potential with non-singular central value, $\kappa \to 2\Omega$ as in a near harmonic potential. Thus all sufficiently eccentric orbits in non-singular stellar systems can be shepherded. In fact, such orbits are downright gregarious in that when a few happen to align their major axes, the combined gravitational field of these orbits generates a torque that precesses other orbits into their alignment. Consequently, a spherical stellar system will be bar-unstable whenever the gregarious characteristic of the low angular momentum orbits outweighs the obstinate individualism of their better rounded fellows.

2.2 Bar Instability and Accreting Galaxies

To understand the implications of this instability for central dominant galaxies, imagine such a galaxy sitting at the centre of a nearly spherical cluster, and accreting material as cold gas in its outskirts condenses into compact objects. To be specific, let the combined potential of galaxy and cluster be

$$\Phi = \tfrac{1}{2} v_0^2 \ln\!\left(r_c^2 + R^2 + z^2/q^2\right) \tag{1}$$

where (R, z) are cylindrical coordinates and v_0, r_c, $q \approx 1$ are constants. We denote by $\rho_\Phi(\mathbf{x})$ the density distribution that generates Φ through Poisson's equation. A useful measure of the elongation of ρ_Φ is the moment

$$Q_{\mathrm{imp}}(r) \equiv \frac{\int \rho_\Phi(r, \Omega) Y_2^0 \, d\Omega}{\int \rho_\Phi(r, \Omega) Y_0^0 \, d\Omega}, \tag{2}$$

where $r \equiv \sqrt{R^2 + z^2}$ and the $Y_l^m(\Omega)$ are the usual spherical harmonics.

We plausibly assume that the newly formed objects, "stars" for short, fall like stones towards the galaxy's centre, thus constantly increasing the radial bias of the galaxy's distribution function. Eventually the galaxy will become unstable to bar formation. How soon does this occur, and what is the elongation of the resultant bar? We answer these questions as follows.

(i) For each of a series of radii $r_i = 2r_c, \ldots, 11r_c$ we numerically integrate the orbits of particles released with uniform number density from the sphere of radius r_i. Let $\rho_{\mathrm{orb}}^{(i)}(\mathbf{x})$ be the time-averaged density generated by all particles released from radius r_i.

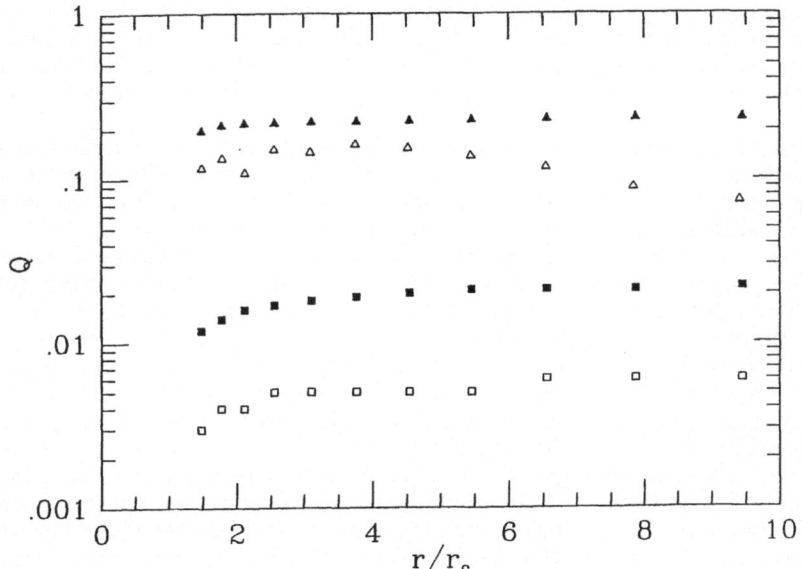

Figure 2. Plots of the quantities Q_{orb} (filled triangles), Q_{imp} (filled squares) and Q_{iso} (open squares) defined by equations (3), (2) and (4) respectively, for the case in which the potential (1) has axis ratio $q = 1.05$ and particles are released from rest. The open triangles show Q_{orb} when particles are released with Maxwellian tangential velocities with one-dimensional dispersion $\sigma = 0.1v_c$.

(ii) We measure the elongation of $\rho_{\text{orb}}^{(i)}$ by calculating

$$Q_{\text{orb}}(\bar{r}_i) \equiv \frac{\int \rho_{\text{orb}}^{(i)}(\mathbf{x}) Y_2^0 \, d\Omega \, r^2 \, dr}{\int \rho_{\text{orb}}^{(i)}(\mathbf{x}) Y_0^0 \, d\Omega \, r^2 \, dr}, \tag{3}$$

where $\bar{r}_i$ is the RMS value of r along the orbit.

Q_{imp} (filled squares) and Q_{orb} (filled triangles) are plotted against r for $q = 1.05$ in Figure 2. Clearly the response density ρ_{orb} is very much more elongated than the density distribution that generates the potential in which the stars move. Consequently, if the potential were entirely in the hands of stars dropped from some spherical shell, any initially small elongation in Φ would amplify on a dynamical time to an axis ratio $q > 1.05$.

In practice is seems probable that most of the galaxy's stars are not on the minority of orbits that can be reached by dropping from a spherical shell, and the density contributed to the galaxy by the majority of stars will respond less enthusiastically to elongation of Φ than does ρ_{orb}. For example, suppose the distribution function of the main mass-bearing population is Maxwellian with the characteristic temperature $\frac{1}{2}v_0^2$ of Φ. Then the density of this population will be $\rho_{\text{iso}}(\mathbf{x}) \propto \exp[-2\Phi(\mathbf{x})/v_0^2]$ and it will generate quadrupole moments

$$Q_{\text{iso}}(r) \equiv \frac{\int \exp[-2\Phi(r,\Omega)/v_0^2] Y_2^0 \, d\Omega}{\int \exp[-2\Phi(r,\Omega)/v_0^2] Y_0^0 \, d\Omega}. \tag{4}$$

The open squares in Figure 2 show Q_{iso}. The fact that these squares lie below the filled squares generated by ρ_Φ betokens the stability of any isothermal system against bar formation.

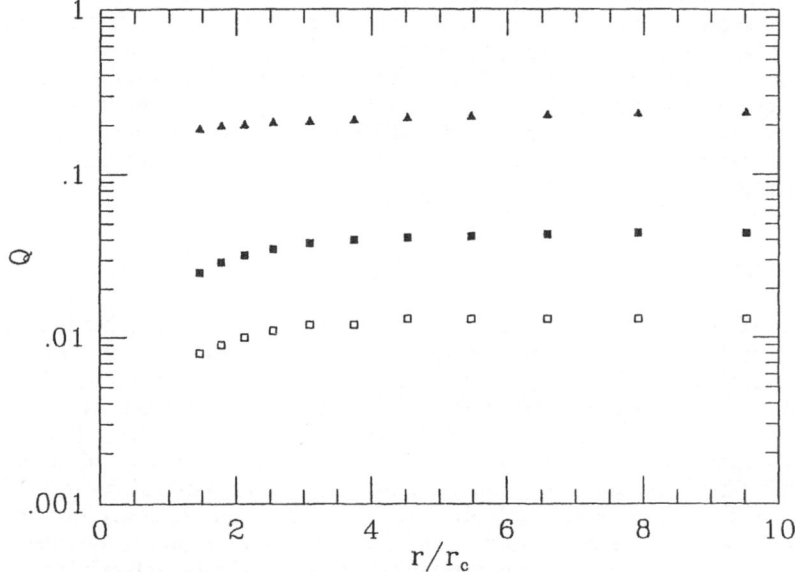

Figure 3. The same as Fig. 2 but for a potential of axis ratio $q = 1.1$.

A system made up of an isothermal background containing most of the mass, and a population of stars released from rest on a series of spheres will be unstable to bar formation if

$$Q_{\text{resp}} \equiv \alpha Q_{\text{orb}} + (1 - \alpha) Q_{\text{iso}} > Q_{\text{imp}}, \tag{5}$$

where α is the fraction of the system's mass in highly eccentric orbits, and the axis ratio q for which the Q's are calculated is any number close to unity. With Figure 2 this criterion suggests that a bar will form provided

$$\alpha \gtrsim \alpha_{\text{crit}} \equiv \frac{Q_{\text{imp}} - Q_{\text{iso}}}{Q_{\text{orb}} - Q_{\text{iso}}} \simeq 0.067. \tag{6}$$

Smaller values of α will lead to the formation of a bar if the mass-bearing background has $\beta \equiv 1 - \sigma_\theta^2/\sigma_r^2 > 0$ since a population with $\beta > 0$ will respond to an incipient bar more enthusiastically than one with $\beta \leq 0$ (eg. May & Binney 1986).

A typical central dominant galaxy might have binding mass $M_{\text{gal}} \approx 10^{12}$ M$_\odot$. According to (6) such a galaxy would be destabilized by the accretion of $0.07 M_{\text{gal}} \approx 7 \times 10^{10}$ M$_\odot$. Thus any accretion rate in excess of $\dot{m} = 5$ M$_\odot$ yr^{-1} should ensure that the galaxy is now barred. Nearly all central dominant galaxies have accretion rates comfortably in excess of this value.

How barred will these galaxies become? Figure 3 shows the same quantities as Figure 2, but for a potential of axis ratio $q = 1.1$. The positions of the filled triangles in the two plots are similar; enhancing the potential's elongation does not increase the elongation of the response density ρ_{orb}. In fact, at small r elongating the potential *decreases* the elongation of ρ_{orb}. By contrast, elongations of both the generating density ρ_Φ and the isothermal response density ρ_{iso} increase roughly in proportion to $q - 1$. Consequently α_{crit}, the critical mass fraction of accreted material required for consistency of the potential with the overall mass distribution, increases with $q - 1$. For $q = 1.1$ I we have $\alpha_{crit} \simeq 0.14$. A fair estimate of the elongation of a system which contains a fraction α' of stars formed from a cooling flow, is probably obtained by finding the value of q for which $\alpha_{crit} = \alpha'$.

How must this discussion be modified if stars form from the cooling flow with non-zero tangential velocities? The open triangles in Figure 2 show how the Q_{orb} values given by the filled triangles are modified if stars are released with Maxwellian tangential velocities of one-dimensional dispersion $\sigma_t = 0.1v_0$. As expected, the elongation of ρ_{orb} is smaller in this case.

3. FIGURE ROTATION

The backbone of a triaxial galaxy is formed by box orbits (Schwarzschild 1979). If the galaxy's potential is time-independent in an inertial frame, box orbits make no contribution to the galaxy's internal streaming motions (Schwarzschild 1982; Merritt 1980). Hence any mean stellar motions observed in a such system must arise from stars that do not form part of the backbone. If, as seems likely, only a minority of stars lie outside the backbone, it follows that only a minor degree of stellar streaming is possible in galaxies whose figures do not rotate (eg. Vietri 1986). To state this more quantitatively, suppose half of all stars are outside the backbone, and these have mean rotation rates uniformly distributed between zero and the circular speed v_c of the monopole component of the galactic potential. Then the overall mean-streaming speed $\bar{v} \simeq \frac{1}{4}v_c \simeq 0.35\sigma$, where σ is the line-of-sight velocity dispersion and I have assumed $\sigma \simeq \sqrt{2}v_c$. Many giant ellipticals show more rotation than this (Davies et al. 1983). Thus it is appropriate to consider the consequences for hot cluster gas of the figures of central dominant galaxies rotating.

Let us assume that the cluster gas does not rotate, or at least did not rotate initially. Then we have to consider the effect upon a giant star-like body of gas of a periodic forcing potential at its centre. Stars are capable of two important classes of oscillation: p modes and g modes (eg. Cox 1980). The p modes are essentially stationary patterns of compressive sound waves, while the oscillations making up g modes are akin to ocean waves; corrugations of the equidensity surfaces propagate around the star just as corrugations on the surface of the oceans move around the World. Clearly periodic forcing by the rotating potential of the central dominant galaxy is liable to excite both p and g waves. Which waves will be most important in practice depends both on the details of the forcing potential and on the forcing frequency. There are two reasons why one might expect the g waves to dominate: (i) their lower frequencies are probably better matched to the central galaxy's pattern frequency; (ii) the fluctuating component of the galaxy's force-field is predominantly tangential, and it is g modes which are associated with mainly tangential motions. However, it is easier to estimate the power radiated in p waves than g waves, so let us obtain such an estimate and treat it as a lower limit on the energy input to all types of oscillation.

What makes p waves easier to handle than g waves is that while g waves cannot be discussed in any context simpler than laminar geometry, p waves survive abstraction of the cluster gas to a homogeneous medium. A treatment based on this abstraction is valid if the gas is approximately homogeneous throughout the near-field of the radiating source—this probably extends to several of the galaxy's corotation radii, say a few hundred kpc. This condition is clearly violated, but not hopelessly so. Therefore we take the background density ρ_0 to be constant and employ the perturbed the flow equations

$$\rho_0 \partial_t \mathbf{v} = -\rho_0 \nabla \Phi - c^2 \nabla \rho \quad ; \quad \partial_t \rho + \rho_0 \nabla \cdot \mathbf{v} = 0. \tag{7}$$

Taking $\mathbf{v}$ to be of the form $\mathbf{v} = \nabla \psi$, these become

$$\nabla(\dot{\psi} + c^2 \nabla \ln \rho + \Phi) = 0 \quad ; \quad \partial_t (\ln \rho) = -\nabla^2 \psi, \tag{8}$$

which yield the single equation

$$\left(\nabla^2 - \frac{1}{c^2} \frac{\partial^2}{\partial t^2} \right) \psi = \frac{\dot{\Phi}}{c^2}. \tag{9}$$

Solving for ψ by the usual technique, we have for $\Phi(\mathbf{x}, t) = \Phi_0(\mathbf{x}) + \Phi_2(\mathbf{x}) e^{-i\omega t}$

$$\mathbf{v}(\mathbf{X}) = -\nabla \int \frac{\exp(ik|\mathbf{X} - \mathbf{x}|)}{4\pi |\mathbf{X} - \mathbf{x}|} \frac{\dot{\Phi}}{c^2} \, d^3 x \simeq \omega k V(\Omega) \frac{e^{i(k|\mathbf{X}| - \omega t)}}{|\mathbf{X}|} \hat{\mathbf{X}}, \tag{10a}$$

where the volume V is defined by

$$V(\Omega) \equiv \lim_{|\mathbf{X}| \to \infty} \left[\frac{e^{-ik|\mathbf{X}|}}{4\pi c^2} \int \exp(ik|\mathbf{X} - \mathbf{x}|) \Phi_2 \, d^3 x \right]. \tag{11}$$

We obtain the power radiated by integrating the flux $\frac{1}{2} c \rho_0 v^2$ over a large sphere centred on the origin:

$$P = \tfrac{1}{4} c \rho_0 |\mathbf{X}|^2 \oint |\mathbf{v}|^2 \, d\Omega = \frac{\omega^4 \rho_0}{2c} \oint |V|^2 \, d\Omega$$
$$= \rho_0 L^3 (\omega L)^2 \omega \left(\frac{\omega L}{4c} \right), \tag{12a}$$

where the length L is defined by

$$L^6 \equiv \oint |V|^2 \, d\Omega. \tag{12b}$$

This drain of rotational energy from the galaxy will profoundly affected the galaxy's structure on the timescale

$$\tau \equiv \frac{\frac{1}{2} I_{\text{gal}} \omega^2}{P} \simeq \left(\frac{\rho_{\text{gal}}}{\rho_0} \right) \left(\frac{R_{\text{gal}}}{L} \right)^5 \left(\frac{2c}{\omega L} \right) \frac{1}{\omega}. \tag{13}$$

Thus the importance of this process depends strongly on the value of L and hence $V(\Omega)$. Writing $\Phi_2(\mathbf{x}) = c^2 \mathcal{R}(r) \Theta(\cos \theta)$ with Θ an even function of its argument, we have

$$V(\theta) = \int_0^\infty \mathcal{R} r^2 \, dr \int_0^1 J_0 \left(\sqrt{1 - \mu^2} \, kr \sin \theta \right) \cos(\mu kr \cos \theta) \Theta(\mu) \, d\mu. \tag{14}$$

232

Since $\int \Theta(\mu)\,d\mu = 0$ and for the frequencies of interest $2\pi/k \gg 2r_e$, the dominant contribution to V comes from points at $r > 2r_e$ at which Φ_2 may be approximated by the external potential of a quadrupole; $\mathcal{R} \propto r^{-3}$. With $\Phi_2 = c^2(b/r)^3 Y_2^0$ I find $L \simeq 0.7b$ independent of k.

Let us apply this result to a concrete example. At large r the vacuum quadrupole contribution to the potential of Schwarzschild's (1979) triaxial model is of the form $\Phi_2(\mathbf{x}) \simeq \sigma(0)^2(4.46 r_c/r)^3 \sum_m \alpha_m Y_2^m$, where σ is the model's one-dimensional velocity dispersion, r_c is its core radius and $\sum_m \alpha_m^2 = 1$. Thus if $\sigma(0) \approx c$, we have $b \approx 4.5 r_c$ and thus $L \approx 3r_c \simeq 3\,\mathrm{kpc}$ in the case of M87. Then taking $\rho_{\mathrm{gal}}/\rho_0 = 100$, $R_{\mathrm{gal}}/L = 3$, $c = 500\,\mathrm{km\,s^{-1}}$ and $\omega L = 80\,\mathrm{km\,s^{-1}}$, we find $\tau \simeq 1.1 \times 10^{13}$ yr. Thus for these parameters the emission of p waves is unimportant.

Two essentially independent factors contribute to the large number of galactic rotation periods required for the emission of significant accoustic energy: the small proportion of the central mass in gaseous form and the large factor by which c exceeds the characteristic pattern rotation speed ωL—this second factor both diminishes L by obliging us to evaluate (14) in the limit of small k, and appears explicitly in equation (13). Clearly the transfer of energy to g waves will also be impeded by the small value of $\rho_0/\rho_{\mathrm{gal}}$. But the excitation of g waves will not be similarly hampered by a large propagation speed, so I think it likely that g waves will prove to play an important role in cluster sources.

4. DISCUSSION

In this article I have assumed that the centres of rich clusters of galaxies are in virial equilibrium. This is undoubtedly a rather dubious idealization; gradually evidence is accumulating from both optical and x-ray studies that many Abell clusters are evolving rapidly, even violently, as they merge with neighbouring clusters and are subjected to a steady rain of infalling galaxies (Geller & Beers 1982, Geller 1984). The restless heaving of the intergalactic medium consequent on this evolution bears on the topics of this paper in two ways: (i) it must contribute to the tangential velocities of stars newly formed from a cooling flow; (ii) it must sustain a quite deafening level of accoustic and g-wave noise by comparison with which the steady note emmitted by a central rotating galaxy must seem like the chirping of a thrush. Thus the analyses of this paper cannot be considered to be of more than exploratory in nature. However, I hope to have convinced the reader that there *are* interesting possibilities of interplay between the internal dynamics of the central galaxy and a surrounding cooling flow, and that these connections merit further investigation.

REFERENCES

Aarseth, S. J. & Binney, J. J. 1978. *Mon. Not. Roy. Astron. Soc.*, **185**, 227.

Antonov, V. A. 1973. In *The Dynamics of Galaxies and Star Clusters*, ed. G. B. Omarov, (Alma Ata: Nauka), p. 139.

Barnes, J., Goodman, J. & Hut, P. 1986. *Astrophys. J.*, **300**, 112.

Bertola, F. & Capaccioli, M. 1975. *Astrophys. J.*, **200**, 439.

Cox, J. P. 1980. *Theory of Stellar Pulsation*, (Princeton: Princeton University Press).

Davies, R. L., Efstathiou, G., Fall, S. M., Illingworth, G. & Schechter, P. L. 1983. *Astrophys. J.*, **266**, 41.

de Zeeuw, T., ed. 1987. *Structure and Properties of Elliptical Galaxies*, (Dordrecht: Reidel).

Geller, M. J. 1984. *Comments on Astrophys.*, **10**, 47.

Geller, M. J. & Beers, T. C. 1982. *Publ. Astron. Soc. Pacific*, **94**, 421.

Illingworth, G. 1977. *Astrophys. J. Lett.*, **218**, L43.

Lynden-Bell, D. 1979. *Mon. Not. Roy. Astron. Soc.*, **187**, 101.

May, A. & Binney, J. J. 1986. *Mon. Not. Roy. Astron. Soc.*, **221**, 13P.

McGlynn, T. 1984. *Astrophys. J.*, **281**, 13.

Merritt, D. 1980. *Astrophys. J. Suppl.*, **43**, 435.

Merritt, D. & Aguilar, L. A. 1985. *Mon. Not. Roy. Astron. Soc.*, **217**, 787.

Palmer, P. L. & Papaloizou, J. 1987. *Mon. Not. Roy. Astron. Soc.*, **224**, 1043.

Polyachenko, V. L. & Shukhman, I. G. 1981. *Astr. Zh.*, **58**, 933 (translated in *Sov. Astr.*, **25**, 533).

Schwarzschild, M. 1979. *Astrophys. J.*, **232**, 236.

Schwarzschild, M. 1982. *Astrophys. J.*, **263**, 599.

Statler, T. 1986. Unpublished Ph.D. thesis, Princeton University.

Statler, T. 1987. *Astrophys. J.*, in the press.

Vietri, M. 1986. *Astrophys. J.*, **306**, 48.

Wilkinson, A. & James, R. A. 1982. *Mon. Not. Roy. Astron. Soc.*, **199**, 171.

COOLING FLOWS IN GALAXIES

P. A. Thomas,
Institute of Astronomy,
Cambridge, CB3 0HA,
U. K.

ABSTRACT. Early-type galaxies contain hot ($\sim 10^7$ K) X-ray emitting gas which is cooling at rates of between a few hundredths and a few solar masses per year. To bind the hot gas to the galaxy the gravitational potential must contain a massive halo component. The dynamical cooling flow equations are discussed in some detail. By considering the energy balance within the flow it can be shown that i) either the usual estimates of the stellar mass loss and supernova rates are too high, or the corresponding mass and energy does not get deposited into the hot gas, and ii) mass deposition occurs throughout the galaxy implying a range of densities at each radius. The cooled gas is not stored in reservoirs of low-temperature gas and does not form stars with a normal initial-mass-function. The interstellar medium in early-type galaxies is complex.

1. INTRODUCTION

This review discusses cooling flows in individual (early-type) galaxies. There is a continuous range of flows from a few hundredths of a solar mass per year in some isolated galaxies up to several hundred $M_\odot$ yr^{-1} at the centres of many clusters. I will concentrate on those systems in which the central galaxy, either through its gravity or mass and energy injection, has an important effect on the structure of the flow. This excludes large cluster flows, of more than a few $M_\odot$ yr^{-1}, in which the main effect of the central galaxy is to provide a focus for the flow.

2. X-RAY OBSERVATIONS

2.1. Review. X-ray emission from clusters of galaxies was first detected with the Uhuru satellite (Bachall & Bachall 1975). Finer imaging with rocket flights established that the brightest emission was associated with the central galaxy (*e.g.* for M 87, Malina, Compton & Bowyer 1976, Gorenstein *et al.* 1977). It was not until the launch of the *Einstein* satellite, however, that Forman *et al.* (1979) found extended emission which they interpreted as from gas at $\sim 10^7$ K, around other galaxies in the Virgo cluster. This was followed by Bechtold *et al.* (1983) for galaxies in A 1367 and Biermann & Kronberg (1983) for NGC 5846 (in a small group). Long & van Speybroeck (1983) surveyed the X-ray emission from normal galaxies and suggested that in general $L_X \propto L_B$ with bright ellipticals perhaps lying a little above this relation. Nulsen, Stewart & Fabian (1984) were the first to find X-ray emission from truly isolated ellipticals and to discuss the resulting cooling flows in detail. Forman, Jones & Tucker (1985, hereafter FJT) published a survey of 55 early-type galaxies and discussed the evidence for gaseous emission. The X-ray surface brightness distribution is more extended than that of the luminous galaxy and in some cases the gas is seen to be interacting with an external

A. C. Fabian (ed.), Cooling Flows in Clusters and Galaxies, 235–244
© *1988 by Kluwer Academic Publishers.*

medium; perhaps the most obvious example is M 86 which is being stripped by motion through the Virgo cluster. The spectra of the (brighter) galaxies are consistent with thermal bremsstrahlung and line radiation from a 0.5 – 2 keV gas, whereas discrete X-ray sources are harder $(kT \gtrsim 5\,\mathrm{keV})$. Finally the high luminosity of early-type compared to late-type galaxies and the correlation of X-ray to blue luminosity, $L_X \sim L_B^{1.5-2.0}$, imply that discrete sources can be important only for the fainter galaxies.

Trinchieri & Fabbiano (1985) have discussed further the relationship between X-ray and optical luminosities. They find $L_X \sim L_B^{1.64\pm0.15}$ for a sample of 29 E and S0 field galaxies. Canizares, Fabbiano & Trinchieri (1987, hereafter CFT) analysed 81 early-type galaxies observed with *Einstein* and obtained $L_x \sim L_B^{1.52-1.98}$. There are a number of upper limits but none of these are very restrictive. These authors, together with FJT, discuss the possible contribution from discrete X-ray sources. This can be extrapolated from the emission from late-type galaxies, which is thought to be dominated by such sources. They rule out M dwarfs but, for faint systems, there can be a significant contribution from low-mass X-ray binaries and from sources in globular clusters which are known to exist in large numbers around bright elliptical galaxies (*e.g.* Hanes & Harris 1986). FJT estimate a normalisation for the discrete source contribution which is one third that of CFT. CFT also investigate the correlation between L_X and $L_B\sigma^2$ which would be directly proportional if, for example, stellar mass loss were to dominate the flow. They find $L_X \sim (L_B\sigma^2)^{1.06-1.38}$. Both this and the above relation between L_X and L_B are steepened if we subtract any discrete source contibution to the X-ray flux. Fabbiano *et al.* (1987) have extended the survey of Trinchieri & Fabbiano (1985) to the radio (see the article by Trinchieri, this volume).

A recent comparison of the optical and X-ray luminosities of all early-type galaxies observed with the *Einstein Observatory*, prepared by Prof. W. Forman, is shown in Figure 1.

2.2. <u>Spectra</u>. Almost all observations of extended X-ray emission from individual galaxies were made using the Imaging Proportional Counter (IPC) of the *Einstein* satellite. This had a spatial point response function with a half-width at half-power of over 1 arcmin (Mauche & Gorenstein 1986). The data are poorly resolved into energy channels but cannot be directly inverted to give the incident spectrum; instead model spectra are folded through the detector response and compared with the data. High Resolution Imager (HRI) data were obtained for some of the brighter galaxies. The HRI was less sensitive to low surface brightness features and had no energy resolution but, as its name implies, had a higher spatial resolution. Further details can be found in Giacconi *et al.* (1979).

Spectral analysis is limited by the quality of the data to the brightest galaxies. FJT have analysed the data from 8 galaxies and Trinchieri, Fabbiano & Canizares (1986, hereafter TFC) 6 galaxies, with 5 in common. They each find that the data are well fit by optically thin thermal emission from an isothermal, hot gas with temperatures between 0.5 and 2.0 keV. There is good agreement between the two groups except in the case of NGC 4649. TFC find that for both NGC 4649 and NGC 4472 the 90 per cent confidence interval for the column density lies above the Galactic value. There is evidence against large quantities of cool gas in these galaxies, however, and the large columns may result from fitting an isothermal model to emission from gas at a range of temperatures. FJT split the emission from NGC 4472 into three radial bins and find that the temperature rises slightly with radius; this is the only spatial information available on the temperature variation in the hot gas found in galaxies.

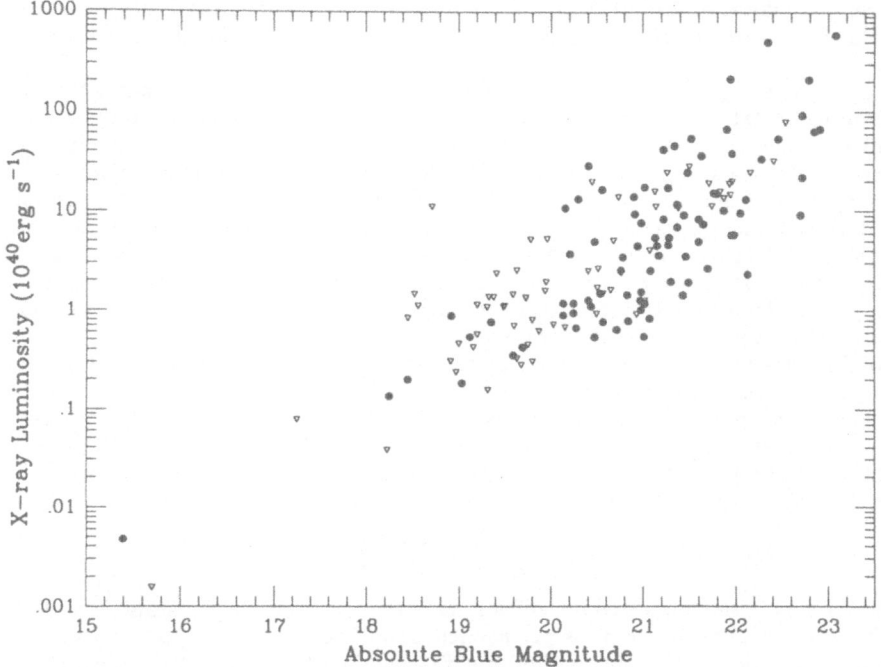

Figure 1. The X-ray luminosity versus absolute blue magnitude of all early-type galaxies observed with the *Einstein Observatory*. The •s are detections and the ▽s upper limits. My thanks to Prof. W. Forman for providing a copy of this figure.

2.3. <u>Amount and distribution of the gas</u>. FJT, Thomas *et al.* (1986, hereafter TFAFJ) and TFC have discussed the distribution of the hot gas. They each binned the the data into circular annuli and fitted models of the form $(1 + (r/a)^2)^{-3\beta+1/2}$ to the surface brightness profile. The IPC model distribution has to be convolved with the spatial point response function (prf, approx. gaussian with $\sigma = 40 - 50$ arcsec) before fitting to the data. The results give $\beta \approx 0.4 - 0.65$ but note that this corresponds to a large range of surface brightness slopes and there is poor agreement between these authors. TFAFJ discussed some of the reasons for this. Firstly, the authors use slightly different forms for the IPC prf, the effect of which is to flatten the core and to steepen the profile at larger radii. Secondly, the deconvolved slopes are sensitive to changes in the core radius, especially for those galaxies with surface brightness slopes, $s = 6\beta - 1$, close to 2, (the total flux is then dominated by emission from the centre of the galaxy), and the core radius is in general poorly determined since it is unresolved by the IPC. Finally, changing the level of background subtraction can alter s by as much as 0.4 in low luminosity sources. TFC also show that the surface brightness can fall at different rates on different sides of the galaxy. In general, the surface brightness distributions of the brightest galaxies are more extended than the optical light. The errors on the low-luminosity sources are large, however, and for these a large contribution to the X-ray flux from discrete sources which follow the galactic profile cannot be ruled out.

    The amount of hot gas inferred to be present in these systems depends on the extent of the X-ray halo. Within the radius at which the data merge into the background the observed quantities are typically 10^9–$10^{10} \, M_\odot$ (FJT), and the total may be many times this as the integrated mass out to large radii is unbounded for $\beta \leq 1$.

2.4. Massive haloes. It has long been known that galaxies contain dark matter. In spirals optical rotation curves and HI measurements at 21 cm enable mapping of the mass distribution out to large radii. Unfortunately there is no test population in the majority of ellipticals (some have external HI disks) with which to trace the mass. The relative spacing of the shells seen around some elliptical galaxies (Hernquist & Quinn 1987) show that the spatial distribution of the mass is more extended than that of the light, but do not provide a quantitative measure. For this we have to use the velocity dispersion, which is only reliably determined in the core, together with an assumed galaxy model. An alternative, model-free measure of the mass is provided by the distribution of the hot gas sitting in hydrostatic equilibrium in the potential well of the galaxy. This has the additional advantage that the observed gas extends to tens of kpc, well beyond the core of the galaxy.

The gas in many of these galaxies is significantly hotter than the virial temperature inferred from the stellar velocity dispersion ($\sigma_\parallel^2 \approx 250 \, \mathrm{km \, s^{-1}} \equiv 4.5 \times 10^6$ K) which implies that there must be a massive halo present in order to confine it. The gas is in hydrostatic equilibrium and so we have

$$M(< r) = -\frac{r}{G}\frac{kT}{\mu m_H}\left(\frac{d\ln T}{d\ln r} + \frac{d\ln \rho}{d\ln r}\right)$$

where $M(< r)$ is the mass within radius r, G is the gravitational constant, k is Boltzmann's constant, μm_H is the mass per particle and T and ρ are the gas temperature and density, respectively. (This method was first applied to M 87 by Fabricant, Lecar & Gorenstein 1980.) Since the emissivity varies as $\epsilon \propto n^2 T^{-0.6}$ the density profile, $n \propto \epsilon^{0.5} T^{0.3}$, can be inferred robustly from the observed surface brightness. However the temperature at large radii, and the temperature gradient, are more uncertain. FJT applied this method to their sample and inferred masses up to $5 \times 10^{12} \, M_\odot$ within 100 kpc, corresponding to M/L ratios of between 7 and $88(M/L)_\odot$. NGC 4472, with its measured temperature gradient, gives $M(r) \geq 3.0 \times 10^{11}(T/10^7 \, \mathrm{K})(r/10 \, \mathrm{kpc}) \, M_\odot$ and has emission detected out to 80 kpc. This is in agreement with the dynamical model of Thomas (1986), using the same data, who found $M(r) = (3.5 \pm 0.8) \times 10^{11}(T/10^7 \, \mathrm{K})$ $(r/10 \, \mathrm{kpc}) \, M_\odot$. TFC have also estimated masses in this way and their results are in general agreement with FJT. They are more conservative, however, in their estimates of the outer radius at which the X-ray emission is reliably detected and hence the required extent of the haloes.

Fabian et al. (1986) have produced a more general limit on the total mass, M_T, of these systems simply by requiring that the observed gas be convectively stable and bound to the galaxy. The steepest allowable temperature gradient, for which the gas is just convectively stable, corresponds to an adiabat and minimises the total mass. If the gas is observed to have temperature T_0 and pressure p_0 at radius r_0 then

$$M_T \geq \frac{5kT_0 r_0 [1 - (p_\infty/p_0)^{0.4}]}{2G\mu m_H (1 - r_0/r_\infty)}.$$

It is assumed that the atmosphere extends to r_∞ where the pressure is p_∞. When applied to the sample discussed by FJT this method gives total masses up to $1.2 \times 10^{13} \, M_\odot$ and M/L ratios of 13–180 $(M/L)_\odot$ — about twice the previous estimates of the mass within r_0. This lower limit suggests that the massive haloes extend far beyond the luminous galaxy.

3. GAS DYNAMICS

3.1. <u>Winds</u>. Winds have long been postulated as an efficient way of removing gas from elliptical galaxies (Johnson & Axford 1971, Mathews & Baker 1971); however they are too effective in this respect. In a steady-state wind the flow becomes supersonic within an effective radius which, for the observed gas densities, leads to mass outflow rates far in excess of those which can be supplied by stellar mass loss (Nulsen, Stewart & Fabian 1984, Thomas 1986). White & Chevalier (1983,4) find that ellipticals can sustain either a steady wind or inflow, but not a partial wind. However Lowenstein & Mathews (1987) in a time-dependent study of mass injection in galaxies with massive haloes found that an initially supersonic wind can be slowed and reversed by the halo potential.

3.2. <u>Cooling flow equations</u>. The dynamical cooling flow equations are discussed below. I first give comoving expressions for conservation of mass, momentum and energy, followed in each case by the simplified form for a steady-state, spherically-symmetric flow. Spherical symmetry implies the absence of angular momentum which might otherwise cause the flow to stagnate at a finite radius (Cowie, Fabian & Nulsen 1980), although viscosity and turbulence can both act to overcome this effect (Nulsen, Stewart & Fabian 1984). Energetically the system may only have attained a steady-state in the inner regions. It is generally assumed to be a good approximation within the 'cooling radius' where the cooling time equals a Hubble time.
Mass:

$$\frac{\partial \rho}{\partial t} + \nabla.(\rho \underline{u}) = \alpha - \beta$$

Here t is the time, ρ the gas density, $\underline{u}$ the gas velocity, and α and β the rates of mass injection and deposition from the flow, respectively. In a steady state and spherical symmetry, with $\underline{u} = -u\hat{\underline{r}}$, this simplifies to

$$\frac{1}{4\pi r^2}\frac{d\dot{M}}{dr} = \beta - \alpha$$

where $\dot{M}(r) = 4\pi r^2 \rho u$ is the mass inflow rate. In a steady state this is equal to $\dot{M}(< r)$, the mass deposition rate within radius r. Note that it is this latter quantity which is derived from the X-ray observations via the energy equation; estimates of the mass deposition rate do not require the steady-state assumption.

The rate of mass injection, α, is not important for clusters, but can be for individual galaxies. In most cases the theoretical rates of stellar mass loss (Faber & Gallagher 1976) are higher than the inferred mass deposition rates (TFAFJ) although this discrepancy will be reduced if the theoretical value of α is too high or if the gas temperature has been overestimated. Thomas (1986) suggests that the mass lost from stars is prevented from mixing with the flow and is confined by the high pressure to cool ($\sim 10^4$ K) blobs. These can survive ram-pressure heating and will be brought to rest relative to the flow without colliding. They may then form low mass stars or flow into the centre of the galaxy carrying dust and heavy elements with them.

β is the rate of mass deposition from the flow. As discussed in §3.4, this term is very important as the observed flows all have $\dot{M}$ rising with radius; a constant $\dot{M}$ carries too much energy into the central regions. β has been ignored in almost all dynamical models however, even though its effect on the flow is large. It makes the temperature more isothermal and moves the sonic radius inwards to within the core of the galaxy. Thomas (1986) included mass deposition in his model of NGC 4472 and White & Sarazin (1987) in their models of clusters.

Momentum:

$$\frac{\partial}{\partial t}(\rho \underline{u}) + \nabla.(\underline{u}\rho\underline{u}) = -\nabla p - \rho\nabla\phi - \beta\underline{u}$$

where p is the gas pressure and ϕ the gravitational potential. This simplifies to

$$\rho u\frac{du}{dr} - \alpha u = -\rho\frac{d\phi}{dr} - \frac{dp}{dr}.$$

The flows are highly subsonic, except maybe in the centre, and the momentum of the injected gas is always negligible so that we may ignore the terms on the left hand side and use the hydrostatic equation:

$$\frac{dp}{dr} = -\rho\frac{d\phi}{dr} = -\frac{GM}{r^2}\rho.$$

To assess the role of gravity we can rewrite this as $d\ln p/d\ln r \approx -3\beta$, where $\beta = \sigma^2/(kT/\mu m_H)$ is the ratio of the galaxy to gas temperatures. Gravity is always important in galactic flows.

Energy:

$$\frac{\partial}{\partial t}\left(\rho(H + \phi + \tfrac{1}{2}u^2)\right) + \nabla.\left(\rho\underline{u}(H + \phi + \tfrac{1}{2}u^2)\right)$$

$$= \alpha(\epsilon_{in} + \phi) - \beta(H + \phi + \tfrac{1}{2}u^2) + h - c$$

where $H = 5kT/2\mu m_H$ is the enthalpy of the gas and ϵ_{in} is the specific energy of the injected material (averaged over supernovae and stellar mass loss).

$$- \rho u\frac{d\epsilon}{dr} = \frac{p}{r^2}\frac{d(ur^2)}{dr} + \alpha(\tfrac{1}{2}u^2 + \epsilon_{in} - \epsilon) - \beta\frac{p}{\rho} + h - c.$$

The first term on the right hand side is the heating by compression, whereas $\beta p/\rho$ is the PdV work done on the cooling gas. Note that gas is removed from the flow with its full kinetic energy and enthalpy. In comparing the model with observations this energy, with the appropriate spectrum for gas cooling to low temperatures, must be added to the emission from the bulk of the flow (*e.g.* Thomas 1986). The cooling component can dominate in the inner regions. $\alpha(\tfrac{1}{2}u^2 + \epsilon_{in} - \epsilon)$ is the energy injection which is critically dependent upon the assumed supernova rate. h represents the heating from other sources and c the radiative cooling rate which is dominated by line radiation as compared to thermal bremsstrahlung from clusters at temperatures above 3×10^7 K.

3.3.   Possible heating mechanisms.   Nuclear activity in these galaxies is generally small. Though compact radio sources have been found in some, they are too weak to provide significant heating of the gas. This would also seem to rule out heating by cosmic rays as has been suggested for the Virgo and Coma clusters by Tucker &Rosner (1983) and Rephaeli (1987). The effect of conduction in M87 has been discussed by Stewart *et al.*(1984) and Bertschinger & Meiksin (1986) who reach opposite conclusions (see also Meiksin, this volume). The latter show that conduction can reduce the mass inflow rate from 10 to 1–3 $M_\odot$ yr^{-1} . However, the density of the gas must be fine-tuned in order to suppress cooling without rendering the gas isothermal (this is less of an objection in galactic than in cluster flows as there are no measured temperature gradients), and even a very weak, tangled magnetic field will reduce conduction to negligible values. The srongest evidence against conduction is provided by spectroscopic data from *Einstein* which show equal quantities of gas cooling down through low temperatures in some clusters of galaxies (see the articles by Canizares and Mushotzky in this volume).

In general the local thermal instability of the gas is such that any heating mechanism, such as supernovae shocks, which induces density variations in the gas may actually promote cooling in the denser regions. There is also the global problem of balancing heating and cooling both in the centre, where cooling is most effective, and in the outer regions. The energy radiated in a large galaxy is typically 3×10^{41} erg s^{-1} which corresponds to one 10^{51} erg supernova per 100 yr, or the thermal energy of 4×10^{10} M$_\odot$ of gas at 10^7 K in a Hubble time.

3.4. Energetics. The supernova rate in elliptical galaxies was estimated by Tammann (1974) to be 2.1×10^{-3} yr^{-1} /10^{10} L$_\odot$. If these supernovae deposit their energy into the hot gas then they supply far more energy than is radiated in X-rays (Nulsen, Stewart & Fabian 1984, Thomas 1986, Sarazin 1987, CFT). It is probable that the supernova rate has been overestimated by a factor of 10 or more (Tammann, private communication) but another possibility is that the supernovae go off in regions of high-mass star formation and deposit their energy into high-density, relatively low-temperature gas.

If the observed cooling gas is transported into the core of the galaxy then the gravitational energy liberated would supply too much energy to the central regions (Thomas 1986, CFT). TFAFJ have applied the deprojection procedure of Fabian et al. (1981) to a sample of 18 early-type galaxies. They considered a variety of models for the gas flow, including successively more terms in the energy equation, and found that in each case the mass deposition rate, $\dot{M}$, increased with r, reaching a total of between 0.02 M$_\odot$ yr^{-1} and more than 1 M$_\odot$ yr^{-1} at the cooling radius. Thus matter matter is being deposited, and dense cooling gas must be present, throughout the galaxy. Thomas (1986) has shown for NGC 4472 that it is consistent to deposit matter as a modified King model $\dot{\rho} \propto (1 + (r/a)^2)^{-3/2}$ but that an isothermal profile $\dot{\rho} \propto (1 + (r/a)^2)^{-1}$ is too extended.

3.5. Multiphase models. The above results suggest that single-phase models for galactic cooling flows are naïve. The interstellar medium must in reality contain gas phases with a wide range of densities (and therefore temperatures) at each radius. These will cool in pressure equilibrium until the cooling time becomes shorter than the sound crossing time of the densest phase which will then cool isochorically. Nulsen (1986) has investigated this situation in detail and concludes that the phases will comove as they flow inwards, although others disagree (Malagoli, Rosner & Bodo 1987). With the co-moving assumption further progress can be made (see contributions to this volume by Nulsen and Thomas). Thomas, Fabian & Nulsen (1987) analysed the data from the centres of clusters using a range of densities and found essentially the same results as in a single-phase analysis. Unfortunately it has not been possible to extend this method to individual galaxies because of the poor quality of the data and the uncertainties in the gravitational profile and heating rate.

3.6. Evolution of cooling flows in ellipticals How large were cooling flows in the past? This is an important question for which there is unfortunately little observational data. Without knowing more about their structure it is not possible to deduce even the future evolution of cooling flows in elliptical galaxies from the present X-ray observations, and all information about the past has been irrevocably erased. Observations of distant galaxies (in other wavebands) do not help unless we can find a way of relating these to cooling flow activity. One diagnostic is the high pressure associated with the hot gas. Fabian et al. (1987) infer in this way a flow of about 100 M$_\odot$ yr^{-1} around the quasar 3C 48, at a redshift of 0.37. This is typical for a central, cluster galaxy but would imply considerable evolution if it later developed into a normal giant elliptical.

In the absence of direct observations, theoretical models abound. The stellar mass loss rate will have been much higher in the past and this has led several authors to speculate that the hot gas may have originated in this way. Indeed, the high ($\sim$ 1/2 solar) abundance of the intracluster medium necessitates processing through some stellar population, though whether Population III, or the first galactic Population II stars, is not known. Cowie & Binney (1977) proposed a 'radiative regulation' model for cluster cooling flows in which mass deposition was balanced by the input of gas stripped from cluster galaxies. A similar mechanism involving mass lost from stars would explain the correlations of X-ray and blue luminosities in individual galaxies, but suffers from the problems that the mass is injected at the centre of the galaxy and at temperatures below that of the hot gas. MacDonald & Bailey (1981) suggested that the inflow of gas at early epochs would be sufficient to power an active nucleus (a similar model has recently been advanced by Kunze, Loose & Yorke 1987) but that the increased importance of (Type I) supernovae relative to mass injection at later times would lead to winds, thus explaining the time evolution of AGN's. We now know, of course, that this has not happened. Loewenstein & Mathews (1987) have carried out time-dependent calculations of gas flows and show that, even if the galaxy is initially devoid of gas and has a supersonic wind, this can be slowed and eventually reversed by a massive halo. The outer gas gradually presses down on the galaxy leading eventually to a pure inflow. Another possibility is that the flow is confined by the external pressure of the intracluster medium. This must be the case in cluster galaxies, some of which have X-ray contours which are seen to be distorted by motion through the ICM. External pressure confinement would weaken any L_X–L_B relation, however, which would then become more dependent upon environment. The distinction between these two models would seem to be whether or not the galaxy is located at the centre of the potential well which contains the hot gas.

It is clear that the currently observed mass flow rates can have little influence on the evolution of the galaxy. Only if the supply of gas were much greater in the past will they be of interest in this respect (see the article by Thomas, this volume).

4. PRODUCTS OF THE COOLING FLOW

4.1. Evidence for low temperature gas.   What happens to the cooled gas? The observations described below, many of which are discussed in more detail elsewhere in this book, show that it does not accumulate in large reservoirs of low-temperature gas.

Caldwell (1984) and Phillips et al. (1986) have searched for emission lines in elliptical galaxies. They each detect ionised gas in about half their sample but at very low levels corresponding to masses of just 10^3–10^4 $M_\odot$ (see the paper by Sadler, this volume). The mass limits from radio observations of HI are less restrictive. The detections and upper limits cover a wide range, 10^6–10^{10} $M_\odot$, of inferred mass and are not well correlated with the optical luminosity, which suggests an external origin for the cool gas (Knapp, Turner & Cunniffe 1985; Knapp, this volume). Sadler & Gerhard (1985) and Sparks et al. (1985) observed dust lanes in one quarter to one half of the ellipticals they looked at, whilst from IRAS observations of early-type galaxies Jura (1986) concludes that at least a third have appreciable amounts of dust. The lifetime for sputtering in a 10^7 K gas is so short that the majority of the dust would be destroyed at its point of injection into the hot gas. However if it is produced by stellar mass loss and does not join the hot phase then it may be preserved.

Thus although it is possible that the mass fraction of cooled gas is large in the cores of these galaxies, the total amount of such gas throughout the galaxy is much less than that which would accumulate at the currently observed cooling rates in the hot gas, and especially from stellar mass loss, if integrated over the galactic lifetime.

4.2. Evidence for star formation   In the absence of large stores for the cooling gas it seems probable that it forms stars of some kind. The evidence for star formation in clusters has been discussed by, for example, Johnstone, Fabian & Nulsen (1987), Fabian (1987), O'Connell (this volume). There is some evidence for normal star formation but at rates much lower than is required to account for all the observed cooling gas. The constraints are much weaker for the lower mass deposition rates found in elliptical galaxies. The main evidence comes from the ultra-violet excess below 2000Å which could either come from old stars on the horizontal or post-asymptotic giant branch, or young OB main-sequence stars (see the review by Bertola, this volume). Infra-red observations show $10\mu m$ emission over and above that from the late-type population, in the nuclei of at least one third of elliptical galaxies (Impey, Wynn-Williams & Becklin 1986). Unfortunately the constraints on the young stellar population are small, though they conclude that the excess emission is not from star-forming regions. Nulsen & Carter (1987) have looked at the optical distortions in M 86 which, as mentioned above, has a plume of X-ray emission extending to the NW. They find an excess in the starlight that follows the X-ray surface brightness distribution and attribute this to star formation from the hot gas. If this is the case then this provides an excellent opportunity for study.

5. CONCLUSIONS

The interstellar medium in elliptical galaxies is complex. It is clear that there is a wide range of densities present causing gas to be deposited from the flow at all radii. In addition, mass lost from stars may be confined in cold blobs by the hot gas thus further increasing the range of densities and metallicities. The inter-relation between the interstellar medium and supernovae, stellar winds/planetary nebulae, radio jets/haloes... is worthy of much more study. Some star formation is certainly taking place although it is not known what form the end products take. We do know, however, that large galaxies contain a substantial fraction of dark matter and, if we believe this matter to be baryonic, then cooling flows would seem to be ideal places in which to form it.

Spiral bulges may have low rates of cooling flow activity; consider, for example, the 'galactic fountain' model of Bregman (1980). In general, however, such activity is much reduced compared to that in bright ellipticals. The main difference between the interstellar medium in early- and late-type galaxies is the pressure. This acts as a confining medium, enhances the cooling of low-temperature gas and may be responsible for the difference in star-formation.

REFERENCES

Bachall, J. N. & Bachall, N. A., 1975. *Astrophys. J. Lett.*, **199**, L89.
Bechtold, J. *et al.* 1983. *Astrophys. J.*, **265**, 26.
Bertschinger, E. & Meiksin, A., 1986. *Astrophys. J. Lett.*, **306**, L1.
Biermann, P. & Kronberg, P., 1983. *Astrophys. J. Lett.*, **268**, L69.
Bregman, J. N., 1980. *Astrophys. J.*, **236**, 577.
Caldwell N., 1984. *Publ. astr. Soc. Pacif.*, **96**, 287.
Canizares, C. R., Fabbiano, G. & Trinchieri, G., 1987. *Astrophys. J.*, **312**, 503. CFT
Cowie, L. L. & Binney, J., 1977. *Astrophys. J.*, **215**, 723.
Cowie, L. L., Fabian, A. C. & Nulsen, P. E. J., 1980. *Mon. Not. R. astr. Soc.*, **191**, 399.
Fabbiano, G., Klein, U., Trinchieri, G. & Wielebinski, R., 1987. *Astrophys. J.*, **312**, 111.
Faber, S. M. & Gallagher, J. S., 1976. *Astrophys. J.*, **204**, 365.
Fabian, A. C., 1987. In *Proc. of 1987 Rencontre de Moriond Meeting 'Starbursts and Star-forming Galaxies'*, ed. Montmerle, T., Editions Frontieres, in press.
Fabian, A. C., Crawford, C. S., Johnstone, R. M. & Thomas, P. A., 1987.

244

Mon. Not. R. astr. Soc., in press.
Fabian, A. C., Hu, E. M., Cowie, L. L. & Grindlay, J., 1981. *Astrophys. J.*, **248**, 47.
Fabian, A. C., Thomas, P. A., Fall, S. M. & White, R. E., 1986.
    Mon. Not. R. astr. Soc., **221**, 1049.
Fabricant, D., Lecar, M. & Gorenstein, P., 1980. *Astrophys. J.*, **241**, 552.
Forman, W., Jones, C. & Tucker, W., 1985. *Astrophys. J.*, **293**, 102. **FJT**
Forman, W., Schwarz, J., Jones, C., Liller, W. & Fabian, A. C., 1979.
    Astrophys. J. Lett., **234**, L27.
Giacconi, R. *et al.*, 1979. *Astrophys. J.*, **230**, 540.
Gorenstein, P., Fabricant, D., Topka, K., Tucker, W. & Harnden, F. R., Jr., 1977.
    Astrophys. J. Lett., **216**, L95.
Hanes, D. A. & Harris, W. E., 1986. *Astrophys. J.*, **309**, 564.
Hernquist, L. & Quinn, P. J., 1987. *Astrophys. J.*, **312**, 1.
Impey, C. D., Wynn-Williams, C. G. & Becklin, E. E., 1986. *Astrophys. J.*, **309**, 572.
Johnson, H. E. & Axford, W. I., 1971. *Astrophys. J.*, **165**, 381.
Johnstone, R. M., Fabian, A. C. & Nulsen, P. E. J., 1987.
    Mon. Not. R. astr. Soc., **224**, 75.
Jura, M., 1986. *Astrophys. J.*, **306**, 483.
Knapp, G. R., Turner, E. L. & Cunniffe P. E., 1985. *Astr. J.*, **90**, 454.
Kunze, R., Loose, H.-H. & Yorke, H. W., 1987. *Astr. Astrophys.*, **182**, 1.
Loewenstein, M. & Mathews, W. G., 1987. *Astrophys. J.*, **319**, 614.
Long, K. S. & Van Speybroeck, L., 1983. In *Accretion Driven Stellar X-ray Sources*,
    eds. Lewin, W. H. G. & van den Heuvel, E. P. J., CUP, Cambridge, England.
MacDonald, J. & Bailey, M. E., 1981. *Mon. Not. R. astr. Soc.*, **197**, 995.
Malagoli, A., Rosner, R. & Bodo, G., 1987. *Astrophys. J.*, **319**, 632.
Malina, R., Compton, M. & Bowyer, S., 1976. *Astrophys. J.*, **209**, 678.
Mathews, W. G. & Baker, J. C., 1971. *Astrophys. J.*, **170**, 241.
Mauche, C. W. & Gorenstein, P., 1986. *Astrophys. J.*, **302**, 371.
Nulsen, P. E. J., 1986. *Mon. Not. R. astr. Soc.*, **221**, 377.
Nulsen, P. E. J. & Carter, D., 1987. *Mon. Not. R. astr. Soc.*, **225**, 939.
Nulsen, P. E. J., Stewart, G. C. & Fabian, A. C., 1984.
    Mon. Not. R. astr. Soc., **208**, 185.
Phillips, M. M., Jenkins, C. R., Dopita M. A., Sadler, E. M. & Binette L., 1986.
    Astr. J., **91**, 1062.
Rephaeli, Y., 1987. *Mon. Not. R. astr. Soc.*, **225**, 851.
Sadler, E. M. & Gerhard, O. E., 1985. *Mon. Not. R. astr. Soc.*, **214**, 177.
Sparks, N. B. *et al.*, 1985. *Mon. Not. R. astr. Soc.*, **217**, 87.
Stewart, G. C., Canizares,. C. R., Fabian, A. C. & Nulsen, P. E. J., 1984. *Astrophys. J.*,
    278, 536.
Tammann, G. A., 1974. In *Supernovae and Supernova Remnants*, ed. Cosmovici, C. B.,
    Reidel, Dordrecht, Holland.
Thomas, P. A., 1986. *Mon. Not. R. astr. Soc.*, **220**, 949.
Thomas, P. A., Fabian, A. C., Arnaud, K. A., Forman, W. & Jones, C., 1986.
    Mon. Not. R. astr. Soc., **222**, 655. **TFAFJ**
Thomas, P. A., Fabian, A. C. & Nulsen, P. E. J., 1987. *Mon. Not. R. astr. Soc.*, in press.
Trinchieri, G. & Fabbiano, G., 1985. *Astrophys. J.*, **296**, 447.
Trinchieri, G., Fabbiano, G., & Canizares, C. R., 1986. *Astrophys. J.*, **310**, 637. **TFC**
Tucker, W. H. & Rosner, R., 1983. *Astrophys. J.*, **267**, 547.
White, R. E., III & Chevalier, R. A., 1983. *Astrophys. J.*, **275**, 69.
White, R. E., III & Chevalier, R. A., 1984. *Astrophys. J.*, **280**, 561.
White, R. E., III & Sarazin, C. L., 1987. *Astrophys. J.*, **318**, 629.

DETAILED SPECTROSCOPIC INVESTIGATION OF EMISSION-LINE NEBULAE ASSOCIATED
WITH COOLING FLOWS

Timothy M. Heckman and Stefi A. Baum
Astronomy Program University of Maryland
College Park, MD

Wil van Breugel and Patrick J. McCarthy
Astronomy Department, University of California
Berkeley, CA

ABSTRACT. High resolution longslit spectra have been used to study the
dynamical, physical, and chemical properties of the emission-line
nebulae associated with seven cluster cooling flows. We find that the
gas is much more centrally concentrated than implied by the radial
accretion profile inferred from the X-ray data. The kinematics show
little evidence for ordered large-scale shears or rotation, but are
consistent with infall and/or turbulence. Gas pressures on scales of
1-10 kpc are about 10 times larger than X-ray pressures on larger scales
and the radial pressure profiles vary significantly among the different
nebulae. We present several lines of evidence favoring shock-heating
over photoionization. In particular, we can conclusively rule out
photoionization by a dilute nuclear continuum source. Several models
for the heating of the nebulae are briefly explored. Two classes of
nebulae with different chemical abundances have been identified.
Finally, we stress the physical relationship between the nebulae and the
coterminous nonthermal radio plasma.

1. INTRODUCTION

The material producing strong optical emission-lines comprises only a
small fraction of the cooling intra-cluster medium (ICM). Nevertheless,
detailed investigation of this material may prove crucial to our quest
to understand cooling flows. In part, this is because (in contrast to
the case of the hotter, more dominant ICM) it is relatively easy to
obtain high quality data which probe the kinematics, physical
conditions, chemical abundances and the morphological/geometrical
configuration of this cool gas. This information, in turn, allows us to
make some important inferences concerning the cooling ICM. Moreover,
understanding the emission-line gas in cooling flows is important
because, as we will discuss below, these nebulae represent significant
channels of energy loss in the flow (particularly at small radii).

A. C. Fabian (ed.), Cooling Flows in Clusters and Galaxies, 245-250.

The following represents a brief summary of the principal results
stemming from our detailed investigation of the nebulae associated with
seven cooling flows in rich galaxy clusters. A much more comprehensive
description and discussion of these results can be found in [1].

2. DATA

We have acquired a set of longslit spectra on the seven cooling flow
nebulae listed in Table 1. Most of these spectra were obtained using
the RC spectrograph with a TI CCD on the 4 m telescope at KPNO, have a
spectral resolution of 3.8Å, and cover the spectral region of the
[OI]λ 6300 Hα, [NII]λ 6584, and [SII]$\lambda\lambda$6717,6731 lines. Some of the
spectra were obtained using the Cassegrain CCD spectrograph on the 3m
telescope at Lick Observatory, and have a spectral resolution of 9Å. In
addition to the spectra, we have also imaged these nebulae in the light
of the Hα + [NII] lines (using several different telescopes+detector
combinations), and have mapped the nonthermal radio continuum emission
using the VLA.

3. RESULTS

Our principal results concerning the morphology of the gas are that the
gas is very centrally concentrated, and generally shows structure
(clumps or filaments) on scales of 1–10 kpc. While the typical nebular
isophotal diameters are tens-of-kpc (Table 1), the radial surface
brightness profiles of the Hα emission are much steeper than the $1/r$
dependence expected if the Hα traces $\dot{M}(r)$ and if $\dot{M}(r) \propto 1/r$, as inferred
from the X-ray data [2,3].
    The kinematics of the gas proved to be especially interesting. The
velocity field usually appears chaotic, with velocities wandering about
the systemic velocity with a typical rms amplitude of 50–100 km/sec.
Lines are broad throughout the nebulae, but in all cases show a marked
radial gradient: line widths fall from 500–1000 km/sec (FWHM) in the
central regions to only 100–300 km/sec at radii of 5–15 kpc. We draw
two conclusions from these results. First, except for NGC 708=A262,
there is little evidence for ordered rotation – suggesting that the flow
is not stalled by centrifugal "spin-up" [4]. Angular momentum can
apparently be transported radially outwawrd in the flow [5]. Second,
the kinematics are consistent with radial infall at about the freefall
velocity, as expected if the emission-line nebulae occur near/inside the
sonic radius of the flow. However, a purely turbulent picture is also
possible as long as the turbulent velocities decrease radially outward.
    Our data are of sufficient quality that we have been able to use
the [SII] lines to measure the electron densities (n_e), and hence to
estimate gas pressure (P_{gas}) in the optical nebulae. At typical radii
of a few kpc, we measure n_e = 200–500 cm^{-3}, corresponding to P_{gas} =
$1-2*10E-09$ dyne/cm^2. These pressures are typically a factor of ten
higher than the pressures in the X-ray gas at somewhat larger radii
(> 10 kpc). In some cases, a radial drop in P_{gas} by factors >3 is

TABLE I

Basic Properties of Cluster Cooling Flows

Cluster	Z	$L_{H\alpha[NII]}$	D	P_{gas}	M_{H+}	FF	W_{nuc}	W_{neb}	Σ	$\dot{M}_x$	L_x	$P_{1.4}$
A262	0.016	3.8	8	2×10^{-9}	4.9×10^4	4×10^{-8}	670	310	150	12	0.16	0.04
Perseus	0.0177	470	>53	$\sim 10^{-9}$	4.1×10^6	1.3×10^{-5}	690	310	90	210	5.5	12
PKS0745-191	0.103	280	23	1.3×10^{-9}	7.9×10^6	2.6×10^{-6}	510	350	20	440	8.0	59
Virgo	0.0043	1.9	14	2.4×10^{-9}	8.3×10^3	1.9×10^{-5}	1000	220	140	9	0.3	5.8
A1795	0.063	82	61	8×10^{-10}	2.6×10^6	1.3×10^{-6}	710	460	90	160	4.6	7.4
A2052	0.035	6.4	15	$\sim 10^{-9}$	3.8×10^4	5.9×10^{-6}	580	240	60	50	0.9	13.5
A2597	0.082	~100	~25	1.6×10^{-9}	$\sim 6 \times 10^6$	3.2×10^{-6}	640	450	60	?	.1.4	2.4

NOTES TO TABLE

Col. 3 – $H\alpha+[NII]\lambda\lambda6548,6584$ luminosity of entire nebula (in units of 10^{40} erg/sec for $H_o = 75$ km/sec Mpc^{-1} and $q_o = 0$)

Col. 4 – The diameter (in kpc) of the nebula, to a limiting surface brightness of $\sim 10^{-16}$ erg/(cm^2sec $arcsec^2$) in $H\alpha+[NII]$.

Col. 5 – The pressure in the optical emission-line gas in dynes cm^{-2} ($P_{gas} \equiv 3n_ekT$, with $T \sim 10^4$ K).

Col. 6 – The mass of ionized hydrogen in the nebula in $M_\odot$.

Col. 7 – The volume filling factor of the ionized hydrogen.

Col. 8 – The FWHM of the emission-lines in the nucleus (in km/sec).

Col. 9 – the FWHM of the emission-lines averaged over the entire nebula (in km/sec).

Col. 10 – The rms velocity variation (relative to V_{sys}) along our slit position angles (in km/sec).

Col. 11 – the mass accretion rate ($M_\odot$/year) estimated from the X-ray data.

Col. 12 – The cluster X-ray luminosity (in units of 10^{44} erg/sec).

Col. 13 – The radio power (in units of 10^{24} Watts Hz^{-1}) at 1.4 GHz.

measured, but in other cases the pressures remain high and level out to
r = 5 kpc., We can also use the values of n_e to derive <u>masses</u> and
<u>filling factors</u> for the emission-line gas. Masses range from 10E04 to
10E07 $M_\odot$, while typical filling factors are 10E-06 to 10E-05.

   While our data are not detailed enough to perform a rigorous
abundance determination, we believe that we have uncovered some
suggestive evidence for two classes of cooling flow nebulae with
markedly <u>different abundances</u>. For each spatially-independent point in
each nebula we have measured the relative intensities of the [OI], $H\alpha$,
[NII], and [SII] lines and have then plotted pairs of line ratios
against one-another. We then find that the seven nebulae (plus two
others from Hu, Cowie, and Wang [6]) fall into two very distinct clumps
in these line-ratio diagrams. For each diagram, the vector joining the
two clumps is nearly orthogonal to the loci of shock models with varying
shock velocity and photoionization models with varying ionization
parameter. This suggests chemical abundance variations. Interestingly,
membership in the two line-ratio classes correlates strongly with the
X-ray and $H\alpha$ luminosities and mass accretion rate for the cluster.

4. DISCUSSION - THE HEATING MECHANISM

Before we can fully exploit data on the emission-line gas to improve our
understanding of cooling flows, it is imperative that we ascertain how
the emission-line gas is actually heated. This is because without such
knowledge, the physical relationship between the emission-line nebulae
and the cooling ICM can not be determined.

   To address this important issue we have taken two complementary
approaches. The first is a statistical one, for which we have assembled
(combining our work with material from the literature) a set of optical,
X-ray, and radio data on 35 Brightest Cluster Galaxies ("BCG's"). The
second approach is a more detailed consideration of the dataset
described above.

   The principal results of the statistical investigation are as
follows:

1. Only those BCG's with well-developed cooling flows (large $\dot{M}$'s and/or
L_x(excess) - [7]) have highly luminous emission-line nebulae. While
this indicates some link to the cooling flow, the scatter in $L(H\alpha)$ at a
given $\dot{M}$ or L_x (excess) is large (a factor of 1000 in the most extreme
cases). Thus the link between the cooling flow and the emission-line
gas is apparently complex.

2. BCG's with cooling flows have greater average ratios of emission-
line to radio luminosities than do "normal" radio galaxies of similar
radio luminosity. Thus, the cooling flow nebulae are special, and not
just the result of an AGN.

3. In no case does the required efficiency for converting the gravi-
thermal energy of the ICM into a powersource for the emission-line
nebulae exceed unity (typical values are 1-10%).

We have then considered in turn several models for heating the emission-line gas. Our data favor shock heating over photoinizaton, but not conclusively. In M87 the [SIII]λλ9069,9532 lines are weak relative to Hα and [SII]λλ6716,6731 — more consistent with shock-heating than photoionization (see [8]). Moreover, for both M87 and NGC1275 the spatial variation in the overall emission-line spectrum can be better modeled as reflecting variations in shock velocity than as variations in local ionization parameter.

More specifically, we can rule out photoionization of the gas by a centrally located continuum source (e.g. an AGN). The relatively flat density profiles we measure (see above) imply that for such a source the ionization parameter in the emission-line clouds should fall like $1/r^2$. The expected strong radial gradients in characteristic emission-line ratios (cf. [9]) are not seen in our data.

As has been frequently noted ([3,5,10]), the simplest models in which the nebular emission is simply that of gas cooling through T=10E04 K fall short of the observed emission-line luminosities by at least one to two orders-of-magnitude. We confirm this result on our large sample of 35 cooling flows. Even models in which repressurizing shocks allow each cooling H atom to recombine several times [4] appear energetically inadequate (see also [3,5]).

One model which should be explored in more detail is X-ray heating of the emission-line filaments by the hot surrounding ICM. It is known that — given the proper ionization parameter — X-ray heating can produce emission-line spectra resembling those of the cooling flow nebulae [11]. We have taken a simple energetics approach and compared the X-ray and total emission-line surface brightnesses (estimated as being about 50 x the Hα surface brightness) in four cases for which the appropriate data exist. In each case, we find that the surface brightnesses of the two gas phases are similar. This suggests that the emission-line clouds may be powered by the X-rays.

Johnstone et al. [3] have reported evidence that formation of massive stars is occuring in the central regions of cooling flows, and suggest that these stars are responsible for photoionization of the emission-line gas. While this mechanism looks feasible on energetics grounds, the emission-line spectra produced by normal OB stars are quite dissimilar to those of the cooling flow nebulae (cf. [12]). We have examined the possibility that the nebulae are shock-heated by supernovae associated with the star-formation, but such a model would require a pathological initial mass function to be consistent with all the existing data.

We have also calculated (using two independent techniques) the kinetic energy flux in the cooling ICM implied by the velocities we have measured in the emission-line gas. Broadly speaking, these "luminosities" are similar to the total luminosities we estimate for the emission-line nebulae. If the ICM is turbulent and/or infalling with velocities of several hundred km/sec, its kinetic energy may be an important powersource for the emission-line gas.

Finally, we have examined our data for evidence of a relationship between the cooling flow nebulae and nonthermal radio plasma. The radio and optical plasmas have very similar size-scales, and in most cases seem to be roughly coterminous. The minimum-energy pressures in the radio sources are comparable (within the order-of-magnitude uncertainty) to both the X-ray and optical gas pressures. We also find that the emission-line widths are significantly broader in regions where the gas overlaps (at least in projection) regions of bright radio plasma. Finally, we find a rough correlation between the radio and Hα luminosities for our larger sample of 35 cooling flows. Further evidence for an interaction between cooling flows and associated radio sources has been discussed elsewhere [6,7,13,14,15,16,17]. However, at this point the mechanism(s) by which the radio source could contribute significantly to the creation, acceleration, and heating of the emission-line clouds is unclear (cloud compression and entrainment, shock-heating, heating by relativistic particles, etc.).

We would like to thank Prof. A. Fabian and the Institute for Astronomy for their hospitality and support. We also thank the staffs of the NOAO, NRAO, and Lick Observatory. TH is supported by NSF grant AST-85-15896 and by an Alfred P. Sloan Foundation Fellowship, and SB is partially supported by the same NSF grant. WvB acknowledges the support of NSF grant AST 84-16177 and PMcC likewise AST-86-14510.

REFERENCES

[1] Heckman, T.M., Baum, S.A., van Breugel, W., and McCarthy, P.J. 1988, in preparation.

[2] Stewart, G.C., Canizares, C.R., Fabian, A.C., and Nulsen, P.E.J. 1984, Ap.J., 278, 536.

[3] Johnstone, R.M., Fabian, A.C. and Nulsen, P.E.J. 1987, M.N.R.A.S., 224, 75.

[4] Cowie, L.L., Fabian, A.C., and Nulsen, P.E.J. 1980, M.N.R.A.S., 191, 399.

[5] Fabian, A.C., Nulsen, P.E.J., and Canizares, C.R. 1984, Nature, 310, 733.

[6] Hu, E.M., Cowie, L.L., and Wang, Z. 1985, Ap.J. Suppl., 59, 447.

[7] Jones, C., and Forman, W. 1984, Ap.J., 276, 38.

[8] Díaz, A.I., Pagel, B.E.J., and Wilson, I.R.G. 1985, M.N.R.A.S., 212, 737.

[9] Ferland, G.J. and Netzer, H. 1983, Ap.J. 264, 105.

[10] Kent, S.M. and Sargent, W.L.W. 1979, Ap.J., 230, 667.

[11] Halpern, J.P. and Steiner, J.E. 1983, Ap.J. Lett., 269, L37.

[12] Heckman, T.M. 1987, in Observational Evidence of Activity in Galaxies, IAU Symp. 121, ed. E.Ye. Khachikian et al.

[13] van Breugel, W., Heckman, T.M., and Miley, G.K. 1984, Ap.J., 276 79.

[14] Baum, S.A. and Heckman, T.M. 1986 in Radio Continuum Processes in Clusters of Galaxies, ed. C.P. O'Dea and J.M. Uson (NRAO), p. 119.

[15] Baum, S.A. 1987, Ph.D. Thesis, University of Maryland.

[16] O'Dea, C.P. and Baum, S.A. 1986, in Radio Continuum Processes in Clusters of Galaxies, ed. C.P. O'Dea and J.M. Uson (NRAO), p. 141.

[17] Feigelson, E.D., Wood, P.A.D., Schreier, E.J., Harris, D.E., and Reid, M.J. 1987, Ap.J., 312, 101.

JET-INDUCED STARBURSTS:
FROM MINKOWSKI'S OBJECT TO DISTANT RADIO GALAXIES

Wil van Breugel, Radio Astronomy Laboratory, UC Berkeley, CA 94720, USA
Patrick J. McCarthy, Astronomy Department, UC Berkeley, CA 94720, USA
Jacqueline van Gorkom, NRAO, P.O. Box O, Socorro NM 87801, USA

1. INTRODUCTION

The occurence of extended optical emission-line regions in many, relatively *nearby* radio galaxies is now well documented (Danziger, Fosbury, these Proc.; Heckman *et al.*, these Proc.; Baum 1986; van Breugel 1986). In most cases the emission-line gas is found along the radio source axes and detailed studies of several sources have shown that many radio and optical properties are intimately related. Often emission-line regions are found adjacent to bright radio knots, and the radio morphologies appear distorted, indicating that jet/gas collisions occur. The low percentages polarization of the radio continuum, and large velocity gradients and line widths of the emission-line gas at these locations furthermore suggest that entrainment of surrounding gas occurs. Recent optical observations of more distant, very powerful (Cygnus A type) radio galaxies show that similar correlations may exist as in the nearby, lower power sources, but that the conditions appear much more extreme (McCarthy *et al.*, these Proc.).

An important question concerns the origin of the gas. In many powerful sources the parent galaxies are peculiar, exhibiting extended loops/filaments suggestive of tidal interactions with gas-rich companions (Heckman *et al.* 1986). In these cases the line emitting gas is probably debris lit up by the radio source. In others the parent galaxies are located at the centers of rich clusters, often surrounded by dense X-ray atmospheres or "cooling flows" (Heckman *et al.*, these Proc.), which would be the likely source of gas. Often neither of these environmental clues to the origin of the gas is readily apparent.

Observations of ionized gas in radio galaxies might yield new information on the presence/properties of mergers, cooling flows and (proto-?) galactic atmospheres. In this paper we focus on a rather dramatic aspect of such jet/gas interactions, which seem to occur as the jets seem to propagate through exceptionally large regions of dense surrounding material: jet-induced starbursts. Evidence for star formation along radio jets is known to exist in two nearby radio galaxies: Centaurus A (eg. Graham and Price, 1981) and Minkowski's Object (van Breugel *et al.* 1985a). We present further observations of Minkowski's Object, and report on the recent discovery (McCarthy *et al.*, 1987a) that star formation may commonly occur along the axes of high redshift radio galaxies.

2. MINKOWSKI'S OBJECT

Minkowski's Object (M.O.) is a small irregular galaxy or extra-galactic HII-region located at the end of a radio jet (PKS 0123-016A), which emanates from the elliptical galaxy NGC 541 in the cluster of galaxies A 194. Its optical morphology

A. C. Fabian (ed.), *Cooling Flows in Clusters and Galaxies, 251–256.*

strongly suggests that it is interacting with the radio jet (Fig. 1). Spectroscopically M.O. resembles starburst galaxies. These observations suggest therefore that the radio jet has entered an extensive region of relatively cool gas, triggering the large scale collapse of clouds and subsequent star formation.

The origin of the gas in M.O. is unclear and may be dense gas which has accreted to center of the cluster, merging debris of NGC 541 and its neighboring pair NGC 545/547, or a chance collision with a pre-existent extra-galactic HII region which was "rejuvenated". Our current optical data does not indicate the presence of an old population of stars, but further observations are warranted.

To search for *cold* gas clouds in the vicinity of M.O. we observed the A 194 field with the VLA in the 21 cm (neutral hydrogen) line mode. The observations, in the D-array, were optimized for good surface brightness sensitivity and large velocity coverage, and were centered at M.O. HI was detected only from M.O., in two channels near its systemic velocity: 3.5 mJy and 3.7 mJy, at 5662 km sec^{-1} and 5619 km sec^{-1} respectively. Although the resolution of the observations ($49''$ $\times$ $45''$) is insufficient to make a very accurate statement, the centroids of the HI emission were located slightly downstream from the bright optical ridge of M.O. and its fainter counter part at the opposite side of the jet (as indicated in Fig. 1). We derive a total neutral hydrogen mass M_{HI} of 3.9×10^8 M_o, which is fairly typical of small irregular galaxies. If we assume that the HI arises from a volume suggested by the optical dimensions of M.O. itself, than we derive a HI column density of 4.0×10^{21} atoms. HI studies of irregular galaxies by Skillman (1987) show that the threshold for massive star formation in HII regions, on scales of ~ 0.5 kpc, is a few times 10^{21} HI atoms. Our HI observations thus provide further evidence for an ongoing starburst in M.O.

The radio and optical morphological evidence strongly suggests an interaction between the radio jet of NGC 541 and M.O. Multiwavelength radio polarization observations, not shown here, support this since they show a rapid increase of Faraday rotation and depolarization in the jet nearby M.O. The optical observations and HI data show that the kinematics is quiescent and that the total mass and column density is typical of "normal" small irregular galaxies. How can this be reconciled with a violent birth scenario of M.O.?

We believe that the radio jet has merely acted as a *trigger* for the starburst. After the onset of star formation the pressure in the HII regions of M.O. has increased rapidly and is now observed to exceed that of the radio jet by an order of magnitude. Evidently the rampressure in the central part of the jet has been large enough to clear a path through M.O. (Fig. 1), but in the outer regions its rampressure may have been insufficient to further penetrate the interstellar medium of M.O. (The interaction has been sufficiently violent though that the jet has become decollimated downstream from M.O.). Both the location of the observed HI and the filametary structure of M.O. downstream from the bright HII ridge indicate that star formation may be continuing in this shielded area.

3. DISTANT RADIO GALAXIES

Considerable progress has recently been made in observations of distant (z $\gtrsim 0.5$), powerful 3CR radio galaxies (Spinrad and Djorgovski 1984a, b; Spinrad *et al.* 1985). One of the most intriguing results is that a large fraction of these distant galaxies have highly elongated or multi-modal shapes (Spinrad and Djorgovski 1984c, Lilly and Longair 1984), with the fraction of elongated galaxies increasing strongly with redshift: 30% for z in the range 0.5 - 1.0, and 50% for z $\gtrsim$ 1.0.

Together with further ongoing observations at several optical observatories and the VLA, these new results have enabled us to explore various radio/optical (emission-line and continuum) correlations as function of radio power, morphology, parent galaxy/environment and redshift. One of the first results of this work is the discovery that the optical *continuum* emission is aligned along the radio source axes in distant radio galaxies (McCarthy *et al.* 1987a; Fig. 2, 3). Using a different radio sample, selected for high redshift parent galaxies using very steep radio spectra, a similar alignment has been found by Chambers *et al.* (1987), reinforcing the significance of this correlation.

Photometry of these distant galaxies shows evidence for large scale, rapid star formation (Lilly and Longair 1984; Spinrad and Djorgovski 1987; Djorgovski 1988). Because of their large redshifts and blue colors the broad band images are dominated by young, massive stars and their elongated structures must be large star-forming complexes along the radio source axes. The simplest, straight forward interpretation of the radio/optical alignments is then that the enhanced star formation has been triggered by outflow associated with the radio sources. Unlike PKS 0123-016A, the distant radio galaxies resemble the classic double source Cygnus A, both in power and morphology. To induce star formation *low* Mach number shocks are probably needed, which may exist near the hotspot (bowshock) and lobes/cocoons (backflow), but not in the actual jets themselves (De Young 1981; Norman *et al.*, 1982). Other possible outflow modes accompanying the radio source (jet) activity, such as nuclear star formation winds surrounding the jets, might act as a trigger as well. The larger fraction of elongated galaxies at high redshifts could mean that the interstellar medium in these (young) distant galaxies is - perhaps not too suprisingly - more filled with potential star-forming clouds when compared to nearby, more mature galaxies.

Perhaps the most dramatic example of a starburst induced by jets in a distant radio galaxy is 3C 326.1 at a redshift of z = 1.825 (McCarthy *et al.* 1987b). In this case the galaxy exhibits surprisingly little stellar continuum, but has extended Ly-α line emission associated with its radio lobes and its emission-line spectrum is consistent with that of large scale star formation.

4. CONCLUSION

As in most other cases of radio galaxies with associated optical emission, also in the distant galaxies the exact origin of the clouds is unclear. One may speculate that while galaxies form, their central regions form first and active nuclei may develop well before the galaxy has finished its formation. Observational evidence for this seems to be found in the above mentioned case of 3C 326.1 , and it has been suggested that certain properties of distant quasars may be attributed to this as well (Miley 1987). Theoretical models of galaxy evolution seem to support this (Silk and Szalay 1987; Silk 1988). Radio jets and other outflowing material from these young Active Galactic Nuclei would then propagate through the cloud-filled (semi-) proto-galaxies. The bowshocks and backflow from the end of the jets would entrain and compress these clouds resulting in the observed star formation.

We would like to thank Prof. A. Fabian and the Institute of Astronomy for their hospitality during the workshop. W.v.B. acknowledges support from NSF grant AST 84-16177 and P. McC. acknowledges support from NSF grant AST85-13416.

REFERENCES

Baum, S. 1986, *B. A. A. S.* **18** 1004.

Chambers, K. C, Miley, G. K, and van Breugel, W. J. M. 1987,*Nature* in press.

De Young, D. S. 1981, *Nature* **293** 43.

Djorgovski, S. 1988, in *Starbursts and Galaxy Evolution*, Th. Montmerle (ed.), Paris: Editions Frontieres, in press.

Graham, J. A., and Price, R. M. 1981, Astrophys. J. **247**, 813.

Heckman, T. M., Smith E. P. van Breugel, W. J. M., Balick, B., Miley, G. K., Bothun, G. D., Illingworth, G. D., and Baum, S. A. 1986 *Ap. J.* **311**, 526.

Lilly, S., and Longair, M. 1984, *M. N. R. A. S.*,**211**, 833.

McCarthy, P., van Breugel, W., Spinrad, H., and Djorgovski, S. 1987*a*, *Ap. J. Lett.* in press.

McCarthy, P., Spinrad, H., Djorgovski, S., Strauss, M. A., van Breugel, W. J. M., and Liebert, J. 1987*b*, *Ap. J.*, **319**, L39.

Miley G. K. 1987, *Observational Cosmology*, proceedings of IAU Symposium 124, ed. G. Burbidge. Dordrecht: Reidel, in press.

Norman, M. L., Smarr, L., Winkler, K.,-H., and Smith, M. D. 1982, *A. and A.* **113**,285.

Silk, J., and Szalay, A. 1987 *Ap. J.*, in press.

Silk, J. 1988 in *Comets to Cosmology*, 3rd International IRAS conference, ed. M. Rowan-Robinson.

Skillman, E. D. 1987, *Star Formation in Galaxies*, NASA Conference Publication 2466, pg. 263, ed. C. J. Lonsdale Persson

Spinrad, H., and Djorgovski, S. 1984a, *Ap. J. Lett.* **280**, L9.

Spinrad, H., and Djorgovski, S. 1984b, *Ap. J. Lett.* **285**, L49.

Spinrad, H., and Djorgovski, S. 1984c, *P. A. S. P.* **96**, 795.

Spinrad, H., and Djorgovski, S. 1987, *Observational Cosmology*, proceedings of IAU Symposium 124, ed. G. Burbidge. Dordrecht: Reidel, in Press.

van Breugel, W. J. M., Filippenko, A. V., Heckman, T. M., and Miley, G. K. 1985*a*, Astrophys. J. **293**, 83.

van Breugel, W. J. M., Heckman, T. M., Miley, G. K., and Filippenko, A. V. 1986, *Ap. J.* **311** 58.

van Breugel, W. J. M. 1986, in *Jets from Stars and Galaxies*, eds. R. N. Henriksen and T. W. Jones; *Can. J. Phys.* **64**, 392.

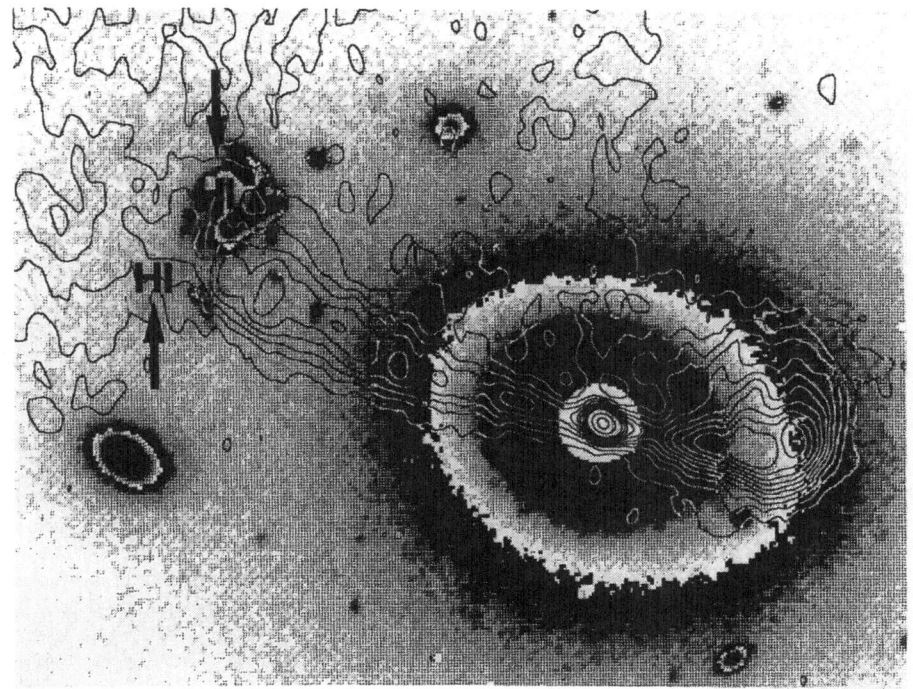

Figure 1: Overlay of a 21-cm radio continuum map (approximately 3 arcsec resolution) on an optical narrow-band image of Minkowski's Object that includes the $H\alpha$ emission-line (from van Breugel *et al.*, 1985). Neutral hydrogen (HI) observations resulted in the two detections as indicated by arrows.

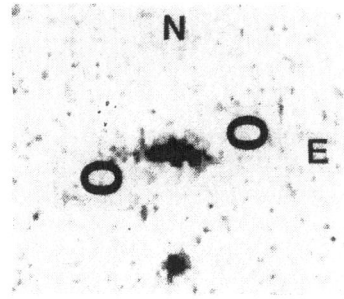

Figure 2: A broad band image of 3C 324 ($z = 1.206$), showing the elongated structure of the galaxy and the positions of the radio hotspots (from McCarthy *et al.*, 1987b).

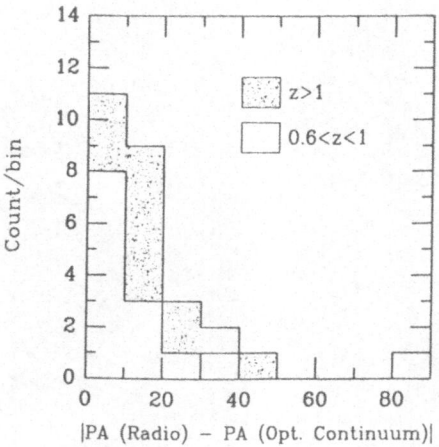

Figure 3: Histogram of the difference between the radio and optical position angles for distant 3CR radio galaxies (from McCarthy *et al.*, 1987*b*).

RADIO JETS AS A PROBE OF THE COOLING FLOW ENVIRONMENT

D M Sumi, M L Norman, and L L Smarr
National Center for Supercomputing Applications
University of Illinois at Urbana-Champaign
152 Computing Applications Building
605 East Springfield Avenue
Champaign, IL U S A 61820

ABSTRACT   We discuss the use of radio jets as a probe of the structure of cooling flows on length scales inaccessible to the currently available X-ray data   The radio structure of one such cooling flow galaxy, IC 1101, suggests a shock disruption of a supersonic beam   This type of radio structure is also found associated with other cooling flow galaxies   The shock disruption is discussed in terms of an interaction of the beam with the cooling flow environment   The current sophistication of these disruption mechanisms and of the radio observations does not compel one to favor one mechanism over another   Each mechanism, however, may be useful in studying the central regions of cooling flows

1   INTRODUCTION

Radio jets associated with the central galaxy of a cooling flow can provide an interesting probe of the structure of this environment   The gas beams associated with radio jets react to changes in the pressure and density of the external medium (see Norman, 1986)   This reaction of the beam is recorded in the observed structure of the jet   In principle, one could then use the structure of the jet to study the structure of the cooling flow

A benefit of studying cooling flows through the radio structure of a jet instead of the X-ray structure of the flow is the higher resolution that can be achieved by radio aperature synthesis arrays   For example, HRI data from the Einstein Observatory has a FWHM resolution of approximately 3"   In contrast, the radio source in M87 (3C 274) has been resolved by the VLA at 15 0 GHz in the A config with a FWHM of approximately 0 15" (Biretta et al , 1983)   Consequently, radio sources would allow the study of changes in the cooling flow structure on scales up to twenty times smaller than the X-ray observations

257

A. C. Fabian (ed.), Cooling Flows in Clusters and Galaxies, 257–261.
© 1988 by Kluwer Academic Publishers.

2 RADIO OBSERVATIONS OF COOLING FLOW GALAXIES

2 1 IC 1101

We have conducted a high resolution study of the radio structure of one such cooling flow galaxy, IC 1101 This galaxy is the dominant central cluster member in Abell 2029 and is believed to have a mass inflow rate onto the galaxy of approximately 100 solar masses per year (Johnstone et al, 1987 for a Hubble constant equal to 100)

Figure 1 shows a 0 6" resolution 4 9 GHz map from a combination of A and B config data taken at the VLA The most interesting feature of this map is the symmetric knots on either side of the nucleus The knots appear to delineate two very different parts of each jet Inward from the knots, the map suggests a weakly radiating flow, with a beam width less than 0 6" and little evidence of bending The lack of bending is suggested by the colinearity of the knots Outward from the knots, the map suggests a strongly radiating flow, with a beam width spread over 2" and observable bends In the southeastern jet, there is evidence of an embedded oscillation superimposed on the larger bend

Figure 2 shows a 1 3" resolution map at 1 5 GHz from A config data taken at the VLA This map is more sensitive to broad low surface brightness structures and shows further extensions of the jets seen in figure 1 There is evidence of strong twists and bends in both jets There are no outer hotspots or radio lobes The emission from each jet simply fades into the noise at the outer edge of the jet in the manner characteristic of FR I type radio sources (Fanaroff & Riley, 1974)

The overall jet structure seen in these maps of IC 1101 suggests a disruption of a supersonic beam into a turbulent subsonic beam by a single shock event (Sumi et al, 1987)

2 2 Other Cooling Flow Galaxies

High resolution observations of three other radio sources associated with cooling flow galaxies, 3C 274 (Biretta et al, 1983), 4C 26 42 (van Breugel et al, 1984), and 3C 84 (Pedlar et al, 1983), all show structure similar to IC 1101 The symmetric knots found in IC 1101 are also found in 4C 26 42 3C 84 and 3C 274, however, both lack a counter knot This may be the result of relativistic beaming or inherent onesideness and not the result of a lack of symmetry All the knots are found close to their radio nucleus In IC 1101, the knots have a projected galactocentric radius of 1 4 kpc In 4C 26 42, 3C 84, and 3C 274, the galactocentric radius of the knots are 0 9, 0 8, and 1 4 kpc, respectively

2 3 Low Resolution Surveys

As a class, radio sources associated with cooling flow galaxies tend to be more compact than other radio sources (Owen et al, 1984) In low resolution surveys, their structure generally consists of a bright unresolved core source with diffuse FR I type jets and/or a diffuse halo (O'Dea & Baum, 1987) This same morphology is observed in IC 1101, 4C 26 42, 3C 84, and 3C 274 at a resolution comparable to these surveys Therefore, it is at least not inconsistent for the diffuse structures in these surveys to be the result of a shock disruption of their beams inside the unresolved radio core Clearly, high resolution observations of more cooling flow galaxies are needed to understand the origin of the diffuse structures in these sources

259

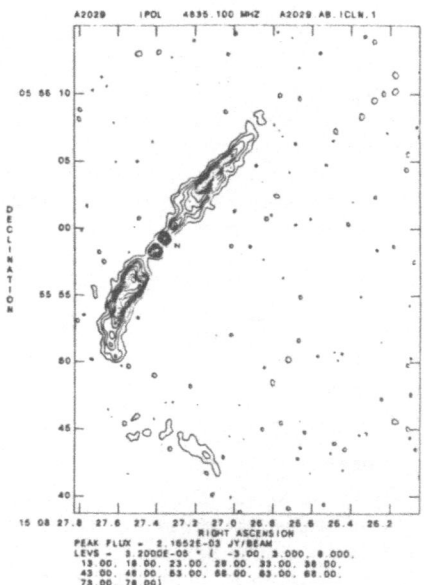

Figure 1. CLEANed untapered 4.9 GHz intensity contour maps of IC 1101 from concatenated A and B configuration data taken at the VLA. The FWHM resolution of the map is 0.58″ x 0.56″ and the RMS noise level is 0.000032 Jy per beam. The nucleus of the galaxy is designated by N.

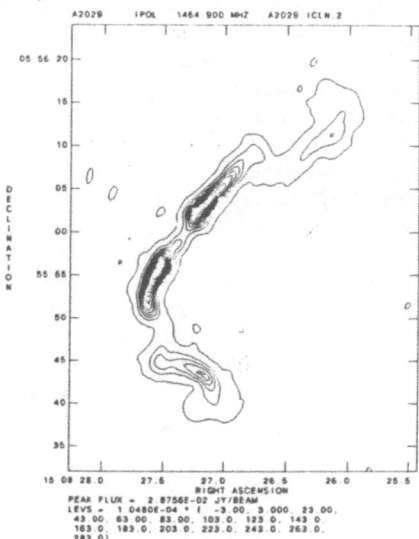

Figure 2. CLEANed untapered 1.5 GHz intensity contour map of IC 1101 from A configuration data taken at the VLA. The FWHM resolution of the map is 1.34″ x 1.27″ and the RMS noise level is 0.00011 Jy per beam.

2 4 Exceptions

Two notable exceptions are 3C 405 (Arnaud et al , 1984) and 3C 295 (Henry & Henriksen, 1986) which have FR II type radio structure (Perley et al , 1984 and Wilkinson,
1982) These sources, however, are much more powerful than the sources previously
mentioned, approximately one thousand times the 1 4 GHz power of IC 1101 High
powered radio sources may have a disruptive effect on the environment This disruption may make them less sensitive to their propagating environment (Norman, 1987)

3 DISCUSSION

The present trend cited above suggests that some factor in the cooling flow environment is responsible for the shock disruption observed in IC 1101 and others Several
possible mechanisms may cause disruptive shocks in beams propagating through a
cooling flow These mechanisms are collisions with dense gas clouds (Blandford &
Königl, 1979), growth of ordinary mode Kelvin-Helmholtz instabilities (Sumi & Smarr,
1986), and pressure walls in the pressure distribution of transonic cooling flows
(Burns, 1986, Burns et al , 1987, and Soker & Sarazin, 1987) In order to explain the
apparent symmetric nature of the knots in IC 1101, and 4C 26 42, the cloud collision
and Kelvin-Helmholtz shock mechanisms require an additional symmetrizing mechanism (see Sumi et al , 1987) The pressure wall mechanism is naturally symmetric in
a spherical cooling flow
     Note that each mechanism would give information about structures or regions in
the cooling flow that cannot be resolved by the available X-ray data A collisional
mechanism gives information on inhomogeneities in the flow Kelvin-Helmholtz instabilities give information on the density and pressure distribution at the center of the
flow A pressure wall would give information about very sharp changes in the pressure distribution at the center of the flow and could indicate the existence of a sonic
point
     At the present time it is not possible to rule out any of these mechanisms
However, the ubiquity of symmetric knots in a large sample of cooling flow galaxies
would help to rule out cloud collisions as the primary shock mechanism Symmetric
collisions in all the sources would be improbable We are in the process of observing
ten cooling flow galaxies with the VLA at a resolution of approximately 0 4" These
observations will allow us to resolve the core structures of the sample at approximately one kpc or better In addition, an investigation of the propagation of beams
in cooling flows using two dimensional beam simulations (Norman et al , 1985) is
currently in progress It is hoped that these simulations will determine the circumstances in which a pressure wall or Kelvin-Helmholtz instabilities can cause beam
disruptions similar to those observed

4 ACKNOWLEDGEMENTS

DS would like to thank Roger Blandford and the Theoretical Astrophysics group at the
California Institute of Technology for their hospitality while this study was being conducted

5 REFERENCES

Arnaud, K A , Fabian, A C , Eales, S A , Jones, C , and Forman, W   (1984), *Monthy Notices R Astr Soc* , **211** , 981

Biretta, J A , Owen, F N , and Hardee, P E (1983), *Astrophys J* , **274** , L27

Blandford, R D , and Königl, A (1979), *Astrophys Lett* , **20** , 15

van Breugel, W , Heckman, T , and Miley, G (1984), *Astrophys J* , **276** , 79

Burns, J O (1986), *Can J Phys* , **64** , 373

Burns, J O , Norman, M L , and Clarke, D A (1987), in *Radio Continuum Processes in Clusters of Galaxies*, Proc NRAO Workshop No 16 (NRAO Green Bank, WV), eds C P O O'Dea and J M Upson, 175

Fanaroff, B , and Riley, J M (1974), *Monthy Notices R Astr Soc* , **167** , 31

Henry, J P , and Henriksen, M J (1986), *Astrophys J* , **301** , 689

Johnstone, R M , Fabian, A C , and Nulsen, P E J (1987), *Monthy Notices R Astr Soc* , **224** , 75

Norman, M L (1986), in *Radiation Hydrodynamics in Stars and Compact Objects*, eds D Mihalas and K-H A Winkler (Springer-Verlag Berlin), 425

Norman, M L (1987), *Bull Amer Astr Soc* , **19** , 651

Norman M L , Smarr, L L , and Winkler, K-H (1985), in *Numerical Astrophysics*, eds J Centrella, J LeBlanc, and R Bowers (Jones and Bartlett Portola Valley, CA)

O'Dea, C P O , and Baum, S A (1987), in *Radio Continuum Processes in Clusters of Galaxies*, Proc NRAO Workshop No 16 (NRAO Green Bank, WV), eds C P O O'Dea and J M Upson, 141

Owen, F N , Burns, J O , and White, R A (1984), in *Clusters and Groups of Galaxies*, eds F Mardirossian, G Givricin, and M Mezzetti (Reidel Dordrecht), 295

Pedlar, A , Booler, R V , and Davies, R D (1983), *Monthy Notices R Astr Soc* , **203** , 667

Perley, R A , Dreher, J W , and Cowan, J J (1984), *Astrophys J* , **285** , L35

Soker, N , and Sarazin, C L (1987), Preprint

Sumi, D M , Norman, M L , Smarr, L L , and Owen, F N (1987), In Preparation

Sumi, D M , and Smarr, L L (1986), in *Physics of Energy Transport in Extragalactic Radio Sources*, Proc NRAO Workshop No 9 (NRAO Green Bank, WV), eds A H Bridle and J A Eilek, 168

Wilkinson, P N (1982), in *Extragalactic Radio Sources* , Proc IAU symp No 97 (D Reidel Dordrecht), eds D S Heeschen and C M Wade, 149

OPTICAL EMISSION LINES IN EARLY–TYPE GALAXIES

Elaine M. Sadler
Kitt Peak National Observatory
P.O. Box 26732
Tucson, AZ 85726
U.S.A.

ABSTRACT. Small amounts of ionized gas are common in nearby E and S0 galaxies, but the origin of this gas, and its relationship to the X–ray coronae, remain unclear. Both the presence of dust within the emission region, and the commonly–observed kinematical decoupling of gas and stars appear to argue against a cooling–flow origin for the gas. Thus if cooling flows operate in individual elliptical galaxies other than those at the centres of clusters, they appear to do so without observable effects at optical wavelengths.

1. INTRODUCTION

Elliptical galaxies have historically been thought of as gas–poor systems, but recent X–ray observations show that many of them contain substantial amounts of gas, and that many early–type galaxies may be as gas–rich as spirals.

Schweizer (1987), in a recent review, describes how our picture of the gas content of elliptical galaxies has changed over the past fifty years with the introduction of more sensitive detectors and the opening–up of new wavelength regions. In general, it now appears that early–type galaxies have a three–phase interstellar medium, comprising 'hot' ($10^6 - 10^7$ K) X–ray gas, 'warm' (10^4 K) ionized gas visible at optical wavelengths, and 'cold' HI or molecular gas observed directly at 21 cm or (more commonly) indirectly via its associated dust. It has long been realised (Sandage 1957; Faber and Gallagher 1976) that the amount of warm and cold gas in these galaxies is far less than that shed from stars within the lifetime of the galaxy, and hence that some mechanism must operate to remove most of the gas.

The discovery by Biermann and Kronberg (1983) and Forman, Jones and Tucker (1985; hereafter FJT) that hot X–ray coronae were a common, and perhaps ubiquitous, feature of early–type galaxies provided new insight into the problem. These authors found that the X–ray coronae contained some 10^9 to 10^{10} $M_\odot$ of gas; roughly the amount expected to accumulate from stellar mass–loss over the galaxy's lifetime. The kinetic energy of the stars, plus a modest amount of supernova heating, are more than sufficient to heat the gas to the observed temperatures (Canizares, Fabbiano and Trinchieri 1987). However, Nulsen, Stewart and Fabian (1984) showed that the cooling time in the inner regions of these galaxies is much shorter than the Hubble time, so that gas should cool and flow to the centre. Pre-

A. C. Fabian (ed.), Cooling Flows in Clusters and Galaxies, 263–272.
© 1988 by Kluwer Academic Publishers.

dicted mass–flow rates for a dozen or so nearby ellipticals observed with the Einstein satellite have been calculated by Thomas *et al.* (1986).

In discussing the general properties of optical emission lines in early–type galaxies, I will draw mainly on results from the recent survey by Phillips *et al.* (1986). Here, I will discuss what is presently known about the $(10^4$ K) ionized gas content of nearby, 'normal' E and S0 galaxies, and how this might be understood in the light of X–ray data which suggest that cooling flows can occur in individual galaxies. In particular, I will attempt to answer two questions:

(1) Is the gas which we see in the centres of nearby ellipticals and S0s a manifestation of cooling flows in these galaxies?

(2) What processes are likely to be important in determining how much gas is observed optically?

2. THE IONIZED–GAS CONTENT OF EARLY–TYPE GALAXIES

2.1 How common are emission lines?

Phillips *et al.* (1986) surveyed a sample of 203 normal E and S0 galaxies at a resolution of 3Å, and detected Hα/[NII] emission in about 55% of these. The equivalent width of the weakest lines detected was less than 1Å, compared to 5–6Å for earlier photographic surveys such as that by Humason *et al.* (1956). Thus the increased detection rate above the value of 15% quoted by Gisler (1978) is due to the improved sensitivity of modern detectors. Caldwell (1984a) has detected [OII] λ3727 emission in about 40% of a sample of E and S0 galaxies, while Keel (1983) found Hα/[NII] emission in 100% of a sample of bright spiral bulges.

Most of the galaxies observed by Phillips *et al.* show weak Hα absorption lines, so a continuum template made from apparently emission–free galaxy spectra was subtracted before measuring the ratio [NII]/Hα. For this reason, and also because [NII] is usually the strongest line, Phillips *et al.* used the presence of [NII] rather than Hα as their detection criterion.

As discussed by Phillips *et al.* and by Sadler (1987), the emission–line spectra seen in low–luminosity ellipticals ($M_B> -18.0$ with $H_o = 100$ km/s/Mpc) differ substantially from those seen in brighter ones. In low–luminosity galaxies, the emission lines are strong and narrow with Hα $>>$ [NII], and the spectra are similar to those of Galactic H II regions. The blue colours of these galaxies also suggest that they are actively forming stars with a normal initial mass function. Recent work by Sadler and Gallagher (1987; in preparation) suggests that current or recent star formation of this kind occurs in at least 20–30% of small elliptical galaxies. A similar effect is seen in spiral bulges (Keel 1983), where the more luminous Sa and Sb bulges show emission–line spectra like those of bright ellipticals, while the bulges of later–type spirals commonly have H II region–like spectra.

However, Phillips *et al.* observed H II region–like spectra only in galaxies fainter than about $M_B -18.0$, and since no galaxies as faint as this have been observed to have X–ray coronae (Fabbiano 1986), they are not relevant to the present discussion. It is interesting, though, that the star formation history of early–type galaxies appears to show a change in behaviour at about the same luminosity where other physical changes occur (for example, a change from rotational support to

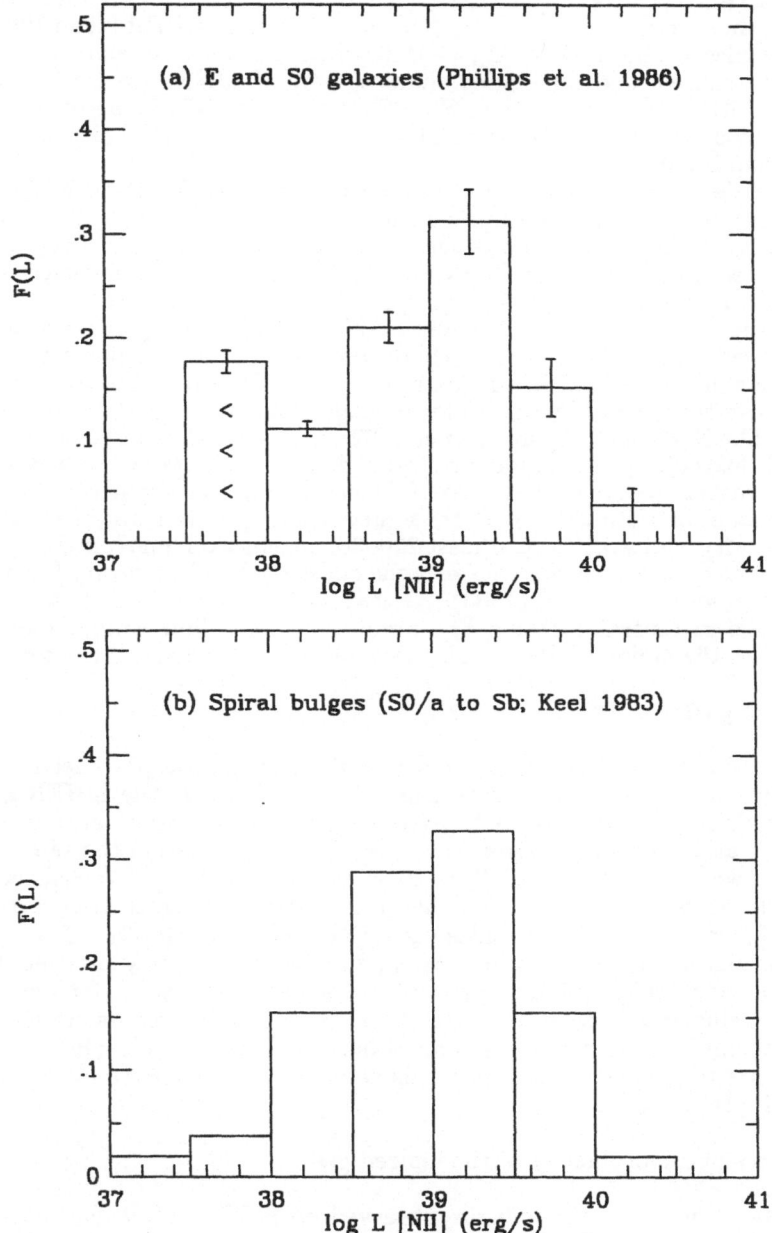

Figure 1. The fractional luminosity function in [NII] for E and S0 galaxies in the Phillips *et al.* sample, derived using a Kaplan–Meier estimator (Feigelson and Nelson 1985). Points below 10^{38} erg/s are upper limits. (b) A similar luminosity function for the bulges of early–type (S0a–Sb) spirals from the survey by Keel (1983), which had a detection rate of 100%.

velocity anisotropy (Davies *et al.* 1983), the onset of central non–thermal radio sources, and the presence of hot X–ray haloes (Trinchieri and Fabbiano 1985)).

All galaxies brighter than $M_B=-19.0$ which showed emission lines in the Phillips *et al.* sample had line ratios characteristic of LINER spectra (Heckman 1980, 1987), with [NII] stronger than Hα. The ratio [NII]/Hα is larger in brighter galaxies, and where the emission region is spatially resolved, [NII]/Hα is larger in the centre than further out.

We can derive an emission–line luminosity function for the Phillips *et al.* sample, using both detections and upper limits for the emission–line luminosity in $\lambda6584$ [NII], L_{em}. Figure 1(a) shows the luminosity function in [NII] for elliptical and S0 galaxies. Although this includes only the luminosity within the spectrograph slit, most of the emission regions were unresolved, so this is likely to be a fair reflection of the total gas content. Not all the galaxies for which spectra are available have a measured value of L_{em}, because conditions were not always photometric, but about 160 galaxies were used in this determination. Note that the distribution is single–peaked rather than bimodal in form, suggesting that well–known emission–line galaxies like NGC 1052 (Fosbury *et al.* 1978) and NGC 4278 (Demoulin–Ulrich, Butcher and Boksenberg 1984) represent the high–luminosity tail of a Gaussian–like distribution, rather than a special class of "emission–line" ellipticals.

The galaxies in Figure 1(a) span three magnitudes (i.e. a factor of about 15) in optical luminosity, and a factor of at least 200–300 in emission–line luminosity. Note the similarity to Figure 1(b), which shows the emission–line luminosity function for spiral bulges observed by Keel (1983). Galaxies of all types from E to Sb show a similar range in central emission–line luminosity, and show similar line–ratios, suggesting that the emission lines in all these galaxies may arise in the same way.

2.2 How much gas?

In order to estimate the mass of ionized gas in these galaxies, we need to make some assumptions about its temperature and density. Many of the LINER galaxies observed by Phillips *et al.* have only [NII] and Hα visible, some also show [SII] $\lambda\lambda6716$, 6731; and a few of the strongest emission–line galaxies in the sample show [OI] $\lambda6300$ as well. With values of $T_e = 10^4$K and $n_e = 10^3$ cm^{-3} (consistent with the observed line ratios), the emission–line luminosities in Figure 1(a) correspond to masses of 10^3 to 10^5 M$_\odot$ of ionized gas (Phillips *et al.* 1986). If a range of densities are present in the emission region, the gas masses may be even smaller than this (Binette 1985), but it is probably reasonable to assume for now that a typical bright elliptical or S0 contains of order 10^4 M$_\odot$ of ionized gas in its central regions. Note that this is much less than either the X–ray or cold (HI/molecular) component (see the paper by Knapp in this volume for a discussion of the cold gas content of ellipticals and S0s).

2.3 Kinematics and morphology of the ionized gas

As stated above, many of the galaxies observed by Phillips *et al.* had unresolved central emission regions. Where the emission region was spatially resolved, it was typically 1–2 kpc in diameter, with the emission always strongly centrally peaked. The kinematics of the extended regions are consistent with the gas being in rapid rotation (with rotation velocities 100–300 km/s). The galaxies with the strongest emission lines usually showed the most extended gas distribution, though some

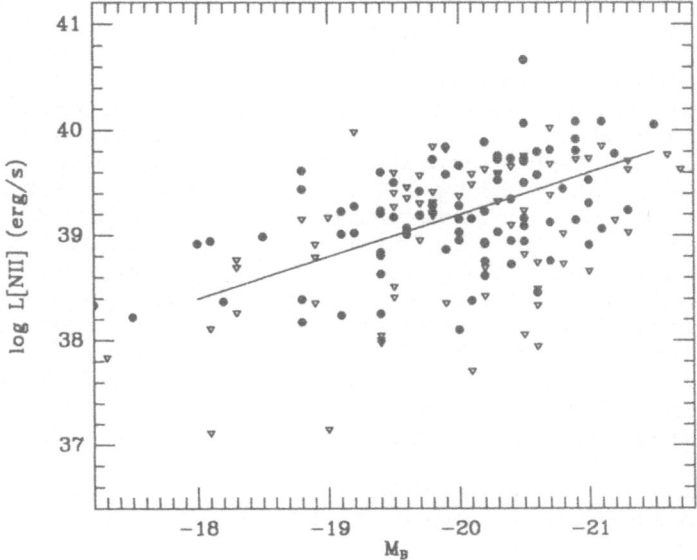

Figure 2. Relationship between optical luminosity and [NII] emission–line luminosity for E and S0 galaxies. Again, $H_0 = 100$ km/s/Mpc. Open triangles show upper limits, and filled circles denote galaxies with detected emission lines. By constructing luminosity functions, one can show that $L_{[NII]} \propto L_B^{1.0}$, in contrast to the relation $L_X \propto L_B^{\sim 1.7}$ found for the X-ray gas by Trinchieri and Fabbiano (1985). Thus the emission–line gas seen in the centre of these galaxies appears to be more closely coupled to the stellar population than to the X-ray halo.

galaxies such as NGC 5090 have strong central emission with no apparent extension.

There have recently been several detailed studies of the kinematics of gas disks in individual elliptical galaxies (Caldwell and Phillips 1981; Sharples *et al.* 1983; Bertola *et al.* 1984; Caldwell 1984b; Demoulin–Ulrich, Butcher and Boksenberg 1984 (hereafter DBB); and Caldwell, Kirshner and Richstone 1986 (hereafter CKR)).

Almost all the galaxies studied show similar behaviour. The gas is in a rotating disk and has a LINER–like spectrum. Most galaxies also have a central compact radio source. The velocity dispersion of the gas is always high in the nucleus ($\sigma_{gas} \sim 200$–300 km/s), but drops sharply with radius. In galaxies where the gas is sufficiently extended that a rotation curve can be measured, the rotation axis of the gas often differs from that of the stars (Bertola *et al.* 1984; DBB; CKR).

2.4 Gas content and other galaxy properties

By constructing emission–line luminosity functions for galaxies of different optical luminosity, it is possible to show that the emission–line luminosity (and thus presumably the *amount* of gas present) correlates with optical luminosity roughly as

$$L_{em} \propto L_B.$$

Figure 2 gives some idea of the scatter in this correlation.

The emission–line luminosity is *not* apparently connected with other galaxy properties such as axial ratio, colour, morphology (as seen above, ellipticals, S0s and spiral bulges all show similar spectra and have similar emission–line luminosities), kinematics or continuum radio power [*]. Gisler (1978) has shown that emission–line galaxies are less common in rich clusters than in the field, but most of the Phillips *et al.* galaxies are in small groups, and there is no clear sign that environment affects the emission–line properties in this sample. Variations between individual galaxy clusters appear to be as large as those between field and cluster galaxies. For example, the Fornax cluster ellipticals show a very low incidence of optical emission lines, but this is not true of galaxies in the Centaurus cluster.

In summary, it appears that essentially *all* early–type galaxies contain 10^3 to 10^5 $M_\odot$ of ionized gas in their central regions. Larger galaxies have more gas on average, but the gas content is not strongly related to any other galaxy properties.

3. X–RAYS AND IONIZED GAS IN FIELD ELLIPTICALS

It now appears that hot X–ray coronae are a common, and perhaps ubiquitous, feature of early–type galaxies (Forman, Jones and Tucker 1985; Canizares, Fabbiano and Trinchieri 1987). Nulsen, Stewart and Fabian (1984) propose that cooling flows from these coronae operate in many elliptical galaxies, including field galaxies as well as those at the centres of rich clusters. In such a picture, the X–ray halo appears to be a natural origin for the small amounts of ionized gas seen in the nuclei of most ellipticals. We therefore investigate the idea that the observed optical emission lines represent the "signature" of a cooling–flow. This investigation falls into two parts; firstly a discussion of the way in which the gas may be ionized, followed by a discussion of its likely origin.

3.1 Correlations between X–ray and optical emission?

The overlap between X–ray and optical observations of 'normal' elliptical galaxies is still very small, and the sample of galaxies for which both sets of data are available is by no means unbiased. X–ray observations of nearby ellipticals with the Einstein satellite (see e.g. FJT; Canizares, Fabbiano and Trinchieri 1987) are strongly concentrated towards northern galaxies, and in particular those in the Virgo cluster, while recent emission–line surveys of elliptical and S0 galaxies (Caldwell 1984a, Phillips *et al.* 1986) have used southern galaxy samples. Phillips *et al.* discussed the relation between X–ray luminosity and emission lines for a sample of 14 galaxies in common with the Einstein data set, and found that most of the galaxies with detected X–ray emission also had optical emission lines. However, one must be cautious about interpreting this as a direct connection between X–ray and optical gas via a cooling flow, because both X–ray luminosity and emission–line luminosity

---

[*] Although, as mentioned in the previous section, radio sources are common in galaxies which show optical emission lines, there is no correlation between the *strength* of the radio source and the amount of gas present.

are strongly correlated with galaxy size. We really need to look at the predicted mass–flow rate from the X–ray cooling flow (e.g. Thomas *et al.* 1986, Johnstone, Fabian and Nulsen 1987) and compare this with the emission–line luminosity in the same region. However, as Thomas *et al.* also point out, there is no straightforward relation between the amount of cooling gas predicted from X–ray observations and the observed emission–line luminosity. For example, NGC 1399 and NGC 1404 are bright elliptical galaxies in the centre of the Fornax cluster, and have high X–ray luminosities and short cooling times in their centres (Thomas *et al.* 1986). Yet these galaxies show no trace of optical emission lines, and the upper limits on the [NII] equivalent width are less than 0.4 Å (Phillips *et al.* 1986). In terms of optical emission–line gas, NGC 1399 and NGC 1404 are extremely gas–poor, yet IC 1459, a less luminous galaxy with a lower X–ray luminosity and longer central cooling time, shows strong optical emission lines in its nucleus.

Thus galaxies in which a cooling flow is expected to operate can show a wide range in optical emission–line luminosities. On the other hand, there are galaxies with very different X–ray properties (bright ellipticals, where the X–ray emission comes from hot gas, and spiral bulges, where it comes from discrete sources; Trinchieri and Fabbiano 1985, Fabbiano 1986) which show remarkably *similar* emission–line spectra, again suggesting that the gas seen in the optical emission lines may be more directly related to the stellar population than to the X–ray corona.

In summary, there is no clear evidence for a correlation between the amount of cooling gas predicted by a cooling flow model and the amount of line–emitting gas seen in the centre of a typical elliptical galaxy. However, it is not clear that such a correlation is even expected, since arguments from energetics (Johnstone, Fabian and Nulsen 1987) suggest that without additional energy input, the cooling gas may simply flow invisibly and undetected through the galaxy.

3.2 Ionization mechanisms for the gas

As discussed above, bright elliptical and S0 galaxies (and early–type spiral bulges) show emission–line spectra characteristic of LINERs rather than H II regions. Any ionization mechanism postulated for these galaxies must produce an emission–line spectrum consistent with that observed, and must also satisfy the energy requirements, i.e. it must be able to ionize the amount of gas which is actually observed.

Heckman (1987) has reviewed recent work on LINERs, and concludes that there are at least four ionization mechanisms which can plausibly produce the observed spectrum. He also suggests that LINERs fall into several superficially similar, but physically distinct classes. Table I summarises various ionization mechanisms which have been proposed to account for the observed line emission. These include:

a. Photoionization by a hot stellar population (i.e. horizontal–branch or post–AGB stars, but *not* OB stars).
b. Photoionization by a central continuum source (active nucleus).
c. Photoionization by X–ray photons from the halo.
d. Shock ionization in cooling flows.
e. Shock ionization from supernovae, collisions, turbulence (in nucleus).

Heckman (1987) also considers LINER emission resulting from shock ionization in galaxy collisions, and from starburst–driven winds in galaxies like M82, but points

TABLE I.   Ionization Mechanisms

Ionizing source	Observed Spectrum?	Energetics?
OB stars	No	Yes
HB/post AGB stars	Yes	No
Photoionization (central source)	Yes	Yes?
X-ray photons	Yes	No
Shocks (nucleus)	Yes?	Yes?

out that these processes are not relevant to the majority of nearby, 'normal' galaxies. Ferland and Mushotzky (1984) show that heating and ionization by cosmic rays may be important in galaxies with active nuclei.

It seems unlikely that a significant fraction of the ionizing flux in most early–type galaxies comes from OB stars formed within a cooling flow, as recently proposed by Johnstone, Fabian and Nulsen (1987), since this would give [NII]/Hα line ratios quite unlike those observed. A population of hot horizontal–branch or post–AGB stars would give the correct spectrum, but Demoulin–Ulrich, Butcher and Boksenberg (1984) show that the ionizing flux from a population consistent with the observed UV spectrum of ellipticals is unlikely to satisfy the energetics. They also show that the kinetic energy of gas falling to the centre of the galaxy is insufficient to produce the required ionization in any but the weakest emission–line galaxies, and they suggest that photionization by a central continuum source is the most plausible source.

Johnstone, Fabian and Nulsen (1987; hereafter JFN) calculate that the Hβ emission luminosity expected from gas which cools from 10^6 K to recombination is of order 10^{37} erg/s for a 1 $M_\odot$/year cooling flow. This corresponds to a luminosity of $\sim 10^{38}$ erg/s in [NII] after changing H_0 to 100 km/s/Mpc for consistency with the present observations. However, the median *observed* [NII] luminosity is $\sim 10^{39}$ erg/s (see figure 1). Thus it appears that an additional source of ionization is needed in the cooling flow picture, not only for the central cluster galaxies and radio galaxies observed by JFN, but for the majority of field ellipticals and S0s.

Photoionization by a central non–thermal source appears at present to offer the simplest explanation for the observed line emission, though shock heating may also play a significant role in the nucleus, where the large velocity dispersion observed in the gas suggests that it is highly turbulent (Gunn 1979). More detailed observations, including a much wider range of line ratios, are required to discriminate reliably between the various mechanisms which can produce a LINER spectrum; and it is also likely that several of the processes listed in table I could operate within a single galaxy.

3.3 Origin of the gas?

Either the gas originates within the galaxy, from stellar mass loss or as the result of a cooling flow, or it has been captured from outside by accretion or merger.

There are two observations which argue against an internal origin. One is that, as mentioned earlier, the gas and stars often rotate about different axes. This has been used (e.g. Schweizer 1987) to argue that the observed gas has been accreted from outside the galaxy. Schweizer also argues that many galaxies with extended gas disks in their centres show other signs of a recent interaction.

A second problem with a cooling–flow origin for the emission–line gas lies in the fact that it is usually mixed with dust. For example, NGC 4696, the prototypical "cooling–flow" galaxy at the centre of the Centaurus cluster (Fabian et $al.$ 1982), contains at least several times 10^4 $M_\odot$ of dust (Sadler and Gerhard 1985; see also the paper by Norgaard–Nielson at this meeting), and dust is frequently seen in the centres of elliptical galaxies (Sparks et $al.$ 1985; Jura 1986; Kormendy and Stauffer 1987; Knapp, this meeting). Dust cannot survive in the X–ray corona, since the destruction time by sputtering is of order 10^7 years; and gas resulting from cooling flows is expected to be dust–free (Fabian, Nulsen and Canizares 1982). Thus the ionized gas seen in the centres of nearby ellipticals does not appear to have cooled out of the X–ray halo recently. It is, however, possible that much of this gas could still have originated within the galaxy, as residual stellar mass–loss which has not been heated to X–ray temperatures.

4. CONCLUSIONS, AND SOME UNANSWERED QUESTIONS

Modest amounts (10^3 to 10^5 $M_\odot$) of ionized gas at 10^4 K are a common feature of nearby early–type galaxies. However, the origin of this gas, and the method by which it is ionized, remain unclear. Several processes can plausibly contribute to the ionized gas content, and these include stellar mass loss, cooling flows from an X–ray corona, and acquisition of gas from outside the galaxy by infall or merger. Similarly, there are several different mechanisms which could plausibly keep all or part of the gas ionized. Active nuclei are common in early–type galaxies, and are perhaps the simplest way of ionizing the gas. Shock ionization may also be important, particularly in the nucleus, where σ_{gas} is large.

X–ray observations have made an important contribution to our understanding of the gas balance in elliptical galaxies, but the overall picture is still not well understood. In particular, it remains difficult to understand why the ionized gas content of early–type galaxies is so small, and why the gas removal processes are both so efficient and so elusive. Within the cooling–flow picture, it would be useful to know the effect of (a) a merger (i.e. input of cold gas) and (b) the turn–on of an active nucleus on a pre–existing cooling flow, since both these processes occur commonly in nearby galaxies. It would also be interesting to know why there is a continuum of emission–line properties (in terms of both ionized gas content and observed line ratios) throughout the Hubble sequence of early–type galaxies from E to Sb, even though these galaxies are believed to have very different X–ray properties.

REFERENCES

Bertola, F., Bettoni, D., Rusconi, L. and Sedmak, G., 1984. $Astron.$ $J.$, **89**, 356.
Binette, L., 1985. $Astron.$ $Astrophys.$, **143**, 334.
Biermann, P. and Kronberg, P.P., 1984. $Astrophys.$ $J.$, **268**, L69.
Caldwell, N., 1984a. $Publ.$ $astr.$ $Soc.$ $Pacif.$, **96**, 287.
Caldwell, N., 1984b. $Astrophys.$ $J.$, **278**, 96.
Caldwell, N., and Phillips, M., 1981. $Astrophys.$ $J.$, **244**, 447.

Caldwell, N., Kirshner, R.P. and Richstone, D.O., 1986. *Astrophys. J.*, **305**, 136.
Canizares, C.R., Fabbiano, G. and Trinchieri, G., 1987. *Astrophys. J.*, **312**, 503.
Davies, R.L., Efstathiou, G., Fall, S.M., Illingworth, G.D. and Schechter, P.L. , 1983. *Astrophys. J.*, **266**, 41.
Demoulin–Ulrich, M.-H., Butcher, H.R. and Boksenberg, A., 1984. *Astrophys. J.*, **285**, 527.
Fabbiano, G., 1986. *Publ. astr. Soc. Pacif.*, **98**, 525.
Faber, S.M. and Gallagher, J.S., 1976. *Astrophys. J.*, **204**, 365.
Fabian, A.C., Atherton, P.D., Taylor, K. and Nulsen, P.E.J., 1982. *Mon. Not. R. astr. Soc.*, **201**, 17P.
Fabian, A.C., Nulsen, P.E.J. and Canizares, C.R., 1982. *Mon. Not. R. astr. Soc.*, **201**, 933.
Feigelson, E.D. and Nelson, P.I., 1985. *Astrophys. J.*, **293**, 192.
Ferland, G.J. and Mushotzky, R.F., 1984. *Astrophys. J.*, **286**, 42.
Forman, W., Jones, C. and Tucker, W., 1985. *Astrophys. J.*, **293**, 102.
Fosbury, R.A.E., Mebold, U., Goss, W.M. and Dopita, M.A., 1978. *Mon. Not. R. astr. Soc.*, **183**, 549.
Gisler, G., 1978. *Mon. Not. R. astr. Soc.*, **183**, 633.
Gunn, J.E., 1979. In: *Active Galactic Nuclei*, ed. C. Hazard and S. Mitton (Cambridge, Cambridge University Press), p. 213.
Heckman, T.M., 1980. *Astron. Astrophys.*, **87**, 152.
Heckman, T.M., 1987. In: *Observational Evidence of Activity in Galaxies, IAU Symposium No. 121*, ed. E.Y. Khachikian, K.J. Fricke and J. Melnick (Dordrecht, Reidel), p. 421.
Humason, M.L., Mayall, N.U. and Sandage, A.R., 1956. *Astron. J.*, **61**, 79.
Johnstone, R.M., Fabian, A.C. and Nulsen, P.E.J., 1987. *Mon. Not. R. astr. Soc.*, **224**, 75.
Jura, M., 1986. *Astrophys. J.*, **306**, 489.
Keel, W.C., 1983. *Astrophys. J. Suppl. Ser.*, **183**, 633.
Kormendy, J. and Stauffer, J., 1987. In: *Structure and Dynamics of Elliptical Galaxies, IAU Symposium No. 127*, ed. T. de Zeeuw (Dordrecht, Reidel), p. 405.
Nulsen, P.E.J., Stewart, G.C. and Fabian, A.C., 1984. *M.N.R.A.S.*, **208**, 185.
Phillips, M.M., Jenkins, C.R., Dopita, M.A., Sadler, E.M., and Binette, L., 1986. *Astron. J.*, **91**, 1062.
Sadler, E.M., 1987. In: *Structure and Dynamics of Elliptical Galaxies, IAU Symposium No. 127*, ed. T. de Zeeuw (Dordrecht, Reidel), p. 125.
Sadler, E.M., and Gerhard, O.E., 1985. *Mon. Not. R. astr. Soc.*, **214**, 177.
Sandage, A., 1957. *Astrophys. J.*, **125**, 422.
Schweizer, F., 1987. In: *Structure and Dynamics of Elliptical Galaxies, IAU Symposium No. 127*, ed. T. de Zeeuw (Dordrecht, Reidel), p. 109.
Sharples, R.M., Carter, D., Hawarden, T.G., and Longmore, A.J., 1983. *Mon. Not. R. astr. Soc.*, **202**, 37.
Sparks, W.B., Wall, J.V., Thorne, D.J., Jorden, P.R., van Breda, I.G., Rudd, P.J. and Jorgensen, H.E., 1985. *Mon. Not. R. astr. Soc.*, **217**, 87.
Thomas, P.A., Fabian, A.C., Arnaud, K.A., Forman, W. and Jones, C., 1986. *Mon. Not. R. astr. Soc.*, **222**, 655.
Trinchieri, G. and Fabbiano, G., 1985. *Astrophys. J.*, **296**, 447.

RADIO AND X-RAY EMISSION IN EARLY TYPE GALAXIES

Ginevra Trinchieri
Arcetri Observatory
Largo E. Fermi 5
50125 Firenze
Italy

ABSTRACT. The radio and x-ray emission in nearby early type galaxies are compared. The results indicate that there is a correlation between the x-ray and the radio luminosities, although with a very large scatter. Part of this scatter might be reduced with better radio data (for the x-ray data, we will have to wait a while), but most of it is probably intrinsic to the data, and indicates a rather complex relation between these two quantities. Sources with low radio luminosity could be effectively confined within the optical galaxy by the hot gas radiating in x-rays. For galaxies with $L_r > 10^{29}$ ergs s^{-1}Hz^{-1} confinement is no longer viable.

1. INTRODUCTION

The x-ray observations of normal, bright elliptical and S0 galaxies have shown that x-ray emission at a level of 10^{38} ergs s^{-1} or above is a property common to all galaxies with L_B above 10^9 $L_\odot$. For the most luminous objects this is most likely due to the emission from hot gas, with temperatures of ~ 1 keV, and total mass of $\sim 10^8 - 10^{10}$ $M_\odot$, while a sizeable contibution from compact, individual sources, such as low mass binary systems in the main body of the galaxy or in the globular clusters, might dominate the emission in the lower luminosity systems (for a review, see Thomas, this volume, and references therein). The central densities and cooling times of the hot gas are shown in figure 1 (from Canizares, Fabbiano and Trinchieri 1987). In most cases the cooling times are shorter than a Hubble time over the entire optical size of the galaxies (Trinchieri, Fabbiano and Canizares 1986, Thomas et al. 1986).

The radio properties of early type galaxies have been the subject of study for a number of years. It has been known for a long time that elliptical galaxies can be very powerful radio sources (for example the 3CR radio galaxies). Since the advent of more sensitive radio observations, normal elliptical galaxies are also detected at the level of a few mJy (typically over factors of 100 lower levels than the Radio Galaxies). These lower radio luminosity sources tend to have on average a compact emission within the optical size of the galaxy, with flat radio spectra, and the radio luminosity is related to the optical luminosity of the galaxies (Hummel, Kotanyi and Ekers 1983, Fabbiano et al. 1987).

The relation between the radio and x-ray emission in E and S0 galaxies has been recently investigated by Fabbiano et al. (1987) using a sample of ~ 30 objects

A. C. Fabian (ed.), Cooling Flows in Clusters and Galaxies, 273–277.
© 1988 by Kluwer Academic Publishers.

observed with the *Einstein* Observatory. These authors find a correlation between the radio and x-ray luminosities which they interpret as consistent with radio nuclear sources being accreting compact objects fueled by the cooling gas. X-ray fluxes (or upper limits) are now available for a larger sample, of ~ 80 galaxies. We can therefore expand the study of the x-ray and radio properties of E and S0 galaxies to this larger, statistically more significant sample. A more complete analysis, which will also be based on new radio data obtained at the VLA, will be given elsewhere (Fabbiano et al. 1988).

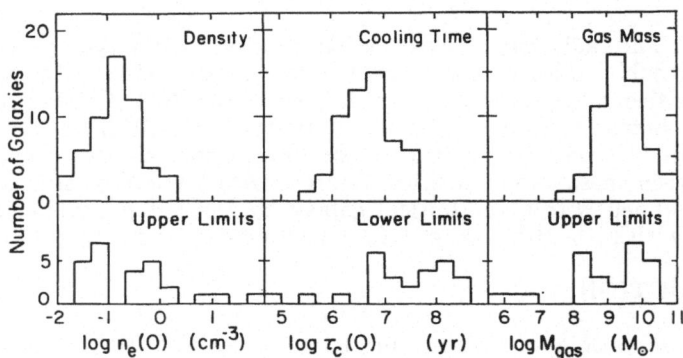

Figure 1. Central densities, cooling times and gas content for E and S0 galaxies.

2. THE SAMPLE

The optical and x-ray properties of this sample of galaxies have been discussed by Canizares, Fabbiano and Trinchieri (1987, see also Thomas, this volume). Of the original sample of 80 'normal'elliptical and S0 galaxies (we have excluded NGC 5128, Cen A), radio data have been published for 76 objects. For 61 the data are at 5 GHz (mostly from Fabbiano et al. 1987, and Birkinshaw and Davies 1985). Table I summarizes some statistics for the radio data. The numbers in parentheses refer to the 5 GHz data.

Table I

	Ell	Lent	Total
Detections	24	17	41(36)
Limits	23	12	35(25)

The radio/x-ray sample therefore contains: 28 detections in both radio and x-ray, 23 radio limits (but x-ray detections), 13 x-ray limits (but radio detections) and 12 double limits.

3. RADIO/X-RAY CORRELATION

The radio and x-ray luminosities of the sample galaxies are plotted in Figure 2a. There is an evident correlation between these two quantities, that ranges continuously for 4 decades in L_x and 7 decades in L_r. The scatter however is quite large, and the number of limits is considerable, thus making the estimate of the slope of the relation extremely uncertain.

We have considered possible biases that could augment the scatter observed in the data. Although there are many upper limits, they are in most cases distributed like the detections, with a few notable exceptions: M 86, with $L_x \sim 4 \times 10^{41}$ ergs s^{-1}, has a very low radio limit at $L_r < 4 \times 10^{26}$ ergs s^{-1} Hz^{-1}, about 2 orders of magnitude lower than suggested by the general trend. Similarly, NGC 1167 and NGC 3894 have $L_r > 10^{30}$ ergs s^{-1} Hz^{-1}, but have not been detected in x-rays at the level of $L_x \sim 10^{41}$ ergs s^{-1}. These will both enhance the spread and weaken the correlation.

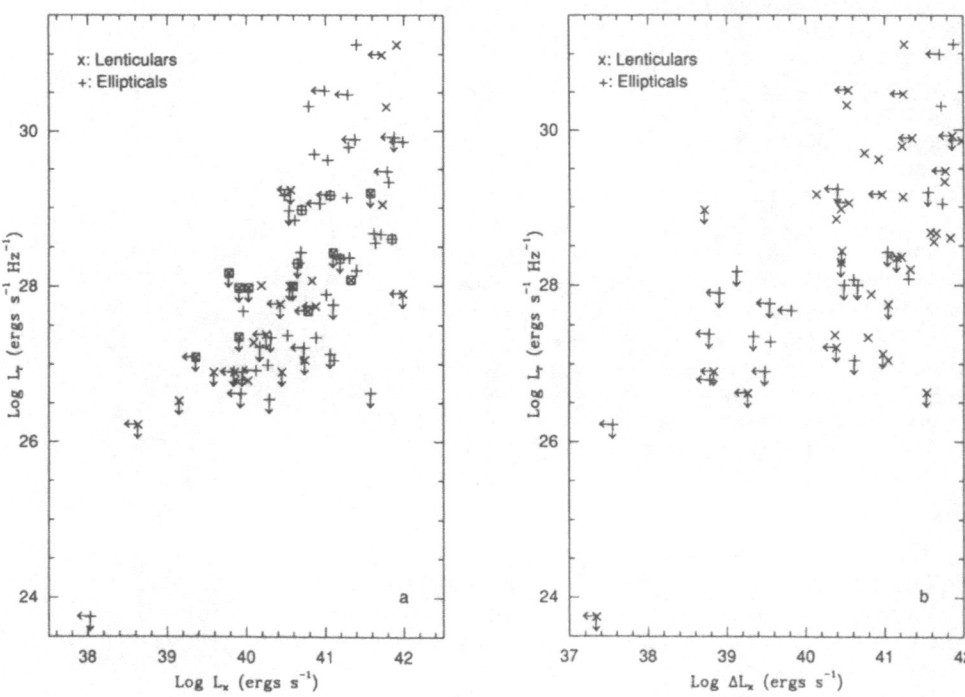

Figure 2. a: Plot of the x-ray and optical luminosity for the early type sample galaxies. Symbols in the box indicate that the radio data are not at 5 GHz. b: Plot of the radio luminosity against the gas component luminosity of the x-ray emission (ΔL_x).

A second possible cause for scatter is the fact that some of the radio data are at frequencies different than 5 GHz (2.3 and 1.4). However, the radio spectra of low luminosity elliptical galaxies indicate a flat spectrum, with a slope close to 0 (Fabbiano et al. 1987, Hummel and Kotanyi 1982). Moreover, Figure 2a shows that these points are in most cases non-restrictive upper limits, therefore their inclusion in the sample should not be responsible for the large spread.

Some spread is certainly introduced by the non-homogeneous radio data, which were obtained by different authors with different instruments. This results in different resolutions in the radio maps, and the radio flux might refer to the total radio emission or in some cases to a more compact central region. These two fluxes could be different by more than a factor of 10, as in the case of M 84, or less than a factor of 2, as in the case of NGC 4472 (Birkinshaw and Davies 1985, Fabbiano et al. 1987, Ekers and Kotanyi 1978).

The presence of two different morphological types does not seem to introduce additional scatter. Both the lenticulars and the elliptical galaxies follow the same trend, and both sample are represented in the full range of both radio and x-ray luminosities (see Figure 2a). Therefore the observed $L_x - L_r$ relation cannot be due to a morphological segregation in the sample.

Finally we have investigated whether the presence of two different components (hot gas and compact binary systems) in the x-ray emission could cause some of the scatter. We have therefore compared the radio emission with the quantity ΔL_x given by Canizares, Fabbiano and Trinchieri (1987), which should represent the hot gas component only in the x-ray emission, without the contribution from the single, compact sources. The results shown in Figure 2b clearly indicate that the scatter is now larger than before, and the relation is less clearly defined.

From the comparison of the x-ray and optical luminosity for these galaxies, only one galaxy is outside of the general correlation and might have a dominating nuclear x-ray source (NGC 3998, see Canizares, Fabbiano and Trinchieri 1987). The other galaxies in the sample contain on average $\leq 10^{10} M_\odot$ of cooling gas (see Figure 1) that can accrete onto the central non-thermal source. However, the large scatter in the correlation, most likely intrinsic to the data, rather than due to the presence of some bias, suggests that, although the cooling gas may provide a suitable fuel for the nuclear engine, neither its presence nor its mass seem to determine the radio emission of these objects. Other properties of early type galaxies (e.g. their 'roundness', Heckman 1983) must play an important role.

4. THERMAL CONFINEMENT

The hot gas can also provide an extremely effective way of confining the radio source. On average, the density of the ~ 1 keV gas ranges from $\sim 10^{-2}$ to 10^{-4} cm^{-3} over the optical size of the galaxy. In the minimum energy conditions (Miley 1980), a gas as tenuous as $n_e \sim 10^{-4}$cm^{-3} at $r \sim 30$ kpc would be effective in confining sources with $L_r \leq 4 \times 10^{29}$ ergs s^{-1} Hz^{-1}, in the most favourable conditions of a flat spectral index $\alpha = 0$, and an axial ratio for the radio source of 1. For a steep radio spectrum and more extreme axial ratios, the gas can confine sources of L_r of a few 10^{27}. Therefore a large fraction of the radio sources in the present sample could be thermally confined by the ambient gas.

A more detailed comparison between the pressure in the radio source U_{min} and the thermal pressure provided by the hot interstellar gas P_{ext} can be done for galaxies for which both good x-ray and radio data (and maps) are available, so that U_{min} and P_{ext} can be derived at the outermost radius observed in the radio structure. The comparison for three low radio luminosity galaxies, NGC 4472, NGC 4636 and NGC 4649, is given in Table 4 in Fabbiano et al. (1987), and shows in fact that $U_{min} < P_{ext}$ at all radii considered.

Fabbiano et al. (1987) have estimated that under simplifying assumptions, L_r is almost linearly proportional to L_x. With large spread, a slope of one could be consistent with the data, but only for the low radio luminosity objects (with L_r lower than $\sim 10^{29}$ ergs s^{-1}). However, about half of the galaxies with $L_x \geq 3 \times 10^{40}$ ergs s^{-1} also have a higher L_r (see Figure 2) and would not fit the relation, but would lie well above that line. The hot gas present in these objects can no longer confine the radio structure, which typically extend outside the optical size of the galaxies. Therefore, confinement could explain the relation between the x-ray and the radio power only for objects with relatively low L_r.

5. REFERENCES

Birkinshaw, M. and Davies, R.L., 1985, Ap. J., **291**, 32
Canizares, C.R., Fabbiano, G. and Trinchieri, G. 1987, Ap. J. **312**, 503.
Ekers, R.D., and Kotanyi, C.G., 1978, Astr. Ap. **67**, 47.
Fabbiano, G., Klein, U., Trinchieri, G. and Wielebinsky, R., 1987, Ap. J. **312**, 111.
Fabbiano, G., *et al.* 1988, in preparation.
Heckman, T.M., 1983, Ap. J. **273**, 505.
Hummel, E. and Kotanyi, C.G. 1982, Astr. Ap. **106**, 183.
Hummel, E., Kotanyi, C.G. and Ekers, R.D. 1983, Astr. Ap. **134**, 207.
Kotanyi, C.G. and Ekers, R.D. 1983, Astr. Astrophys. **122**, 267.
Miley, G. 1980, Ann. Rev. Astr. Ap. **18**, 165.
Thomas, P.A., Fabian, A.C., Arnaud, K.A., Forman, W., and Jones, C. 1986, MNRAS, 222, 655.
Trinchieri, G., Fabbiano, G. and Canizares C.R. 1986, Ap. J. **310**, 637.

ECOLOGY AND EVOLUTION OF ELLIPTICALS

William G. Mathews
Lick Observatory and Board of Studies in Astronomy and Astrophysics,
University of California, Santa Cruz

1. INTRODUCTION

The X-ray luminosities of elliptical galaxies indicate a large cooling rate, ~ 1 $M_\odot$ yr^{-1}, similar in magnitude to the current rate of mass ejection from evolving stars. The mass loss rate and cooling rate were almost certainly larger in the past when more massive stars were leaving the main sequence, so the total accumulated mass from cooled-off gas could be extremely large, perhaps even larger than the total stellar mass of these galaxies. Time-dependent evolutionary gas dynamical models of the hot gas in ellipticals have been largely successful in accounting for the soft X-ray luminosity and surface brightness profiles in present day ellipticals (Loewenstein and Mathews 1987). Typically these calculations indicate that the total amount of cooled-off gas is 5 to 10 times that of the current mass in hot gas and is from 5 to 200 times the stellar core mass of the galaxies, depending on the size of the core radius. The total amount of cooled off gas is insensitive to the distribution of dark halo matter that is responsible for binding the hot gas to the galaxies.

Since the total mass of gas that has cooled off is so large it is important to inquire about possible dynamical effects that this inflow may have had on the distribution and properties of visible stellar mass in these galaxies. The central core regions of ellipticals, where most of the cooling is occurring (Phillips *et al.* 1986), is most vulnerable to massive inflows. If low mass stars form preferentially from the cooling gas, galactic colors may be only very slightly affected, but observed central mass to light ratios, velocity dispersions and optical brightness distributions must be consistent with the large amount of matter that has precipitated in elliptical cores.

To address these issues briefly here, we consider an "isolated" elliptical which formed during an episode lasting only ≲0.1 of the Hubble time and which has received no large quantities of additional mass since, although some mass may have been lost in early winds. A single phase, hot interstellar medium is assumed to have been established at an early date and maintained thereafter. In the following we discuss limits on the amount of dark (stellar) matter that may have been deposited near elliptical cores by cooling flows, limits imposed by the contraction of the observed surface brightness profiles or set by observed mass to light ratios.

A. C. Fabian (ed.), Cooling Flows in Clusters and Galaxies, 279–284.

2. DISTORTION OF SURFACE BRIGHTNESS

It is well known that a stellar system will shrink if some of its mass is caused to move closer to the center of the system as in a cooling flow. As a simple galactic model, assume a King-type distribution of stellar mass $\rho(r) = \rho_0[1 + (r/r_c)^2]^{-3/2}$ for $r < r_t$, where r_c is the stellar core radius. When evolving stars shed mass which eventually cools to the galactic core, $\rho(r)$ will compress. For cooling flows, the mass accumulated in each dynamical (orbital) time is small, so the stellar orbits evolve according to an adiabatic invariant; for circular orbits this requires $r \propto M(r)^{-1}$. Stellar orbits in the core region of ellipticals are probably random as expected from N-body collapse calculations. Purely radial orbits are unstable to bar formation and do not allow density profiles flatter than $\rho(r) \propto r^{-2}$. A conservative approximation to the change in $\rho(r)$ due to the inflow can be found by assuming all orbits in the central core of the galaxy are circular; this is conservative because the orbit-averaged r decreases faster than $M(r)$ for radial than for circular orbits in the King-type potential. Results of such calculations demonstrate that, while only a tiny fraction ($M_{in}/M_{*c} \lesssim 0.05$) of the cooling flow can process to $r = 0$ without destroying the flat projected core, observed $\Sigma_*(R)$ are consistent with $M_{in}/M_{*c} \lesssim 3$ if the dark matter is distributed exponentially with a scale height of one core radius.

3. EFFECT OF COOLED MASS ON CENTRAL MASS TO LIGHT RATIOS

Mass to light ratios in elliptical galaxies are determined from measurements of the line of sight stellar velocity dispersion σ_ℓ and surface brightness. Values of M/L characteristic of the galactic core $(M/L)_c$ and out to the effective radius $(M/L)_e$ have been determined by Lauer (1985). It is remarkable that for 13 well-resolved ellipticals Lauer finds an average $(M/L)_c/(M/L)_e = 1.2 \pm 0.2$, providing no apparent evidence of a substantial buildup of dark stellar matter in the galactic cores. However, the procedures used in determining M/L values are notoriously uncertain, particularly for values $(M/L)_e$ characteristic of the galaxy as a whole and when only the central velocity dispersion is available, as for Lauer's galaxies.

To examine the sensitivity of M/L determinations in elliptical galaxies, I have constructed model galaxies described by a de Vaucouleurs density profile (but containing cores and outer cutoffs) and a quasi-isothermal distribution of dark halo matter containing 90 percent of the total galactic mass. The models are assumed to have velocity ellipsoids which are spherical near the galactic core, but become progressively more prolate at large radii (as in N-body collapse models). Then an additional component of dark stellar matter is assumed to be distributed exponentially near the galactic core. By examining the central velocity dispersion in these galactic models, it is possible to determine "mass to mass" ratios by putting observable properties of the models through the same equations that observers use to determine $(M/L)_c$ and $(M/L)_e$. By this means it is possible to gain quantitative insight into the sensitivity of published (M/L) values to large central depositions of dark matter.

These studies indicate that observed mass to light ratios are not responsive to large central depositions of dark matter. For one particularly relevant model galaxy – having $R_e/r_c = 30$ and with a total dark (stellar) component of $10M_{*c}$ distributed $\propto e^{-r/r_{ds}}$ with $r_{ds} = r_c$ – the value of $(M/L)_c/(M/L)_e$ is identical to the same model with no central dark component. Therefore radial variations in M/L values may provide very little information about the total amount of matter deposited by cooling flows. However, as the total amount of diffuse central dark matter is increased, the central velocity dispersion eventually becomes too large compared to observations and $\sigma_\ell(R)$ decreases too fast with projected radius to match the observations (Davies and Illingworth 1983). By this means it can be determined that the total amount of dark matter near the center cannot exceed $\sim 10 M_{*c}$ if $r_{ds} = r_c$, or $\sim 30 M_{*c}$ if $r_{ds} = 3r_c$.

It appears that the distortion of the stellar surface brightness $\Sigma_*(R)$ constrains the total diffuse dark matter near the galactic cores more than the M/L values do. But the limits on M_{in}/M_{*c} set by observations of $\Sigma_*(R)$ fall short by an order of magnitude of the total amount of gas that is expelled from stars during the entire lifetime of these galaxies. In the following we consider three possible resolutions of this "inflow crisis": (1) perhaps matter cools into low mass stars at large distances from the galactic cores, contrary to our initial assumption above, (2) perhaps an initial mass function for the galactic stars can be found such that relatively little mass was ejected during the Hubble time, or (3) a galactic wind may have been present in the early history of ellipticals, drastically reducing the total amount of cooled off gas.

4. FORMATION OF OPTICALLY DARK STARS AT LARGE RADII

To address this possibility it is necessary to carefully examine the sources of entropy variations in the hot interstellar gas which could lead to thermal instabilities; the first and necessary step in star formation are regions of low entropy. For an unmolested and isolated elliptical, there are only three sources of entropy inhomogeneities: supernovae, mass ejection from red giants, and magnetic field fluctuations.

Our hydrodynamic models of the evolution of supernova blast waves in cooling flows demonstrate that only regions of higher entropy result; ultra-hot bubbles are generated in $\sim 10^5$ years and undergo buoyant mixing in $\sim 10^6$ years – no star formation can result from supernova blasts. Winds and planetary nebulae are obvious sources of relatively cool, low entropy gas, but are unlikely to remain coherent long enough to be consolidated into low mass stars. Since the stars are generally moving about sonically relative to the hot interstellar medium, the cooled-down ejected shells are both compressed and rammed by the ambient gas, becoming Rayleigh-Taylor unstable. Essentially all the kinetic energy of the fragments goes into heating the hot interstellar medium; no contiguous low entropy regions of stellar (or sensibly substellar) mass remain which have appreciably shorter cooling times than the overall flow. Finally, our studies of the evolution of magnetic fluctuations indicate that if they are large enough to cause significant changes in the cooling time, the magnitude of the field is so large (even in the relatively low field regions) that thermal instabilities are arrested by magnetic stresses long before stellar densities are reached. In summary, low mass star formation, if it occurs at all, must occur near the galactic core, as we have assumed above, where the cooling is rapid, even catastrophic.

5. TOTAL STELLAR MASS LOSS AND THE INITIAL STELLAR MASS FUNCTION: AN ARGUMENT FOR EARLY EPOCH GALACTIC WINDS

Here we examine the possibility that an initial mass function (IMF) can be found for which the total amount of ejected gas is reduced, alleviating the "inflow crisis" described above. To this end, we have considered all possible power law IMFs, $\Psi(m) = Am^{-(1+\beta)}$ for $m_\ell < m < m_u$. We have followed the evolution of these stars and determined the total amount of gas ejected over the lifetime of the galaxy. The solutions are required to agree with stellar mass to light ratios observed at the present time, $M/L_V \approx 6 \pm 1$, a ratio which is remarkably constant among all observed ellipticals. The adjustable parameters are m_ℓ, m_u, β, and the current age of a typical galaxy, t_{now}. We evaluate the ratio of total gas ejected by the present time t_{now} to the total current stellar mass, M_{gas}/M_*. This ratio is evaluated from a knowledge of the rate that stars leave the main sequence and the total fraction of mass ejected from stars of a given initial mass; this last information is based on the Tinsley-Renzini Fuel Consumption Theorem (Renzini and Buzzoni 1982).

Figure 1 shows the results of such a calculation for $t_{now} = 10^{10}$ years, $m_u = 50$ and for four values of m_ℓ. All masses are in solar units. There are three sets of curves in Fig. 1; the uppermost set is the current value of $log(M_*/L_V)$ as a function of β, the set labeled "no wind" are values of $log(M_{gas}/M_*)$ if all the ejected gas is kept within the galaxy, and the lowest set of curves are values of $log(M_{gas}/M_*)$ for which all gas produced before 1.5×10^9 years is assumed to have been removed from the galaxy as a wind. The results shown in Fig. 1 are similar to those for other reasonable values of t_{now} and m_u and are not unique to the assumption of single power law IMFs.

The first interesting result illustrated in Fig. 1 is the relatively small range of β which allows a match with the current mass to light ratio, $log(M_*/L_V) \approx 0.8 \pm 0.1$. Discounting the curve for $m_\ell = 0.3$, which seems too close to the current turnoff mass $m_{to} \approx 0.85$, observed mass to light ratios can be reached only for $\beta \approx 0.75 \pm 0.20$; the Salpeter value $\beta_{sp} = 1.35$ clearly gives mass to light ratios that are too large, unless absolute M/L determinations from observations are incorrect by $\gtrsim 2$. The strong constraining influence of the observed mass to light ratios on the slope β of the IMF was unexpected.

Solutions for the normalized gas production M_{gas}/M_* must now be viewed only in the vicinity of $\beta = 0.75$. The "no wind" family of curves indicates gas to stellar ratios M_{gas}/M_* that exceed unity; more gas has been produced by stellar evolution than exists in stars today! Main-sequence stars, giant stars, and stellar remnants all contribute to M_*. The "no wind" values of M_{gas}/M_* are also two orders of magnitude greater than the relative gas content in ellipticals observed today, $log(M_{gasnow}/M_*) = -1.7 \pm 0.3$.. This may indicate a breakdown in the single starburst model for galactic evolution, i.e., raising the possibility that some of the large quantity of ejected gas was incorporated into subsequent generations of stars; this possibility of continued star formation must have occurred to some extent in ellipticals to produce the observed radial metallicity gradients. However, it also seems likely that much of the gas produced at early times – by far the bulk of all liberated gas – was actually ejected from the galaxy by supernova evolving from stars having $m \gtrsim 5$. It is easy to show that this is energetically possible.

However, M_{gas}/M_* values computed (but not shown in Fig. 1) excluding all gas lost from supernova-producing stars of mass $m > 5$ are still about ten times larger

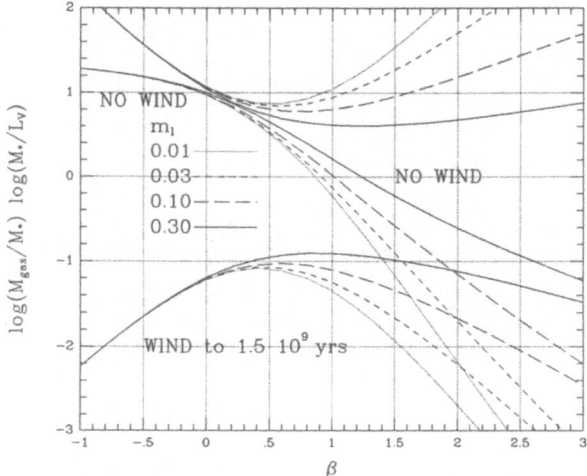

Figure 1. Upper set of curves are values of $log(M_*/L_V)$ as a function of β. Each curve in the set refers to a lower cutoff mass m_ℓ for the IMF. Middle set of curves labeled "no wind" represent values of $log(M_{gas}/M_*)$, the total amount of gas produced in terms of the total remaining stellar mass now. Lower set of curves are values of $log(M_{gas}/M_*)$ if all mass expelled from stars during the first 1.5×10^9 years is assumed to have been lost in a galactic wind. All curves refer to $t_{now} = 10 \times 10^9$ years and $m_u = 50$.

than the total amount of gas observed today. Depending on the core mass for the galaxy of interest, the total cooled-off gas generated over the galactic lifetime even by stars with mass $m < 5$ could easily violate the limits discussed above for the distortion of $\Sigma_*(R)$. Winds that last beyond several 10^8 years must be generated by continued massive star formation (and subsequent supernovae) due to internal processes or to mergers with other gas-rich companion galaxies. As an extreme case Fig. 1 shows the results for M_{gas}/M_* when a wind is assumed to last for the first 1.5×10^9 years of the galactic lifetime. The results in this case indicate that the currently observed amount of gas ($log(M_{gasnow}/M_*) \approx -1.3$) is accounted for if $m_\ell \lesssim 0.10$. The total amount of gas produced per unit stellar mass during galactic evolution must not exceed ($\sim 3M_{*c} + M_{gasnow})/M_*$, according to the surface brightness distortion argument. However, for ellipticals having the smallest core radii, $3M_{*c} \ll M_{gasnow}$, so the current gas content must represent most of the gas produced at all times or the nuclear regions in ellipticals having the smallest core radii should be systematically more distorted by inflow; this latter possibility may not be ruled out by present observations. In any event, winds that last for $\sim 10^9$ years have many astronomical implications which cannot be considered here except to note that a relatively long-lasting galactic wind can also account for the large masses of metal-enriched gas observed in clusters of galaxies today.

The author is pleased to acknowledge support from the California Space Institute Grant CS-16-87 and from a UCSC Faculty Research Grant.

REFERENCES

Davies, R. L., and Illingworth, G. 1983, *Ap. J.*, **266,** 316.
Lauer, T. R. 1985, *Ap. J.*, **292**, 104.
Loewenstein, M., and Mathews, W. G. 1987, *Ap. J.*, (in press).
Phillips, M. M., Jenkins, C. R., Dopita, M. A., Sadler, E. M., and Binette, L. 1986, *A. J.*, **91**, 1062.
Renzini, A., and Buzzoni, A. 1986, in *The Spectral Evolution of Galaxies*, eds. C. Chiosi and A. Renzini, (Dordrecht:Reidel), p. 195.

Warm gas in radio galaxies

R A E Fosbury*
Space Telescope–European Coordinating Facility
European Southern Observatory
Karl-Schwarzschild-Straße-2
D-8046, Garching bei München, FRG

Abstract. Many active early-type galaxies exhibit extended distributions of ionized gas which are detected by optical spectroscopy and narrow-band imaging. A summary is presented of the basic observational properties of these extended emission line regions (EELR) with a view to establishing their connection with the "cooling flow" phenomenon

1. Observed properties

EELR are found around early-type galaxies with currently active nuclei — as evidenced by a strong, narrow emission line spectrum. Their common properties can be summarised as follows:

a) *Size and shape.* The spatial extent is typically a few tens of *kpc* with the largest objects having a projected radius of 100 *kpc* or more. The morphology is sometimes reminiscent of a distorted disk but can be fairly chaotic (see, e.g., Fosbury *et al.* 1982; Danziger *et al.* 1984; Fosbury 1986). There can be associations with radio structures but this is by no means always the case.

b) *Kinematics.* The observed velocity fields divide into two apparent classes: those with a well-defined and systematic "rotation curve" with a half amplitude ranging up to $\sim 350\,km\,s^{-1}$ and a more chaotic type with lower total velocity amplitude (Tadhunter 1987). The *specific angular momentum* associated with the former systems is large — $\sim 10^4\,km\,s^{-1}\,kpc$ — and is comparable with that found in the disks of large spirals. It is possible that the second type have not yet been observed sufficiently completely for us to have found a kinematic major axis if one exists.

---

* Affiliated to the Astrophysics Division,
Space Science Department, European Space Agency

A. C. Fabian (ed.), Cooling Flows in Clusters and Galaxies, 285–287.
© *1988 by Kluwer Academic Publishers.*

c) *Luminosity.* The [O III] λ5007 line luminosities from the extended regions in the radio galaxies range up to a few by $10^{42}\ erg\ s^{-1}$ — with $H_0 = 50\ km\ s^{-1}\ Mpc^{-1}$ — (di Serego Alighieri, personal communication) and thus overlap the values found for the QSO EELR (Stockton & MacKenty 1987).

d) *Optical spectra.* A wide range of ionization is present in the EELR; those with the highest surface brightness tend to have the highest degree of ionization. He II λ4686 is a common feature and ions as high as Fe^{9+} have been seen (Tadhunter *et al.* 1987). The low ionization "filaments" have *LINER-like* spectra.

e) *Ionization sources.* The gas appears to be *photoionized* by a source with a mean ionizing photon energy of between 30 and 50 eV (Robinson *et al.* 1987 and Robinson, this volume). The source of these photons is likely, in the majority of cases, to be the active nucleus although there is evidence in a few objects for *in situ* ionization (van Breugel *et al.* 1985; van Breugel *et al.* 1986; Tadhunter *et al.* 1987). In some of the radio galaxies, the apparent nuclear luminosity — as seen directly by us in the optical and when extrapolated into the ultraviolet with a power-law spectrum — would provide insufficient ionizing flux to illuminate the EELR. Possible solutions to this dilemma are: (i) a beamed (but with a large opening angle) nuclear radiation field and/or, (ii) an ionizing source with a *thermal* rather than a power-law spectrum.

f) *Chemical composition.* From the results of comparing emission line measurements with photoionization models (Robinson *et al.* 1987), a rather narrow range in EELR composition is allowed—approximately $0.2 \leq Z_{EELR} \leq 2\,Z_\odot$, i.e., the gas consists of processed material.

2. Conclusions

Two classes of possible origin of the EELR can be proposed:

1. The gas is the debris from tidal interactions or merger events. There is evidence for companion galaxies in some, but not all, of the observed cases.

2. The material is condensing from a hot $(T > 10^6\ K)$ halo and has reached a *warm*, $10^{-4}\ K$ phase where it can be re-ionized by the nuclear activity. Within this "cooling flow" scenario, there are again two distinct and observationally testable hypotheses:

   The hot gas from which the EELR is cooling is a hydrostatic atmosphere with a low specific angular momentum and originates from stellar processes within the central elliptical galaxy.

   - The hot gas has a specific angular momentum comparable to that of large spiral galaxy disks and could be part of the (continuing) process of galaxy formation.

In at least a subset of the observed EELR, we know that the gas has a large specific angular momentum and so the 'static' cooling flow picture can be excluded. It remains to distinguish between the disk formation and accretion/merger possibilities. Whatever conclusion is reached about their origin, the EELR phenomenon is widespread and shows a number of surprisingly uniform properties.

Acknowledgments. Several people have been involved in this study, I should like to thank, in particular, Sperello di Serego Alighieri, Luc Binette, John Danziger, Peter Quinn, Andrew Robinson and Clive Tadhunter.

References

Danziger, I.J., Fosbury, R.A.E., Goss, W.M., Bland, J. & Boksenberg, A., 1984. *Mon. Not. R. astr. Soc.*, **208**, 589

Fosbury, R.A.E., Boksenberg, A., Snijders, M.A.J., Danziger, I.J., Disney, M.J., Goss, W.M., Penston, M.V., Wamsteker, W., Wellington, K.J. & Wilson, A.S., 1982. *Mon. Not. R. astr. Soc.*, **201**, 991

Fosbury, R.A.E., 1986. In: *Structure and Evolution of Active Galactic Nuclei*, Trieste, April 10-13 1985, Eds. Giuricin, G. *et al.* Reidel, pp297-308.

Robinson, A., Binette, L., Fosbury, R.A.E. & Tadhunter, C.N., 1987. *Mon. Not. R. astr. Soc.*, **227**, 97

Stockton, A. & MacKenty, J.W., 1987. *Astrophys. J.*, **316**, 584

Tadhunter, C.N., 1987. 'Emission line radio galaxies'. *D. Phil. Thesis*, University of Sussex.

Tadhunter, C.N., Fosbury, R.A.E., Binette, L., Danziger, I.J. & Robinson, A., 1987. *Nature*, **325**, 504

van Breugel, W.J.W., Miley, G.K., Heckman, T.M., Butcher, H. & Bridle, A.H., 1985. *Astrophys. J.*, **290**, 496

van Breugel, W.J.W., Heckman, T.M., Miley, G.K., & Filippenko, A.V., 1986. *Astrophys. J.*, **311**, 58

He II λ4686/Hβ AS A CONSTRAINT ON THE IONIZING CONTINUUM

L. Binette*
European Southern Observatory
D8046 Garching bei München, FRG

*Canadian Institute of Theoretical Astrophysics
University of Toronto, Ont. M5S 1A1, Canada

ABSTRACT. Working on the assumption that the emission line filaments associated with radio-galaxies are photoionized, the relative reliability of the He II λ4686/Hβ line ratio as a test of the hardness of the continuum ionizing distribution is reviewed. It may provide a useful discriminant for determining whether or not young stars are responsible for exciting the emission line filaments.

1. INTRODUCTION

The excitation mechanism of the emission line ionized gas observed in the environment of cooling flows remains an open question. Excitation by means of interstellar shocks appears insufficient to account for the observed recombination line luminosities (*cf.* Johnstone, Fabian and Nulsen, 1987). Although photoionization appears by default favoured, is it caused by (young) stars or by an (central) active nucleus? The possibility that the excitation proceeds from nuclear activity –a possibility indicated by Robinson (this workshop)– does not rule out star formation as such but would simply indicate that cooling flows do not favour the formation of giant H II regions as we know them. Furthermore, the fact that in specific cases nuclear activity has so far been seen only in the radio domain does not necessarily preclude the possibility of photoionization by the nucleus, assuming that the ionizing shape, as argued by Binette, Robinson and Courvoisier (1988), is not simply described by a power law but rather by an energy distribution which could peak above $13\,eV$ allowing for a weak yet undetected optical counterpart to the ionizing spectrum. [Photoionization by a partly covered nuclear UV source is also a possibility].

2. USEFULNESS OF THE He II λ4686/Hβ RATIO

In computing photoionization models under a variety of ionizing continuum energy distributions, Stasińska (1984) pointed out the sensitivity of He II/Hβ to the relative amount of ionizing flux in the $100\,eV$ region. Fig. 1a, shows the behaviour of He II/Hβ in the case

289

A. C. Fabian (ed.), Cooling Flows in Clusters and Galaxies, 289–291.
© 1988 by Kluwer Academic Publishers.

of different black body ionizing continua temperatures (T_{bb}) ranging from 6.0×10^5 K to 8.0×10^6 K. The calculations were carried out using the multi-purpose modeling code *MAP-PINGS* (Binette, Dopita & Tuohy 1985) and adopting the simplest possible geometry, that of a plane parallel slab of gas assumed to have a constant hydrogen density $n_H = 2\,000\,cm^{-3}$ with one face illuminated by ionizing photons emitted by a distant continuum source. The emitting cloud is ionization bounded, that is, optically thick to the ionization radiation. The ionizing parameter [1] is $U_t \approx 10^{-3}$ for $T_{bb} < 10^{5.6}$ K (varying somewhat with T_{bb} due to the normalization procedure adopted –see Binette, Courvoisier and Robinson, 1987) and is not a critical parameter in the determination of He II/Hβ as indicated below.

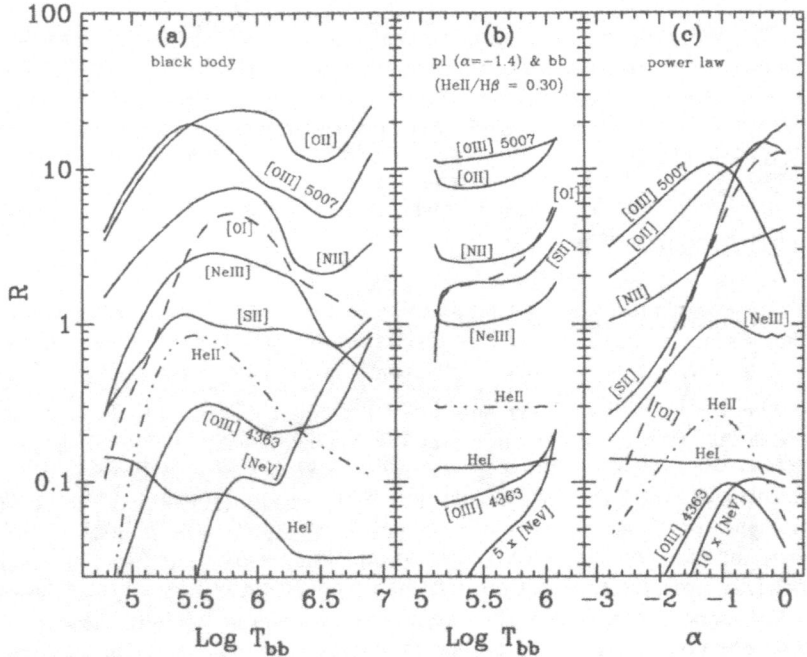

Fig. 1: Calculated line intensities (relative to Hβ) as a function of (a) temperature, T_{bb}, for pure black body ionizing continua; (b) temperature, T_{bb}, of the black body component added to an underlying power law of slope $\alpha = -1.4$ (the contribution of the black body is such as to result in He II/H$\beta = 0.30$); (c) spectral index, α, for power law ionizing continua (no energy cut-off below 5 keV). The relative intensities are given for the following lines: [O I] $\lambda6300$, [O II] $\lambda\lambda3726,29$, [O III] $\lambda5007$ and [O III] $\lambda4363$, He I $\lambda5876$, He II $\lambda4686$, [N II] $\lambda6583$, [S II] $\lambda\lambda6716,31$, [Ne III] $\lambda3869$, [Ne V] $\lambda3426$.

---

[1] $U_t = \frac{Q_t}{4\pi r^2 n_H c}$ where c is the speed of light and Q_t the total number of ionizing photons emitted by an UV ionizing source imagined to be at a distance r from a cloud of density n_H

The He II/Hβ ratio resulting from photoionized gas has some advantageous properties generally not shared by other ions line ratios: firstly, the ratio is insensitive to the exact abundance of He; secondly, the He II/Hβ ratio, because it involves for the most part recombination processes rather than collisional processes is not sensitively dependent on the electronic temperature in contrast to diagnostics involving forbidden lines; thirdly, the ratio is found to be constant over quite a wide ionization parameter range: we have verified that with the $T_{bb} = 130\,000\,K$ and $500\,000\,K$ ionizing continua, for instance, He II/Hβ varies by less than 15% within the range $10^{-1} \geq U_t \geq 10^{-3}$. (He II/H$\beta$ is more sensitive to U_t at higher temperatures, however.)

The following approximate expression can be used to determine He II/Hβ as a function of T_{bb} (with a precision of $\leq 6\%$ and a validity limited to $T_{eff} \leq 300\,000K$ *and* black body-like ionizing energy distributions):

$$\text{He II/H}\beta \simeq 2.2\,\frac{Q_{4-6.2}}{Q_{1-4} + Q_{6.2-\infty}} = 2.2\,\frac{Q_{4-6.2}}{Q_t - Q_{4-6.2}}$$

where Q_{1-4}, $Q_{4-6.2}$, $Q_{6.2-\infty}$ represent the number of ionizing photons whose energies are between 1.0 and 4.0 Ry, 4.0 and 6.2 Ry, and 6.2 and ∞ Ry respectively, Q_t being, by definition, the sum of all three quantities.

The sensitivity of He II/Hβ to the hardness of the ionizing spectrum is apparent in Fig. 1a where it is seen to rise steeply with T_{bb} for moderate black body temperatures. Therefore, if young stars of effective temperature < 60 000K were photoinizing the filaments, we could reasonably expect He II/Hβ < 0.1. Any measurement that would indicate a higher value would point to a non-stellar ionizing continuum.

The possible complication of He II/Hβ being emitted by Wolf-Rayet stars appears unlikely in those cases where no sign is found in the underlying continuum of the presence of these very hot stars. Another weakness of the He II diagnostic is the possibility that the gas clouds may be optically thin to the ionizing radiation (*cf.* Stasińska, 1984) . This appears quite unlikely in the case of the cooling flow filaments which often present a strong [O I] line (see Johnstone, Fabian and Nulsen, 1987) which is emitted by partially ionized gas beyond the fully ionized H$^+$ zone.

To conclude, He II/Hβ potentially constitutes a powerful test to determine whether young stars are responsible for exciting the gas or not. Its application to low excitation emission line filaments, however, may lie behond the sensitivity of current detectors given the practical difficulty of measuring any line significantly weaker than Hβ .

REFERENCES.

Binette, L., Courvoisier, T.J.-L., Robinson, A.: 1987, *A & A* (in press)
Binette, L., Robinson, A., Courvoisier, T.J.-L.: 1988, submitted to *A & A*
Binette, L., Dopita, M.A., Tuohy, I.R.: 1985, *Ap. J.* **297**, 476.
Johnstone, R.M., Fabian, A.C., Nulsen, P.E.J.: 1987, *Mon. Not. R. astr. Soc.*, **224**, 75.
Stasińska, G.: 1984, *A & A* **135**, 341.

Photoionization of extended emission line regions

A. Robinson[1]
European Southern Observatory
Karl-Schwarzschildstr.-2
D8046 Garching bei München
Federal Republic of Germany

Abstract. Extensive regions of optical line-emitting gas are commonly found around quasars, radio galaxies and also the central galaxies of clusters in which cooling flows are thought to be occuring. The spectra of these extended emission line regions show a remarkable continuity in excitation which, when analysed in the context of photoionization models, can be explained by an ionizing continuum source characteristic of active galactic nuclei. The location and nature of the continuum source are discussed and it is suggested that ionizing radiation generated by nuclear activity may play an important role in exciting the emission lines associated with cooling flows.

1   Introduction

In the conventional picture, the characteristic emission line spectrum of active galactic nuclei (AGN) is emitted by gas confined to within a few kiloparsecs of the central continuum source (e.g., Osterbrock 1984). It is now becoming apparent, however, that emission lines often extend to much greater distances, tens or even hundreds of kiloparsecs, from the nucleus (Fosbury 1986 and references therein). Although extra-nuclear line emission has also been discovered in some Seyfert galaxies (Unger et al., 1987), extended emission line regions (EELR) are most prominent around objects such as radio-loud quasars, broad-line and narrow-line radio galaxies which have both powerful extended radio sources and relatively strong nuclear emission lines (Fosbury 1986, van Breugel 1986, Stockton & MacKenty 1987).

The origins of this gas are not yet understood but one possibility is that the line-emitting clouds have condensed out of X-ray haloes similar to those recently detected around nearby elliptical galaxies (Forman, Jones & Tucker 1985). If this is the case, EELR will provide important tracers of the properties of distant cooling flows. A question which must first be addressed, however, is that of the nature of the excitation mechanism. In this paper I wish to summarize some recent work on this problem which shows that photoionization by an AGN-like ionizing continuum can both explain the emission line spectra of the high

---

[1]Current address: Institute of Astronomy, Cambridge, UK

A. C. Fabian (ed.), Cooling Flows in Clusters and Galaxies, 293–303.

excitation nebulosities surrounding relatively distant radio galaxies and quasars and may also play a role in powering the emission line filaments associated with cooling flows around the central galaxies of X-ray luminous clusters.

2 Observations and properties of extended emission line regions

Over the last few years, searches for extra-nuclear line emission have been carried out for a variety of reasons by a number of groups using different samples of galaxies. The following discussion is based on a subset of galaxies with EELR for which measurements of enough emission lines exist to allow them to be usefully analysed in line ratio diagrams. The largest and most homogeneous set of observations has been collected by Danziger, Fosbury, Tadhunter and co-workers for a sample of Southern Hemisphere radio galaxies (SRG sample). Several other radio galaxies have been studied in detail by van Breugel, Heckman, Miley and co-workers as possible sites of radio jet–cloud interactions. In addition to these objects, the well-known emission line filaments associated with NGC 1275, Cen A and M 87 are also well documented (see Table 1 for sources).

Symbol	Extended emission line regions	Sources
■	Southern radio galaxies (SRG sample)	1
●	Radio galaxies from other sources	2 – 10
▲	Optical filaments in X-ray luminous clusters	11,12
	Nuclear narrow emission line regions	
□	Host galaxies (SRG sample)	as above
○	Host galaxies (other sources)	as above
□	Seyfert 1.5, Seyfert 2, Narrow-line Radio Galaxies	13,14,15
◇	Liners and other emission line galaxies	13
⋆	Starburst and H II region galaxies	13

Table 1: Sources of data and key to the symbols used in the figures. References: (1) Robinson *et al.*, 1987; (2) Ford & Butcher 1979; (3) Kent & Sargent 1979; (4) Osmer 1978; (5) Phillips 1981; (6) van Breugel *et al.*, 1985a; (7) van Breugel *et al.*, 1985b; (8) van Breugel *et al.*, 1986; (9) van Breugel *et al.*, 1984; (10) Heckman *et al.*, 1982; (11) Hu *et al.*, 1985; (12) Johnstone *et al.*, 1987; (13) Veilleux & Osterbrock (1987) and references therein; (14) Cohen 1983; (15) Cohen & Osterbrock 1981.

These galaxies have been studied essentially because they are strong radio sources. However, spatially extended emission lines have also been discovered around the central galaxies of X-ray luminous clusters in which cooling flows are thought to be occuring (Heckman 1981). Extended emission line regions appear to be quite common in these systems and are expected to arise naturally within the cooling flow picture as thermally unstable condensations (*e.g.*, Cowie, Fabian & Nulsen 1980). Spectroscopy of optical filaments associated with cooling flows has been obtained by Hu, Cowie & Wang (1985) and Johnstone, Fabian & Nulsen (1987) but, in any case, the two categories of EELR (*i.e.*, 'radio selected' and

'X-ray selected') are not mutually exclusive: several of the radio selected EELR, notably NGC 1275 and M 87, have been interpreted as cooling flow filaments.

Many other EELR have been discovered for which adequate spectrophotometry does not yet exist. In particular, radio-loud quasars with extended, steep spectrum radio sources commonly have EELR (Stockton & MacKenty 1987) which are comparable in extent and [OIII]λ5007 luminosity to those found around the radio galaxies (di Serego Alighieri, private communication).

The observational properties of EELR will be discussed in more detail elsewhere in these proceedings (see contributions by Danziger, Fosbury, McCarthy and van Breugel) but it is worth summarizing some relevant aspects here. The line-emitting gas typically extends to projected distances $\sim 20 - 50\,kpc$ from the nucleus of the host galaxy but in some objects may extend as far out as $100\,kpc$ or more. The morphology is varied and irregular, individual structures typically have scale sizes of $10\,kpc$ or less but long filamentary or shell-like features also occur. With the exception of some distant 3CR radio galaxies, the extended emission line features do not, in general, appear to be associated with an underlying stellar continuum.

The EELR seem to occur preferentially in galaxies which have strong extended radio sources and in some galaxies, notably those studied by van Breugel and co-workers, there is a detailed correspondance between features of the radio source (e.g., jets, lobes etc.) and the emission line gas while in others the EELR extend roughly along the radio source axis. However, it is important to note that this is not always the case: emission line gas is also commonly found at large distances from the radio source and in some galaxies there is no clear association between the optical and radio emitting gas.

Since it is usually difficult to measure the appropriate diagnostic line ratios accurately, our knowledge of the detailed physical conditions within the extended nebulosities is somewhat limited. When measured, the temperature-sensitive [OIII]$\lambda\lambda(4959 + 5007)/4363$ ratio indicates electron temperatures ($10^4 \leq T_e < 3 \times 10^4\,K$) which are significantly lower than would be expected for the O^{++} zone in a shock-ionized gas and therefore favour photoionization as the dominant excitation process. Electron densities have been measured from the [SII]λ6716/6731 ratio in a few objects and range from 10^2 to about $10^3\,cm^{-3}$.

3 Photoionization and line ratio diagrams

Any model for the excitation mechanism must be capable of explaining both the measured relative intensities of the emission lines and the luminosity radiated. The question is, does photoionization fulfill these conditions for the EELR (including those we might term cooling flow filaments) and if so, what is the nature and location of the required photon source? Perhaps the best way of approaching this problem, other than attempting to model each individual galaxy in detail, is to construct line ratio diagrams (see for example Baldwin, Phillips and Terlevich 1981) in which any relationships within and between the spectra of various groups of objects can be identified and compared with computed models.

The simplest model of a photoionized cloud is that of a plane parallel slab of gas of constant hydrogen density, n, which is illuminated by a flux, F, of ionizing photons emitted by a distant source. The ionization and thermal structure of the slab and hence the emergent line emissivities can be calculated using standard procedures (Davidson & Netzer 1979). Given the geometry, the predicted emission line intensities are, in general, governed by five parameters: the ionization parameter U ($= F/c\,n$), the spectral energy distribution, $S(\nu)$, of the ionizing continuum, the density, the element abundances, Z and the column density, N_c, of the slab.

The photoionization models shown in the diagrams presented here have been calculated using the multi-purpose modeling code *MAPPINGS* (Binette, Dopita & Tuohy 1985) and are described in more detail in Robinson *et al.*, (1987). We will be mainly concerned with the effects of varying U and $S(\nu)$ while keeping other parameters fixed. In all calculations the density is $n = 100\,cm^{-3}$ and the clouds are assumed to be radiation-bounded in the sense that almost all photons with energies $< 2\,keV$ are absorbed in photoionizing the gas. Solar abundances are assumed for most calculations.

The photoionized slab can crudely be thought of as consisting of two zones: a 'fully-ionized zone' (FIZ) where all H is ionized (*i.e.*, $n(H^+)/n(H) = 1$) and a 'partially ionized zone' (PIZ) where H is mostly neutral ($n(H^+)/n(H) \leq 0.1$). The FIZ is ionized and heated by low energy photons ($13.6 \leq h\nu \leq 200\,eV$) and for photon fluxes and gas densities of interest, it is cooled principally by [OIII]λ5007. The PIZ, on the other hand, is heated by X-ray photons ($h\nu > 200\,eV$) and emits low ionization lines such as [OI]λ6300 and [SII]λ6725. The [OI]λ6300/Hα ratio thus provides a useful way of discriminating between gas photoionized by 'hard' continua with strong X-ray components which create a large PIZ and stellar photoionized H II regions in which the PIZ is absent (see for example Veilleux & Osterbrock 1987). Several other useful line ratios will be used in constructing the diagrams. The recombination ratio HeIIλ4686/Hβ depends, to a good approximation, only on the fraction of ionizing photons effective in ionizing He$^+$ and therefore, also provides a useful indicator of continuum 'hardness' (*c.f.*, Binette, Courvoisier & Robinson 1987). The [OIII]λ-5007/Hβ ratio is sensitive to U and since [OIII]λ5007 is the principal coolant of the FIZ, also measures the efficiency with which this zone is heated and hence also depends on $S(\nu)$. Finally, because [OIII]/Hβ increases while [OI]/Hα decreases with increasing U, the [OI]/[OIII] ratio is a more sensitive measure of this parameter than [OIII]/Hβ alone.

4 Spectra of EELR

The emission line spectra of EELR are strikingly similar to those of the *nuclear* narrow emission line regions (NNLR) of active galaxies. This is demonstrated in Fig. 1 where almost all EELR lie within the region occupied by the NNLR. Only a few stray towards the area occupied by the the starburst and H II-region galaxies, the most notable of these being Minkowski's Object which van Breugel *et al.*, (1985b) suggest is a starburst triggered by a collision with a radio jet. Similarly, the upper limit on [OI]/Hα for the high velocity system of NGC 1275 is also suggestive of stellar photoionization, consistent with the interpretation of Kent & Sargent (1979).

A notable feature of Fig. 1 is that, as measured by the [OIII]/Hβ ratio, both the NNLR and the EELR show a trend in excitation which follows the power law photoionization model locus. Amongst AGN this trend is well known and explained as a decrease in U reflecting the decrease in the ionizing continuum luminosity of the active nucleus from Seyfert galaxies to Liners (*e.g.*, Ferland & Netzer 1983, Halpern & Steiner 1983, Binette 1985). The correspondence between EELR and NNLR extends down to low [OIII]/Hβ, the cooling flow filaments apparently forming a low ionization tail comparable to the Liner nuclei.

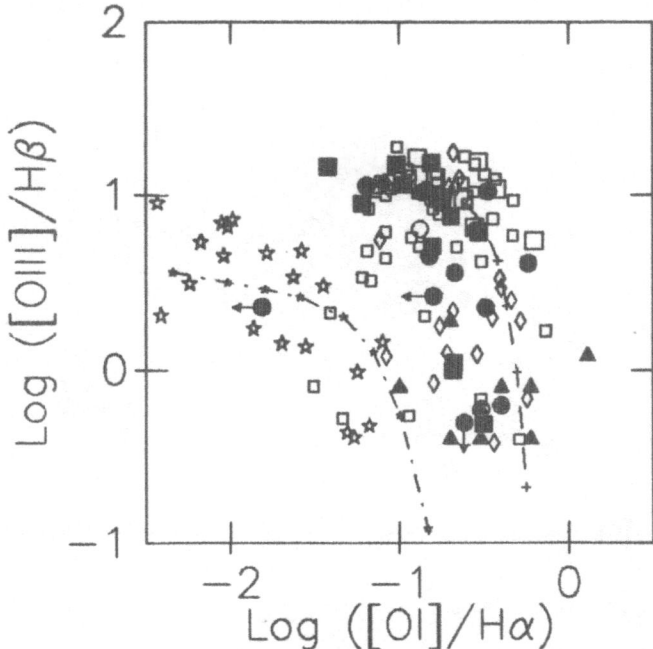

Figure 1: Comparision of extended emission line regions (large filled symbols) with the nuclear narrow emission line regions of the host galaxies (large open symbols), other active galactic nuclei, starburst galaxies and H II region galaxies (small open symbols). See table 1 for key and sources. Upper limits to [OI]/Hα are for Minkowski's object and the high velocity system of NGC 1275 (see text). Models: theoretical ionization parameter loci corresponding to power-law (spectral index $\alpha = -1.5$; dashed line) and stellar atmosphere (effective temperature $T_{eff} = 5 \times 10^4$ K; dot-dashed line) ionizing continua; U increases from bottom to top along the loci.

This effect can be seen more clearly in Figs 2a and 2b where [OIII]λ5007/Hβ and [OI]λ6300/Hα, respectively, are plotted against an excitation axis defined by [OI]/[OIII] for the EELR, the corresponding NNLR and several sequences of photoionization models. The distribution of the EELR in these diagrams is strongly suggestive of a trend runing from very high excitation EELR characterised by [OIII]/Hβ> 10 and [OI]/[OIII]≈ 0.01 to cooling flow filaments for which [OIII]/Hβ≤ 1 and [OI]/[OIII]≥ 1. Similar diagrams can be constructed for [SII]/Hα and [NII]/Hα (Robinson et al., 1987).

The theoretical U locus corresponding to a stellar atmosphere continuum, which represents an an extreme case for normal stars (effective temperature $T_{eff} = 5 \times 10^4$ K), falls well below the general trend. Furthermore, almost all stellar photoionized H II regions have [OI]λ6300/[OIII]λ5007< 0.01 (Evans & Dopita 1985) and most would fall beyond the left hand edges of Figs 2a and b. It appears, therefore, that photoionization by normal stars can be ruled out as a general excitation mechanism for EELR.

Most of the EELR cluster about the U-loci corresponding to power law (spectral index $\alpha = -1.5$) and hot blackbody ($T_{bb} = 1.3 \times 10^5$ K) ionizing continua. These loci are almost

298

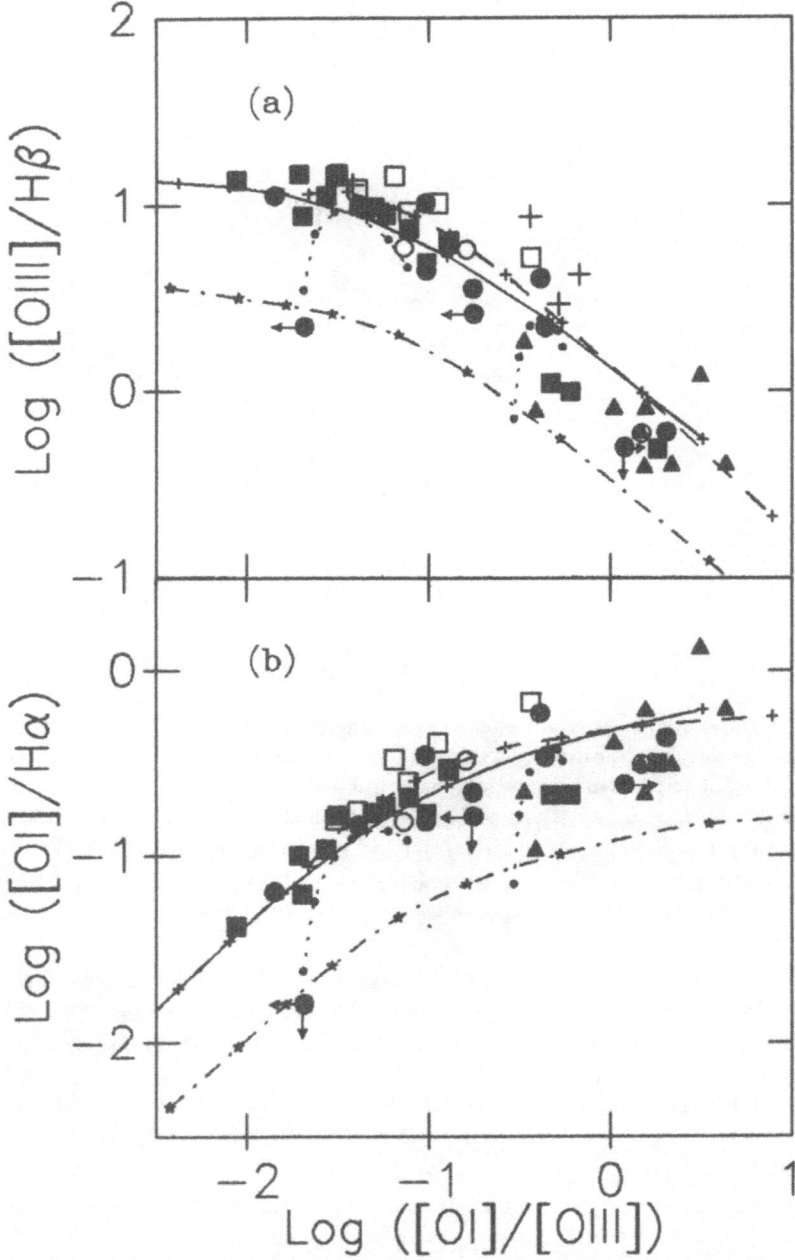

Figure 2: Line ratios of EELR and the host galaxy NNLR: (a) [OIII]λ5007/Hβ and (b) [OI]λ6300/Hα plotted against [OI]λ6300/[OIII]λ5007. All data are corrected for reddening derived from Hα/Hβ except for NNLR (crosses) in which Hα is contaminated by a broad component or or blended with [NII]. Other symbols as for Fig. 1. Models: U-loci corresponding to photoionization by stellar atmosphere ($T_{eff} = 5 \times 10^4$ K; dot-dashed line), power-law ($\alpha = -1.5$; dashed line) and blackbody ($T_{bb} = 1.3 \times 10^5$ K; full line) ionizing continua (U increases from left to right along loci); abundance-loci (dotted lines) for blackbody continuum at two values of U (abundances run from 0.1 to 2× solar).

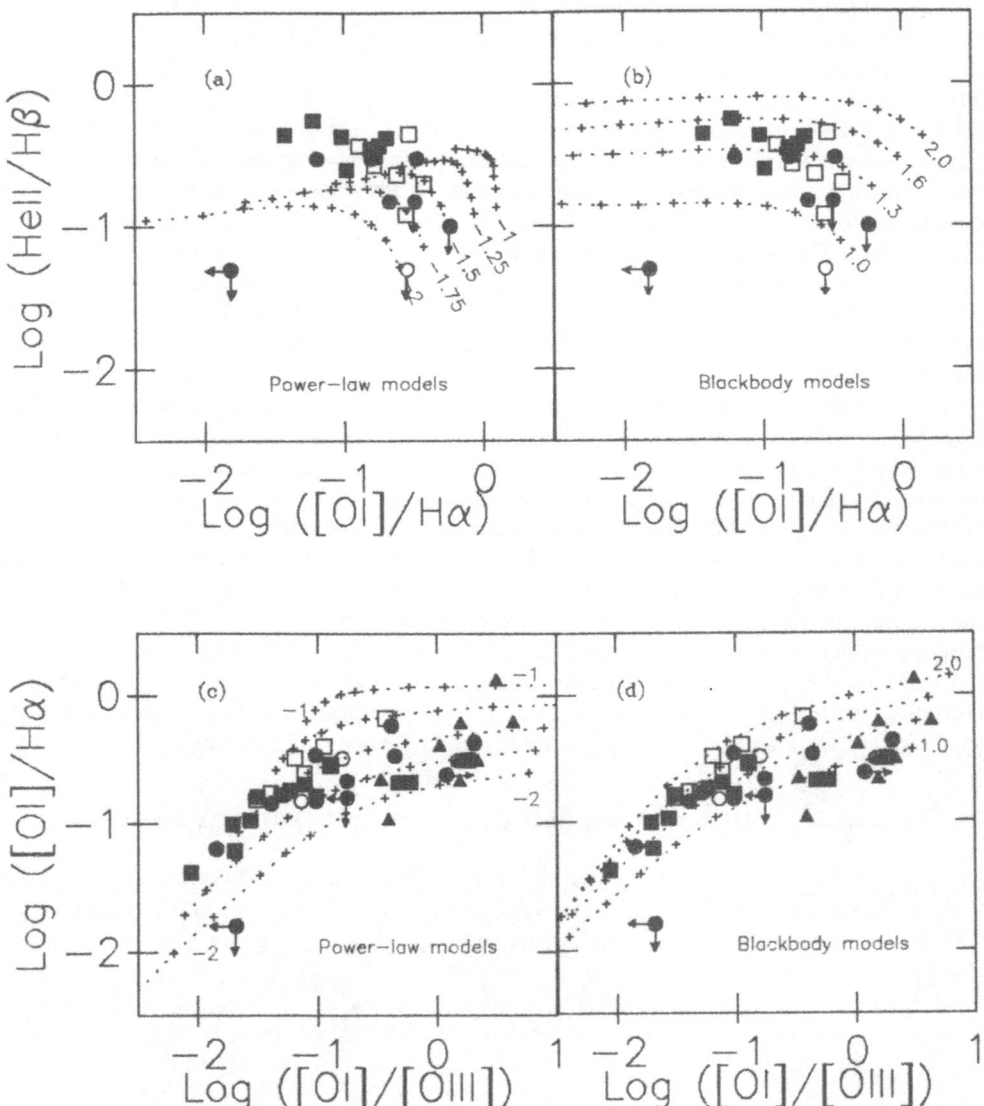

Figure 3: Comparison of EELR and corresponding NNLR with theoretical U-loci for photoionization by power-law and blackbody continua: (a) HeII/Hβ versus [OI]/Hα for power-law continua of spectral index $\alpha = -2, -1.75, -1.5, -1.25, -1$; (b) as (a) but for blackbody continua of temperature $T_{bb} = (1.0, 1.3, 1.6, 2.0) \times 10^5$ K; (c) [OI]/Hα versus [OI]/[OIII] for power law models; (d) as (c) but for blackbody models. Data as Fig. 2.

indistinguishable but if we assume that all scatter in the overall trend is due to differences in the spectral energy distribution, $S(\nu)$, we can attempt to set limits on α and T_{bb} for power law and blackbody photoionization models separately. It is apparent from Fig. 3a, in which HeII/Hβ is plotted against [OI]/Hα, that power law models cannot easily account for those EELR which have the highest observed values of HeII/Hβ. Making the model clouds matter bounded does not help; reducing N_c has the effect of truncating the PIZ while the He^{+2} zone at the illuminated edge of the cloud is unaffected with the result that the points move horizontally towards smaller [OI]/Hα. In the [OI]/Hα versus [OI]/[OIII] diagram (Fig. 3c), however, most EELR lie between the power law model loci corresponding to $\alpha = -1.75$ and -1.25. On the other hand, the high HeII/Hβ values do not pose any problem for the blackbody models. The observed spread in HeII/Hβ suggests a range $T_b = 1.0 - 1.6 \times 10^5$ K (Fig. 3b) and this is consistent with the scatter about the general trend in the [OI]/Hα diagram (Fig. 3d).

As with AGN, therefore, the trend from high to low excitation can be interpreted in terms of photoionization at progressively smaller ionization parameter. However, it is impossible to say whether the spectral energy distribution of the ionizing continuum for any given emission line region can be characterized by a power law or a high T_{bb} blackbody, except perhaps for the few EELR where high HeII/Hβ favours the latter. The required ionizing continuum is, nevertheless, AGN-like in the sense that it contains sufficient high energy photons to maintain a significant PIZ and therefore produce [OI]λ6300. In this context, a useful measure of the 'hardness' of the spectral energy distribution is the mean ionizing photon energy, $\bar{\epsilon}$, which for the ranges in α and T_b derived from Fig. 3 is $30 \leq \bar{\epsilon} \leq 50\,eV$.

Sequences of calculations in which metallicity Z is varied from 0.1 to $2 \times Z_\odot$ for the hot blackbody ionizing continuum at two values of U are also plotted in Figs 2a and b. These loci show that abundances in the EELR are certainly $> 0.1Z_\odot$ and probably $> 0.25Z_\odot$.

5   Energetics and the location of the continuum source.

The most straightforward interpretation of these results is that the EELR are photoionized by the radiation field of the active galactic nucleus and that as seen by the EELR this is weaker, on average, for the low-excitation objects. This explanation has the virtue of being the simplest that can account for the observed distribution of the emission line ratios considered here but we cannot exclude the possibility that the nuclear and extended emission line regions of the same galaxy are photoionized by physically distinct continuum sources. The ionizing photon luminosity, Q_N, which the active nucleus would have to radiate in order to photoionize the surrounding EELR can be estimated from the observed Hβ luminosity, $L_{H\beta}$. Assuming all ionizing photons are absorbed, $L_{H\beta} \approx 5 \times 10^{-13} \Omega\, Q_N$ erg s^{-1}, where Ω is the fraction of the sky covered by the EELR system as seen from the nucleus. Thus, if Ω can be estimated from line images, for example, $L_{H\beta}$ gives a direct measure of Q_N and this can, in turn, be related to the continuum flux density at an observable wavelength by making a reasonable guess as to the form of the spectral energy distribution (for simplicity, this is usually approximated as a power law).

In quasars (Stockton & MacKenty 1987) and broad-line radio galaxies (Tadhunter 1987) there is no evidence that any ionizing source other than the central continuum is necessary to account for the observed EELR line luminosities. However, in at least three of the narrow-line radio galaxies of the SRG sample, a simple power law extrapolation of the comparatively weak non-stellar optical continuum would imply an ionizing photon luminosity too small to account for the observed EELR Hβ luminosities (Tadhunter 1987). This in itself does not necessarily imply that additional continuum sources are required. Recent UV and soft X-ray observations (*e.g.*, Arnaud *et al.*, 1985, Elvis *et al.*, 1986), suggest the presence of a 'big bump' dominating the AGN continuum in the far UV. As already described, an ionizing continuum radiating like a hot blackbody can reproduce the observed emission line ratios and if such a component is present, simply extrapolating a power law from the observed optical continuum would seriously underestimate Q_N. Another possibility is that the nuclear radiation field is anisotropic in such a way that the ionizing continuum recieved by the EELR is stronger than inferred from observations.

However, in one case, a highly ionized cloud in the outskirts of the radio galaxy PKS 2152–69, there is direct evidence from line intensity gradients and the presence of a localized optical continuum source that the ionizing radiation is generated within the cloud itself, probably as a result of a collision with a radio jet (Tadhunter *et al.*, 1987). Similarly, van Breugel *et al.*, (1985a) and van Breugel *et al.*, (1986) argue that the EELR around, respectively, 3C 277.3 and 4C 29.30 are photoionized *in situ* by ionizing radiation generated as a by-product of the interaction between the extended radio source and the ambient gas.

In at least a few cases, therefore, the ionizing continuum appears to be generated locally by mechanisms associated with the extended radio source. These local sources are still powered, albeit indirectly, by the active nucleus and furthermore, the line ratio diagrams show that the mean ionizing photon energy must be similar to that of the *nuclear* continuum. Another process which might be important in the vicinity of the radio sources is direct heating and ionization by relativistic particles which can significantly increase the emission line luminosities without greatly affecting the relative intensities (Ferland & Mushotzky 1984). However, it is unlikely that this, *in situ* generation of ionizing radiation or indeed any other process associated with the extended radio source could power all EELR because emission line gas is often found at large separations from the radio structures. In general, therefore, it is difficult to avoid the conclusion that most EELR are predominantly photoionized by the radiation field of the active nucleus.

6 Implications for cooling flow filaments

It is clear from the line ratio diagrams that photoionization by an AGN-like continuum can explain the relative intensities of the emission lines associated with cooling flows. Moreover, the distribution of EELR in the diagrams suggests a continuity from the highly excited nebulosities, which are almost certainly powered (directly or otherwise) by nuclear activity, to the low ionization cooling flow systems.

In at least some of the central galaxies, the radiation field of the active nucleus may be strong enough to produce the observed $L_{H\beta}$. Thus, photoionization by the AGN continuum seems plausible for all three of the radio galaxies in the SRG sample (PKS 1404–267, PKS 2152–69[2] and PKS 0521–36) which have low ionization EELR (Tadhunter 1987). Similarly, the UV luminosity (at 1500Å) of the unresolved continuum sources detected in 4C 26.42 (Abell 1795) and NGC 1275 by Nørgaard-Nielson, Jørgenson & Hansen (1984) suggests that,

---

[2]That is, the low ionization filaments; as discussed above the high ionization cloud requires a local continuum source.

in each case, the ionizing continuum of the active nucleus is powerful enough to produce the line luminosities of the surrounding EELR. Even if the active nucleus is so weak that its presence is marked only by the existence of an extended radio source it is possible, as discussed in the previous section, that the emission line filaments are powered by ionizing radiation generated locally by the radio source itself.

In general, of course, other ionization mechanisms such as shocks or other sources of ionizing photons which are intrinsic to the cooling flow phenomenon, such as hot stars and the ambient thermal plasma, will be relatively more important when nuclear activity is weak. It seems, however, that none of these mechanisms acting alone can adequately account for the line emission. Thus, although shocks would produce line ratios similar to those observed, each blob of cooling gas would have to undergo 50–100 shocks in order to reconcile the observed $L_{H\beta}$ with mass inflow rates deduced from X-ray data (Johnstone et al., 1987). Similarly, it is unlikely that photoionization of the cooling gas by the diffuse X-ray emission of the hot component could produce the observed $L_{H\beta}$. On the other hand, star formation may generate a sufficent ionizing luminosity but as we have seen, photoionization by normal stars alone cannot reproduce the strong [OI]λ6300 which is observed. Hence it may be more appropriate to consider a combination of excitation mechanisms and in this context the evidence outlined above shows that it is important to take account of the possible role of nuclear activity.

7 Conclusion

Extended emission line regions are commonly found around quasars and active galaxies which have extended radio sources. The spectra of these EELR show a remarkably well defined trend in excitation which is very similar to that shown by the narrow emission line regions of AGN and to which emission line filaments associated with cooling flows apparently form a low ionization tail. Comparison with theoretical ionization parameter loci suggests that the trend can be explained in terms of photoionization by an 'AGN-like' ionizing continuum characterized by a mean ionizing photon energy of between 30 and 50 eV. In most cases the source of the ionizing radiation is related to nuclear activity, the EELR being photoionized directly by the radiation field of the active nucleus itself or occasionally, by ionizing radiation produced in situ by mechanisms associated with the extended radio source. The situation is more complex for low ionization 'cooling flow' EELR but it is at least plausible that ionizing radiation generated by nuclear activity plays a significant role.

References

Arnaud, K.A., Branduardi-Raymont, G., Culhane, J.L, Fabian, A.C., Hazard, C., McG-lynn, T.A., Shafer, R.A., Tennant, A.F., & Ward, M.J., 1985. *Mon. Not. R. astr. Soc.* **217**, 105.

Baldwin, J.A., Phillips, M.M. & Terlevich, R., 1981. *Pub. astr. Soc. Pacific*, **93**, 5.

Binette, L., 1985. *Astron. Astrophys.*, **143**, 334.

Binette, L., Courvoisier, T.-L. & Robinson, A., 1987. *Astron. Astrophys.*, in press.

Binette, L., Dopita, M.A., & Tuohy, I.R., 1985. *Astrophys. J.* **297**, 476.

Cohen, R.D., 1983. *Astrophys. J.*, **273**, 489.

Cohen, R.D. & Osterbrock, D.E., 1981. *Astrophys. J.*, **243**, 81.

Cowie, L.L., Fabian, A.C. & Nulsen, P.E.J., 1980. *Mon. Not. R. astr. Soc.*, **191**, 399.

Davidson, K. & Netzer, H., 1979. *Rev. Mod. Phys.*, **51**, 715.

Elvis, M., Green, R.F., Bechtold, J., Schmidt, M., Neugebauer, G., Soifer, B.T., Matthews, K. & Fabbiano, G., 1986. *Astrophys. J.* **310**, 291.

Evans, I.N. & Dopita, M.A., 1985. *Astrophys. J. Suppl.*, **58**, 125.

Ferland, G.J. & Mushotzky, R.F., 1984. *Astrophys. J.*, **286**, 42.

Ferland, G.J. & Netzer, H., 1983. *Astrophys. J.*, **264**, 105.

Ford, H.C. & Butcher,H., 1979. *Astrophys. J. Suppl.*, **41**, 147.

Forman, W., Jones, C. & Tucker, W., 1985. *Astrophys. J.* **293**, 102.

Fosbury, R.A.E., 1986. In *Structure and Evolution of Active Galactic Nuclei*, Trieste, April 10–13 1985, Eds. Giuricin,G. *et al.*, Reidel, pp297–308.

Halpern, J.P. & Steiner, J.E., 1983. *Astrophys. J.*, **269**, L37.

Heckman, T.M. 1981. *Astrophys. J.* **250**, L59.

Heckman, T.M., Miley, G.K., Balick, B., van Breugel, W.J.M. & Butcher, H.R., 1982. *Astrophys. J.* **262**, 529.

Ford, H.C. & Butcher, H.R., 1979. *Astrophys. J. Suppl.* **41**, 147.

Hu, E.M., Cowie, L.L. & Wang, Z., 1985. *Astrophys. J. Suppl.*, **59**, 447.

Johnstone, R.M., Fabian, A.C. & Nulsen, P.E.J., 1987. *Mon. Not. R. astr. Soc.*, **224**, 75.

Kent, S.M. & Sargent, W.L.W., 1979. *Astrophys. J.*, **230**, 667.

Nørgaard-Nielson, H.U., Jørgenson, H.E. & Hansen, L., 1984. *Astron. Astrophys.*, **135**, L3.

Osmer, P.S., 1978. *Astrophys. J.*, **226**, L79.

Osterbrock, D.E., 1984. *Q. Jl. R. astr. Soc.*, **25**, 18.

Phillips, M.M., 1981. *Mon. Not. R. astr. Soc.*, **197**, 659.

Robinson, A., Binette, L., Fosbury, R.A.E. & Tadhunter, C.N., 1987. *Mon. Not. R. astr. Soc.*, **227**, 97.

Stockton, A. & MacKenty, J.W., 1987. *Astrophys. J.*, **316**, 584.

Tadhunter, C.N., 1987. *D.Phil. Thesis*, University of Sussex.

Tadhunter, C.N., Fosbury, R.A.E., Binette, L., Danziger, I.J. & Robinson, A., 1987. *Nature*, **325**, 504.

van Breugel, W.J.W., 1986. *Can. J. Phys.*, **64**, 392.

van Breugel, W.J.W., Filippenko, A.V., Heckman, T.M., & Miley, G.K., 1985b. *Astrophys. J.*, **311**, 58.

van Breugel, W.J.W., Heckman, T.M., & Miley, G.K., 1984. *Astrophys. J.*, **276**, 79.

van Breugel, W.J.W., Heckman, T.M., Miley, G.K., & Filippenko, A.V., 1986. *Astrophys. J.*, **311**, 58.

van Breugel, W.J.W., Miley, G.K., Heckman, T.M., Butcher, H. & Bridle,A.H., 1985a. *Astrophys. J.*, **290**, 496.

Unger, S.W., Pedlar, A., Axon, D.J., Whittle, M., Meurs, E.J.A., & Ward, M.J., 1987. *Mon. Not. R. astr. Soc.*, **228**, 671.

Veilleux, S. & Osterbrock, D.E., 1987. *Astrophys. J. Suppl.*, **63**, 295.

THE EVOLUTION OF EARLY-TYPE GALAXIES

Richard S Ellis
Physics Department
University of Durham
England

ABSTRACT: I review evidence from a number of optical searches for recent star formation in rich cluster early-type galaxies. High quality spectra of nearby ellipticals show an additional component of young starlight when compared to metal-rich globular clusters. Studies in intermediate redshift ($z \sim 0.4$) clusters show that early-type galaxies, selected as those obeying the optical colour-luminosity relation, often reveal ultraviolet excesses and Balmer absorption lines indicative of recent star formation. At higher redshift ($z \sim 0.8$) although some galaxies are found to have completed their star formation, the fraction that have done so remains unclear because of the complex selection effects involved in studying such remote objects. A consistent picture is emerging whereby, at all epochs studied so far, early-type galaxies suffer time-dependent bursts of star formation that are probably environmental in origin.

1. INTRODUCTION

If cooling flows are an important ingredient in the story of galaxy formation and evolution, it is important to consider evidence on the history of star formation with look-back time and see whether the form of such evolutionary changes can be naturally understood in the context of cooling flows. Ideally this should be done in a number of different environments, but the subject of galaxy evolution is observationally very young and here I will mostly concentrate on early type galaxies in rich clusters, since it is in this population that most of the X-ray studies of relevance have been made.

The traditional picture accounting for the mean stellar population and colour of the early type (E/S0) galaxy adopts a current star formation rate (SFR) substantially lower than the past time average. Such a trend is needed to account for the low present day gas fraction and the dominant evolved stellar population at optical and near ultraviolet wavelengths. This could arise in one of three ways:

- An initial burst of SF, whereby substantially all the gas was consumed in the

305

A. C. Fabian (ed.), Cooling Flows in Clusters and Galaxies, 305–313.
© *1988 by Kluwer Academic Publishers.*

first Gyr or so; passive evolution then reddens the stellar population.

- A declining SFR where the rate at any time is somehow governed by a smooth function e.g. Bruzual's (1983) exponential decay models.

- A series of star formation bursts whose intensity or frequency decreases with time.

A decreasing sequence of bursts might be regarded as a more realistic version of the smooth SFR decay model, but I separate these here since a burst has to be induced by some local process (c.f. Scalo 1988) whereas a smooth SFR decline is more indicative of some long-term phenomenon, although both might be driven by environmental effects like cooling flows.

To distinguish between the three forms of past activity, one can scrutinise the present day fossil stellar populations in elliptical galaxies via evolutionary synthesis (Rose 1985), make comparisons between hopefully similar types of objects at various redshifts (Ellis 1984, Djorgovski *et al* 1986), and search for the luminous phase associated with the initial burst of star formation (Baron and White 1987). Each of these topics is reviewed in detail more often than progress deserves. However, it is worth pointing out those weak assumptions made in previous analyses.

In interpreting and predicting evolutionary changes, we assume the model ingredients are correct. Until recently, metallicity changes were largely ignored in synthesis models (e.g. Bruzual 1983), but recent work shows these to be crucially important in modelling ellipticals (Arimoto and Yoshii 1987). Indeed, as observers are now able to gather *spectra* of distant objects rather than simply colours, there is a need to widen the stellar libraries to include metallicity changes, as well as to understand the later stages of evolution where infrared changes can also be monitored (Lilly 1987). At high redshifts we assume (often without evidence) that we can select representative objects in an unbiassed way in order to make comparisons over a range of redshifts. If this cannot be checked, the evolution inferred could easily be a distortion of the truth. Finally, whether there is an absence of luminous primeval galaxies or not depends on how readily we would recognise a primeval galaxy whose properties did *not* match those searched for. In this respect the safest way to place a limit on the number of primevals is to do spectral surveys to fixed apparent magnitude limits. Although the limits so far probed are not very deep (B $\sim$ 22 - Ellis 1987), no anomalously luminous objects have yet been found.

2. NEARBY GALAXIES

First we will review the observational limits on the proportion of star formation that occurred in typical early type galaxies over the last 3 - 5 Gyr. To do this we must first find the most sensitive probe of recent star formation.

Figure 1 shows the optical and uv spectral evolution of a normal passively-evolving (i.e. single burst) elliptical galaxy which suffers a second era of star formation amounting to 10 % by mass consumed within a brief 1 Gyr burst. The panel displays the subsequent spectral evolution and that in rest frame B - V colour is

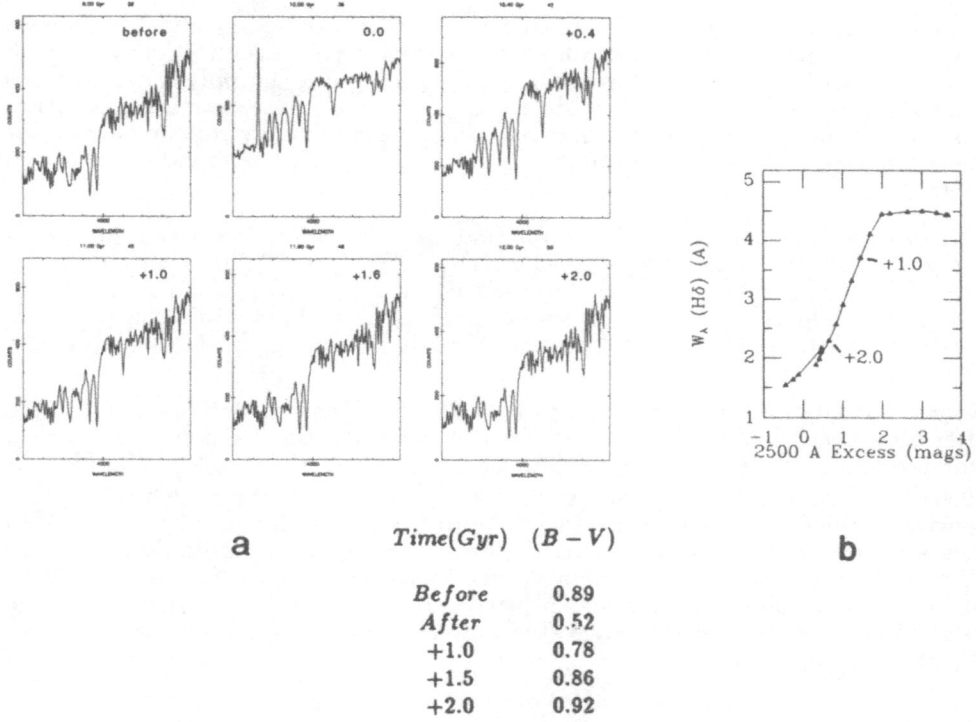

Time(Gyr)	$(B-V)$
Before	0.89
After	0.52
+1.0	0.78
+1.5	0.86
+2.0	0.92

Figure 1: Probes of star formation in otherwise passively evolving stellar systems: (a) Optical spectra during and after a 10% by mass burst of star formation superimposed on a 9 Gyr old initial burst. (b) Ultraviolet spectral energy distributions for the same model. In both cases, ages marked refer to times after the end of the burst.

tabulated alongside.

We note the following:

- Optical colours are only sensitive to this second burst during the actual star formation period. The homogeneity of optical colours for E galaxies of fixed luminosity therefore merely indicates the absence of on-going activity.

- Spectral indicators of the main sequence turn off temperature (Ca II K/ H + Hϵ, Hδ) and the 4000 Å break index, D_{4000}, remain sensitive up to about 1.5 Gyr after the burst has subsided, after which time the A stars responsible for these features have left the main sequence. High quality spectra are required to examine these features quantitatively. The model spectra here are gathered from the Jacoby *et al* (1985) library at 2 Å resolution.

- Ultraviolet colours at 2500 Å remain sensitive for somewhat longer because the continuum light here is sensitive to stars cooler than those dominating the Balmer spectrum. Such changes may also be observationally easier to detect at high redshift. Unfortunately, at low redshift the IUE satellite has only surveyed some 20-30 ellipticals and much of the data is of poor signal/noise. Furthermore, there is the added complication of contamination in the continuum from evolved stellar populations, such as post asymptotic giant branch stars, horizontal branch stars etc.

- Gravity sensitive indicators have been identified by Rose (1985) and applied to high signal/noise optical spectra of nearby E/S0s. He claims giant ellipticals in Virgo and nearby groups are dwarf-enriched compared to globular clusters of comparable metallicity. If the interpretation is correct, these studies suggest there was some activity in E/S0 galaxies over the last 5 Gyr (see also O'Connell 1980).

Rose's indicators not only promise to be valuable in sensing activity a long time after any recent burst, but also in terms of detecting small amounts of *current* star formation. In present day E/S0s, he claimed less than 2 % of the light at 4000 Å could come from a young stellar population; interestingly this is a *tighter* constraint than that imposed by the uv upturn analysed by Bertola *et al* (1985) and recall that latter analysis is critically dependent upon the assumption that the IUE short wavelength radiation is dominated by young starlight. Rose's comparison with metal rich globulars suggests, at least in some environments, that early type galaxies have an additional population whose age is substantially less than that of the primary population.

3. INTERMEDIATE REDSHIFT GALAXIES

A population of age 3-5 Gyr inferred today should have produced observable star formation at redshifts within reach if H_o is low. The problem in searching for the counterpart is that a 10 % burst superimposed on an otherwise uninteresting early type galaxy would transform its appearance and without morphological information such a galaxy would be difficult to distinguish from a normal spiral. However, by using spectral/colour indices in combinations that yield information on the ratio of *current-to-average star formation*, and probing in rich cluster environments where spirals are normally rare, this problem can be overcome.

Couch and Sharples (1987) have completed a thorough spectroscopic investigation of 3 z = 0.31 rich clusters, each of which from photometric studies (Couch 1981) showed significant blue excesses, compared to present day counterparts of comparable central concentrations, in the sense originally claimed by Butcher and Oemler (1978). Their spectral indices (principally Hδ) and colours show that the majority of galaxies bluer than early types (forming the excess) do *not* share the properties of normal spiral galaxies - their current star formation rate is either unusually high (and emission lines are visible) or was high very recently. The latter systems show deep Balmer absorption lines characteristic of main sequence A stars with ages $\sim$ 1 Gyr first identified in other distant clusters by Dressler and Gunn (1983).

At z $\sim$ 0.4, the rest-frame ultraviolet emerges into the range of optical CCD detectors and MacLaren *et al* (1987) have used this advantageously to probe similar

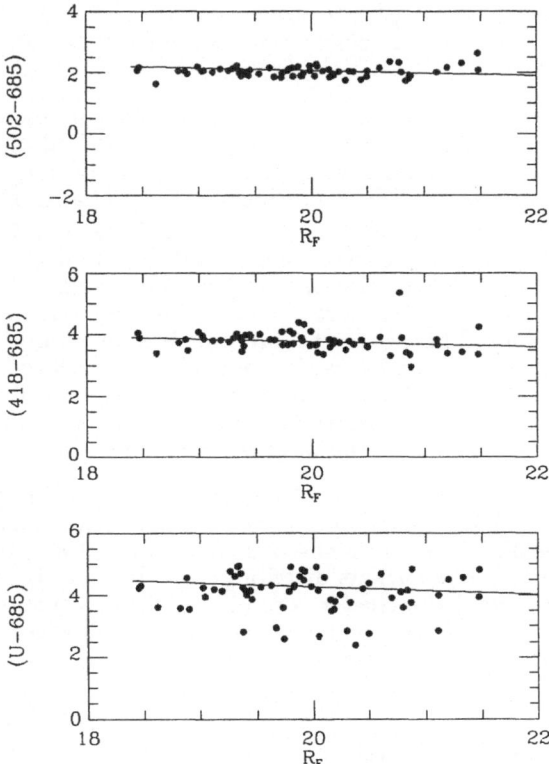

Figure 2: Colour-luminosity relations for objects classed as early-type members in Abell 370 (z = 0.37) from the multi-passband analysis of MacLaren et al (1987)

clusters to rest wavelengths of $\sim$ 2500 Å. They find enhancements of uv flux, but only in *some* optically red galaxies (Figure 2). Limited spectroscopy of these "UVX" E/S0s, suggests post-starburst features such as Hδ are unusually common in this subset of red galaxies.

MacLaren *et al* discuss a unifying picture whereby some cluster members suffer a burst of star formation for some reason, become optically active (blue B-O galaxies) and decay via the post-starburst phase to become optically red galaxies with uv excesses. About 3-5 Gyr after the burst, it requires careful detective work to find any evidence at all for this activity. In their analysis of Abell 370 (z = 0.37) they find evidence for various stages of this cycle simultaneously in one cluster, strongly suggesting the bursts are locally induced.

On the other hand *some* early-type cluster galaxies at high redshift are found to be indistinguishable from present day cluster E/S0s. Couch and Sharples analysed a composite spectrum of these and concluded < 6 % of the light at restframe 4000 Å could come from young stars - almost as tight a constraint as that reached by Rose nearby. However, at this redshift, as discussed in §2, this only pushes back the epoch of star formation activity in these galaxies to a redshift of z ~ 0.8. The application of this technique to higher redshift clusters will be an important constraint (c.f. Hamilton 1985 and §4).

It is appealing therefore to consider that the activity seen in clusters at z ~ 0.3-0.4 is related to the intermediate age population. Couch and Sharples specifically suggested their star-forming galaxies were the precursors of today's S0 galaxies but preliminary studies by Rose, Ellis and Sharples (in preparation) using a large database at 2 Å resolution in the nearby cluster Abell 2670 (z = 0.07, Sharples, Ellis and Gray 1987), failed to find any difference in Rose's line indices between Es and S0s. Furthermore, the UVX phenomenon found by MacLaren et al increases the proportion of objects undergoing this sort of activity beyond just those observed to have strong Hδ absorption. What we see at any time is a lower limit to the number that have undergone this cycle unless we are especially priveleged. Therefore, it is tempting to consider the possibility that we are seeing some phenomenon that affects *most* early-type galaxies rather than simply a path for producing present day S0s. Morphological information would, of course, help to resolve this question.

The origin of these bursts must surely be of importance for cooling flow advocates. Alternative ideas proposed for the enhanced star formation include infalling spirals shocked in collisions with the cluster's gaseous wall (Gunn 1988) or galaxy interactions (Thompson 1987). The curious distribution of UVX objects in Abell 370, showing a preference to avoid the cluster core (Figure 3) is a further clue perhaps supporting the infall picture. However, the objection to both these explanations is the dearth of similar activity in present day clusters.

4. HIGH REDSHIFT GALAXIES

Beyond z ~ 0.5 we can no longer see *typical* M* elliptical galaxies - we only observe what we are allowed to see. Even rich clusters where luminous galaxies are in abundance are subject to complex selection difficulties. The largest catalogues of distant clusters are contrast-selected from wide field prime focus photographs (c.f. Gunn et al 1986, Ellis et al 1988). Elementary considerations (c.f. Couch et al 1984) reveal that a cluster of richness comparable to Coma should *not* be seen beyond z > 0.7 unless its galaxies are more luminous than present-day counterparts (due to evolutionary effects) or it overlaps a foreground cluster.

The first problem is dangerous for those who attempt to compare the properties of cluster galaxies at different redshifts. If luminosity evolution is even partially induced by local processes that occur over a wide range of redshift, the bias would introduce a false evolutionary trend; at high z we only see those clusters exhibiting strong luminosity and colour evolution. The second problem of overlapping clusters makes it essential to gather as many spectroscopic redshifts as possible for very distant clusters. In the AAT programme, Couch, MacLaren and I have now gathered typically 4 redshifts/cluster for approaching 75 % of our strictly contrast-limited IIIa-J and IIIa-F distant cluster survey. The average interloper field redshift in these areas is larger than expected for our spectroscopic limit of R ~ 22 confiriming that foreground contamination by other z~ 0.4-0.5 *clusters* occurs quite often.

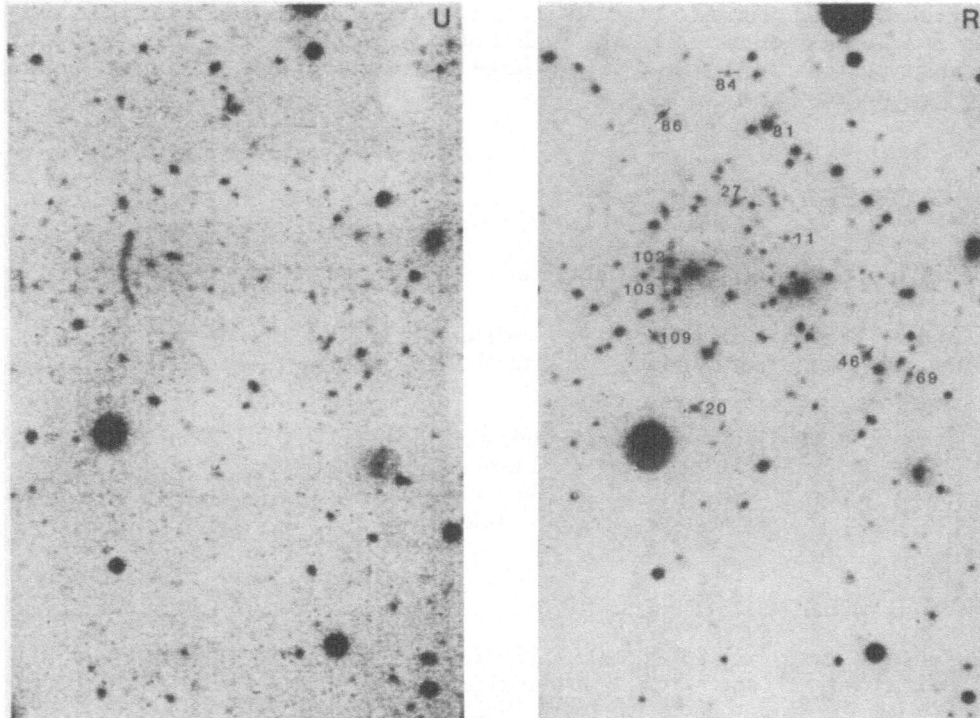

Figure 3: Deep U and red CCD frames for Abell 370 (z = 0.37) from the study of MacLaren et al(1987); the U band probes to 2500 Å in the cluster restframe. Marked galaxies are red objects with strong ultraviolet excesses consistent with star formation over the last 2-3 Gyr.

Assuming the disentangling can be done reliably, there are two simple tests to perform with distant cluster samples. The first is to examine some spectral index such as the 4000 Å break index, D_{4000}, to see if most cluster galaxies were star forming at large z. Hamilton (1985) selected the *reddest* objects in his deep field analysis of D_{4000}. Whilst he found examples of old systems at high z, it is difficult to know what proportion of objects share this property. Star formation could still be predominant at these epochs but sporadic in form. This would guarantee that *some* passive objects could always be found.

Controlled cluster samples could quantify the proportion of bursting objects at high z. Preliminary results have been discussed by Gunn (1988). The important point is to ensure that the spectroscopic targets are chosen by well-understood photometric criteria. Other workers (Persson 1988) refer to a "red envelope" effect similar to that observed by MacLaren *et al* which seems a clear indication of locally affected galaxy evolution.

The second technique is to use multicolour photometry to extract colour-luminosity sequences for members and to simultaneously examine mean colours and their scatter at various wavelengths for the *reddest* objects selected in bands unaffected by young stars (c.f. Ellis *et al* (1985). At large z the addition of K photometry would be advantageous here.

5. CONCLUSIONS

Recent work at a variety of redshifts leads us to two principle conclusions:

- Nearby E/S0 spectra cannot be reconciled with a single initial burst of star formation at some remote epoch; a substantial amount of more recent star formation is needed.

- At moderate redshifts, evidence for this extended star formation can be found. It manifests itself in various short-lived forms and strongly suggests such secondary bursts of star formation are driven by local events in rich clusters. The enigma to solve is why such activity is not prevalent in present day clusters.

ACKNOWLEDGEMENTS

Thanks are due to my collaborators Warrick Couch, Ray Sharples and Iain MacLaren on various aspects of the distant cluster work. I acknowledge useful discussions with Jim Gunn, David Koo, Gus Oemler and Jim Rose.

REFERENCES

Arimoto, N and Yoshii, Y 1987 *Astron. Astrophys.* 173, 23.

Baron, M and White, S D M 1987 preprint.

Bertola, F, Capaccioli, M and Oke, J B 1982 *Ap. J.*, 254, 494.

Butcher, H and Oemler, A 1978 *Ap. J.*, 219, 18.

Bruzual, G 1983 *Ap. J.*, 273, 105.

Couch, W J 1981 Ph.D., Australian National University.

Couch, W J and Sharples, R M 1987 *MNRAS*, in press.

Couch, W J, Ellis, R S, Kibblewhite, E J, Malin, D F and Godwin, J 1984 *MNRAS* 209, 307.

Djorgovski, S, Spinrad, H and Marr, J 1986 "New Aspects of Galaxy Photometry", ed Nieto, J-L, Springer Verlag.

Dressler, A and Gunn, J E 1983 *Ap. J.* 263, 533.

Ellis, R S 1984, "Spectral Evolution of Galaxies" ed. Gondhalekar, P, RAL Lecture Series RAL 84-002

Ellis, R S 1987, *IAU Symposium 124*, "Observational Cosmology", eds. Hewitt et al, D. Reidel Publ. Co., 367.

Ellis, R S, Couch, W J, MacLaren, I and Koo, D 1985 *MNRAS* 217, 239.

Ellis, R S, Couch, W J, MacLaren, I and Malin, D F 1988, in preparation

Gunn, J E 1988 in "Towards Understanding Galaxies at High Redshift", eds. Kron and Renzini.

Gunn, J E, Hoessel, J and Oke, J B 1986 *Ap. J.* 306, 30.

Hamilton, D 1985 *Ap. J.* 297, 31.

Jacoby, G H, Hunter D A and Christian C A 1986 *Ap. J. Suppl.* 56, 278.
Lilly, S J 1987 *MNRAS*, in press.
MacLaren, I, Ellis, R S and Couch, W J 1987 *MNRAS*, in press.
O'Connell, R 1980 *Ap. J.* 236, 436.
Persson, E 1988 in "Towards Understanding Galaxies at High Redshift", eds. Kron
        and Renzini.
Rose, J A 1985 *Astron J* 90, 1927.
Scalo, J 1987 in "Towards Understanding Galaxies at High Redshift", eds. Kron
        and Renzini.
Sharples, R M, Ellis, R S and Gray, P M 1987 *MNRAS*, in press
Thompson, L 1987 *Ap. J.* in press

SOME IMPLICATIONS OF DISTANT COOLING FLOWS

A.C. Fabian
Institute of Astronomy
Madingley Road
Cambridge CB3 OHA
U.K.

ABSTRACT. There are cooling flows in many nearby clusters and X-ray evidence for them out to redshifts of $\sim$ 0.5. Cooling flows would have been more common at earlier epochs if clusters are built up in a hierarchical fashion out of subclusters. Disruption of the cooling flows as the subclusters merge provides a mechanism for changing the accretion flow on to active nuclei and for the recent evolution observed in powerful radio galaxies and quasars. Much of the dark matter around large elliptical galaxies may be due to early strong cooling flows, which also play a significant part in their formation.

1. INTRODUCTION

Cooling flows appear to be relatively common at the present epoch. The mass deposition rate in some systems is sufficiently high that the total mass accumulated over a Hubble time is comparable to the luminous mass of the underlying galaxy. It is therefore of some interest to consider whether cooling flows are long-lived (see also White 1988) and so are part of the formation process of galaxies. Furthermore, if cooling flows were both more common and of greater magnitude in the past, then they could contribute significantly to the dark matter and to the formation and evolution of many galaxies.

I review here the relevant properties of nearby cooling flows before outlining how more distant ones could be observed and, indeed, whether they are already detected. Many nearby flows contain an active nucleus or radio source which may be powered by accretion from the cooling gas. It is possible that the evolution of some quasars and radio galaxies is related to the evolution of cooling flows through the hierarchical growth of clusters. The existence of many more luminous active nuclei in the past then argues for many more powerful cooling flows.

Finally, I discuss some of the outstanding problems to our understanding of cooling flows.

2. NEARBY COOLING FLOWS

2.1 Observational evidence for cooling flows

A cooling flow is defined here to be a region of quasi-hydrostatic gas in which the radiative cooling time, t_{cool}, lies between the age of the system, t_{sys}, and the dynamical free-fall time t_{ff}, i.e. $t_{sys} > t_{cool} > t_{ff}$. Strong heat sources are assumed to be absent. Cooling leads to

315

A. C. Fabian (ed.), Cooling Flows in Clusters and Galaxies, 315–324.

an increasing gas density as pressure is maintained to support the weight of the overlying gas. A rising density implies inflow. In order that the gas is distributed throughout the galaxy or cluster in the first place and is not just a small puddle at the centre, its sound speed must be of the same order as the local velocity dispersion; i.e. the gas temperature $T \sim 10^6 - 10^8$ K. The above timescale condition then means that the gas density, $n \gtrsim 10^{-4} - 10^{-2}$ cm^3, respectively, for a cooling flow to occur.

Most of the gas is then at X-ray emitting temperatures and the major observational evidence for cooling flows is derived from X-ray studies. X-ray images showing a highly peaked X-ray surface brightness distribution are usually indicative of a cooling flow (see Fig. 1 of Fabian, Nulsen & Canizares 1984). Stewart et al. (1984) and Arnaud (1985, 1988) have found that at least 50 per cent of clusters of galaxies show this property. It is also apparent in some poor clusters (Schwartz et al. 1980; Canizares et al. 1983) and in elliptical galaxies (Nulsen et al. 1984; Thomas et al. 1986).

Clear evidence for low temperature components in the X-ray spectrum of the brightest flows has been obtained from medium and high resolution X-ray spectroscopy (Canizares et al. 1979; Mushotzky et al. 1981; Lea, Mushotzky & Holt 1982; Canizares et al. 1982; also contributions by Mushotzky (1988) and Canizares (1988) to this Meeting). In some cases such as the Perseus cluster, the cooling gas has lost at least 90 per cent of its initial thermal energy. A strip scan of that cluster reported by Ulmer et al. (1987) shows the decreasing central temperature well. It is stressed several times in these Proceedings that the emission measures of the X-ray cool gas components agree with the simple cooling interpretation and would be extremely difficult to account for if there is strong heating or conduction.

Finally, there is evidence at ultraviolet (Fabian, Nulsen & Arnaud 1984; Norgaard Nielsen, Jorgensen & Hansen 1984), optical (Kent & Sargent 1979; Heckman 1981; Fabian et al. 1982; Cowie et al. 1983; Hu, Cowie & Wang 1985; Johnstone, Fabian & Nulsen 1987 and several contributions to these Proceedings) and radio (21 cm; Crane, Van der Hulst & Haschick 1982; Bregman 1988) for some cooled gas at temperatures below 10^3 K. Also, radio emission is common in and around the central galaxy of a cooling flow (e.g. 3C84; 4C26.42; Cygnus A; PKS 0745-191).

2.2 Mass deposition

The mass deposited as cooled gas can be very substantial:

$$\dot{M} \approx \begin{cases} 1\,M_\odot\,\mathrm{yr}^{-1} & \text{(isolated ellipticals)} \\ 10 - 100\,M_\odot\,\mathrm{yr}^{-1} & \text{(poor clusters)} \\ 10 - 1000\,M_\odot\,\mathrm{yr}^{-1} & \text{(rich clusters)} \end{cases}$$

The total mass accumulated within a Hubble time can thus easily exceed $10^{12}\,M_\odot$ in the higher $\dot{M}$ cases. The mass deposition profile inferred from X-ray imaging data indicates that $\dot{M}(r) \propto r$, approximately (Arnaud 1985). This means that $\dot{M}(r) \sim 1\,M_\odot\,\mathrm{yr}^{-1}\,\mathrm{kpc}^{-1}$ in the cores of many rich clusters of galaxies.

We (Thomas, Fabian & Nulsen 1987) have refined our analysis approach to the X-ray data by considering the emission to originate from a comoving inflow of a multiphase gas. The number of phases used equals the number of radial zones into which the data are divided. An example of the mass deposition profile is shown in Fig. 1a for A496.

The volume filling factors of the various phases determines the way in which matter is deposited Nulsen (1986) and is shown in Fig 1b. If the gas were homogeneous at each radius then it would cool into a singularity at the centre. The spread of densities shown

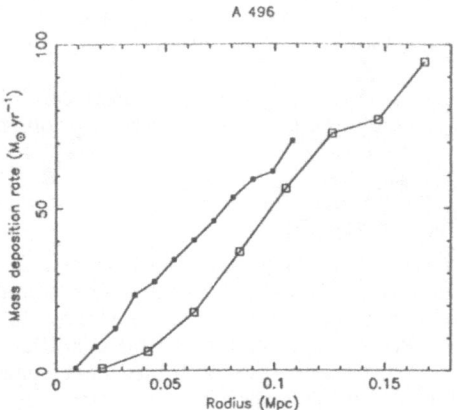

Figure 1a. Mass deposition profile of A496, from Thomas *et al.* (1987).

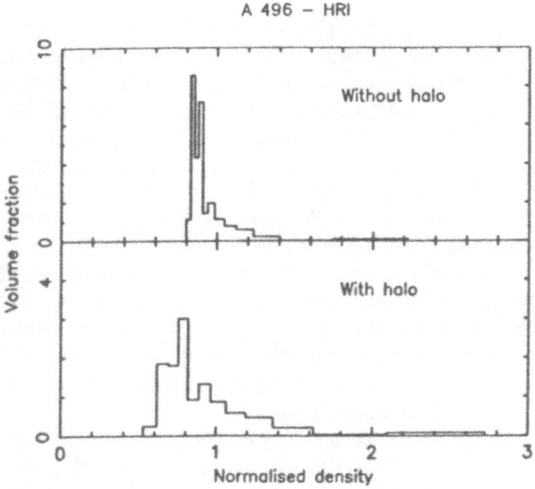

Figure 1b. Volume filling fraction of gas at 100 kpc in A496 (Thomas *et al.* 1987).

in Fig. 1b at a radius of 100kpc allows gas to cool and drop out of the flow throughout that region. The densest gas cools out at the largest radii whilst the rarest gas flows to the centre. The high density tail is consistent with radiative cooling.

What the bulk of the cooled gas forms remains a mystery. If stars, then the magnitudes and colours of the underlying galaxy require them to be predominantly of low mass (Cowie & Binney 1977; Fabian, Nulsen & Canizares 1982; Sarazin & O'Connell 1983). Between less than one and about ten per cent of the cooled gas forms 'normal' stars, similar to those found in the Solar Neighbourhood (Johnstone, Fabian & Nulsen 1987, Romanishin

1986 and O'Connell 1987 and in these Proceedings).

2.3 Implications from nearby Cooling Flows

The masses of stars formed from gas clouds depend upon local conditions such as the pressure and dust content, which are high and low, respectively, in cooling flows relative to our Galaxy. We infer

$$\frac{\dot{M}(\text{normal stars})}{\dot{M}(\text{low} - \text{mass objects})} \approx \frac{M_{visible}}{M_{dark}}$$

in the central galaxy. Therefore, cooling flows produce dark matter around cD and gE galaxies in the form of low-mass objects (Fabian et al. 1986a,b). The formation of high mass stars may require the presence of Giant Molecular Clouds or clumps of mass $> 10^3 \, M_\odot$, which may be absent from cooling flows except in the special conditions that create the observed optical filaments, where there may be some diffuse heat source. Most of the cooled gas forms into low-mass objects for which the mass deposition profiles inferred from X-ray data are consistent with isothermal dark haloes ($\rho \propto r^{-2}$). Consequently, larger flows in the past could generate much of the dark matter now seen around large galaxies (Fabian et al. 1986a,b).

The gas in the intracluster medium is expected to be multiphase and blobby. This is then the environment of any central black hole in an early-type galaxy and may affect the accretion rate.

The relatively large cooling flows in some poor clusters (e.g. MKW3s; 3A0335 + 096) and the absence of 'focussed' cooling flows in those clusters with two dominant galaxies (e.g. Coma and CA 0340-53) can be explained, in a picture of hierarchical clustering, by assuming that the poor clusters represent the conditions in the subclusters that merged to form the current rich clusters. This means that cooling flows were more common in the past when subclusters were more common. Disruption, mixing and heating during a subcluster merger can then lead to no cooling flow, an unfocussed flow or even an enhanced flow depending upon whether the merger took place between equal subclusters or between a subcluster and a rich cluster, and the impact parameters (Stewart et al. 1984; McGlynn & Fabian 1984). Subcluster mergers are also a good way to create an inhomogeneous intracluster medium, which is necessary for the distributed mass deposition. The implications of this picture are that much of the gas has been heated up in merger collisions and that the only substantial flows to have survived are either those in poor clusters that have not yet merged with any similar-sized or larger cluster, or those originally at the centres of large subclusters that have only accreted smaller clusters.

As emphasized elsewhere in these Proceedings by Nulsen, the good agreement between the spectroscopic data over a range of temperatures, and thus cooling times, with the imaging data indicate that cooling flows are long-lived and have ages $\gtrsim 10^{10}$ yr. The large fraction of clusters that contain flows also supports this view. Cooling flows are a major energy flow and involve 10^{62} ergs per cluster (~ 100 times the minimum energy of the radio source Cygnus A), so any heat source that significantly affects cooling flows has to be an important one.

3. DISTANT COOLING FLOWS

3.1 How are Distant Cooling Flows observed?

The most obvious means of detecting distant cooling flows is through their X-ray emission.

Extended soft X-ray emission from regions of 10 to 20 arcsec is expected. We must wait for the launch of ROSAT, AXAF and XMM before making much progress in that way. In the meantime, the Einstein Medium Sensitivity Survey (Gioia *et al.* 1984) may be turning out some flows at moderate redshifts ($z < 0.5$). We know that cooling flows exist at the distance of the 3C295 cluster ($z \sim 0.5$; Henry & Henriksen 1985). The ROSAT Sky Survey will contain many clusters of galaxies and should allow us to determine the fraction containing cooling flows with much more precision than is possible now. More detailed pointed observations, especially with AXAF, will reveal the distant flows. As argued here and by Fabian *et al.* (1986a), many of these flows will contain fairly powerful active galaxies so high spatial resolution may be needed to distinguish the point-source component from the intracluster one. This may have an important affect on the source counts (and thus redshift- dependent luminosity functions) of clusters. In a picture of hierarchical clustering without powerful central active galaxies we would expect the space density of clusters to increase with redshift and (ignoring spatial resolution problems and assuming that the gas temperature keeps most of the luminosity in the observed waveband) the slope of the source counts to be steeper than expected from no density evolution (a slope of 1.5 in Euclidean space). With an increasing probability of a powerful central active galaxy (quasar) at larger redshifts, together with a large cooling flow, there will be an increasing chance of missing the cluster emission, which will be swamped by that of the active galaxy. There is already some indication of a flattening of the cluster source counts (Gioia *et al.* 1984) which may be due to this effect. One implication of this possibility is that distant cooling flows should be sought around active objects.

Diffuse optical emission lines similar to, or enhanced above, those observed around NGC 1275 and PKS 0745-191 are a candidate indicator of cooling flows. Actual observational evidence for such emission is discussed in the next Section, but it is noted here that if density diagnostics in the line ratios indicate a high gas density ($\gtrsim 10\,\mathrm{cm}^{-3}$ at $T \sim 10^4\,\mathrm{K}$ so that $P = nT \sim 10^5\,\mathrm{cm}^{-3}\,\mathrm{K}$) then the high implied pressure suggests confinement by surrounding hot gas. This in turn can imply cooling flow conditions (Fabian *et al.* 1986a). We have used that argument to infer a cooling flow surrounding the radio-loud quasar 3C48 (Fabian *et al.* 1987; Crawford 1987). Hintzen & Romanishin (1986) have earlier emphasized the similarity between the extended emission-line region around the quasar 3C275.1 and nearby cooling flows. These indications are consistent with an increasingly strong link between cooling flows and active galaxies (or at least radio galaxies) with redshift.

Another way in which distant cooling flows may be observed is by absorption. The distributed deposition of cooled gas in cooling flows provides a significant covering fraction of cold gas that absorbs the light of background quasars (Crawford *et al.* 1987). The low-ionization, metal-rich intervening absorption line systems commonly seen in quasar spectra are strong candidates. The column density of CI and CII can be $10^{14}\,\mathrm{cm}^{-2}$. Different blobs suspended in the hot intracluster gas would give rise to the multiple absorption systems sometimes observed.

Evidence for extensive blobs of gas would also support the concept of a multiphase medium. Gas blobs may well need to be confined in some way and the thermal pressure of a surrounding diffuse gas at the virial temperature is the most likely explanation. This may eventually be observed with high-spatial and -spectral resolution X-ray instruments but at the present time less direct approaches are required. The manner in which radio jets propagate is of interest here. The report from Miley (1987) that radio sources are more bent at $z > 1.5$ is some evidence for a clumpy environment.

3.2 Are Distant Cooling Flows already detected?

The observational work of Spinrad & Djorgovski (1984a,b), McCarthy *et al.* (1988, this

Table 1

Line strengths in Cooling Flows and Radio Galaxies

	z	EW Ly-α	OII	FWHM km s^{-1}	Luminosity Ly-α	OII
N1275	0.018	> 86			> 6.10^{42}	1.10^{43}
P0745	0.1		170	660	< 1.10^{43}	4.10^{43}
A1795	0.06	200		480		> 2.10^{42}
3C265	0.81			600		
3C368	1.132			1000		6.10^{43}
3C326.1	1.82			1000	4.10^{44}	
3C324	1.206		755	650		6.10^{42}
3C13	1.4	~ 300				
3C266	1.272					8.10^{43}
3C256	1.819	> 340				
3C239	1.781	340				
P1614A	3.2	200		900		

meeting) and others has shown that many distant 3CR radio galaxies ($z \gtrsim 1$) have extended, low-ionization,emission lines of strengths and widths similar to those of the largest nearby flows. Their evidence is summarized in Table 1. Such emission lines are indirect, in terms of X-ray gas, but strong cooling flows in the past could well have associated powerful radio sources. The relationship with detectable star formation is not clear. Perhaps the radio emission induces both extensive optical emission and more massive stars. For example, the injection of relativistic plasma may increase the pressure and make the cooling gas more thermally unstable and promote the formation of large blobs of cold gas that form into massive stars. This is of course pure speculation without some theoretical guide to star formation, but the formation of several hundred solar masses of massive OB stars per year would create a spectacular object. We need a much clearer idea of what triggers which mode of star formation. Many nearby flows would look like the most distant 3CR galaxies if their cooled gas went into massive stars rather than quietly into low-mass stars.

I assume that where dense clumps of gas are observed they are pressure-confined by hotter, less dense gas. We only see the densest gas at optical wavelengths. The less dense gas is probably at the local virial temperature and constitutes the cooling flow and its outer atmosphere. $10^{11} - 10^{14}$ M$_\odot$ of hot gas can happily surround most distant galaxies, provided the potential well is deep and extensive enough, without being detectable. The optically-observed gas is only a small fraction of that which has cooled, or is cooling, and totals only $10^8 - 10^{10}$ M$_\odot$.

3.3 Some Implications for Distant Cooling Flows

If much of the mass of a present-day large galaxy was virialized, then $t_{sys} > t_{cool} > t_{ff}$ and its formation would resemble a cooling flow. The masses of elliptical galaxies typically exceed 10^{13} M$_\odot$ (Fabian *et al.* 1986) and, if much of the dark matter is baryonic and from a cooling flow,

$$\dot{M} \approx 10^4 \, \text{M}_\odot \, \text{yr}^{-1} \,,$$

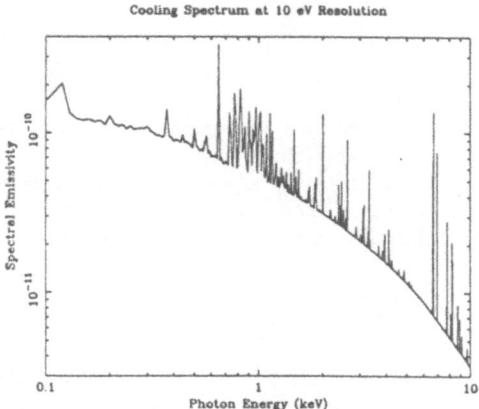

Figure 2. The soft X-ray spectrum of a cooling flow.

assuming a formation time $\sim 10^9$ yr. This implies a cooling luminosity

$$L_{cool} \approx \frac{5\dot{M}kT_{virial}}{2m}$$

$$\approx 10^{45}T_7 \, \text{erg s}^{-1} ,$$

where $T_{virial} = 10^7 T_7$ K. Most of this is radiated (Fig. 2) into the unobservable ultraviolet (now).

If hierarchical clustering has occurred (see e.g. Geller 1984) then, as already discussed, mergers of subclusters would be an important process for disrupting cooling flows. This mechanism provides a relationship between the evolution of radio galaxies and radio-loud quasars, which appear to be often in subclusters (Yee & Green 1984), with the recent evolution of clusters. The late decline in radio galaxies is due to the recent growth of rich clusters. Disruption of the radio soures may lead to radio haloes and wide-angle-tail sources. An alternative view on the evolution of cooling flows is given in these Proceedings by Valentijn (1987, 1988).

The mergers of subclusters will also mean that any spiral galaxies experience very large pressure changes. (The gas pressure in a cluster is 10 - 1000 times greater than that in a spiral such as our Galaxy.) Through induced star formation, this may help to explain the prevalence of blue galaxies in some distant clusters.

Finally, mergers of clusters will help to spread the dark matter formed through a cooling flow over the core and further in the new cluster. In this way it is possible that much of the cluster dark matter could be baryonic from cooling flows.

4. SOME PROBLEMS WITH COOLING FLOWS

I have summarized in earlier Sections the strong evidence for cooling flows and have suggested that they could have been more important in the past. Nevertheless there are some outstanding problems to be faced.

First, there is the problem of what is formed out of the cooled gas and how it can be detected. The gas probably cools into sheets rather than spherical blobs (cf. supernova remnants, Schwarz *et al.* 1972). Compression and acceleration of cosmic rays and magnetic fields in these sheets could contribute to the halo radio emission such as seen around 3C84. The high gas pressure in a cooling flow does reduce the Jeans mass,

$$M_J \approx 10^6 T_4^2 \left(\frac{P}{10^6 \, \text{cm}^{-3} \, \text{K}} \right)^{-1/2} M_\odot,$$

but to obtain sub-solar-mass objects requires that the gas cools below 10K. A simple application of the formula indicates that impossibly low temperatures may be needed to obtain Jupiters or still lower mass clumps directly. Fragmentation is presumably necessary.

The properties of cold, dust-free, metal-rich gas at high pressure have not been investigated in any detail yet. For example, does molecular hydrogen form? Can small blobs form by acting as 'cooling centres'. By this I mean that a cooled gas blob does not evaporate but absorbs and radiates the thermal energy of surrounding hotter gas. McKee & Cowie (1977) point out that this can occur if a cloud is large enough. If conduction is suppressed by a large fraction in the intracluster gas, then the size of clouds onto which gas will condense rather than evaporate will be much reduced.

The length scale on which conduction is important is relevant here. It appears that conduction is inhibited over large length scales, but its effects on smaller scales are unknown. The length scale on which the gas is turbulent and the scale sizes of gas blobs are also of interest. The optical data on emission line gas in nearby flows and around the distant 3CR galaxies show velocity FWHMs of up to $1000 \, \text{km s}^{-1}$ (Table I). Whether this is motion of both the hot and cool gases together or just of the cool gas is not yet known. It is possible that the scale sizes vary depending upon the subcluster merging rate and radio source activity in the cluster.

It is also interesting to speculate on the likelihood of cooling flows in the Coma cluster, which may be disrupted, in the Local Group and in the bulges of spiral galaxies. Widespread thermal gas at temperatures $< 2 \times 10^6 \, \text{K}$ around and in spiral galaxies would be very difficult to detect at the present time.

5. CONCLUSIONS

X-ray data show that cooling flows are common now and existed out to a redshift of at least 0.5 (3C295 cluster).

The hypothesis that higher mass-flow-rates were more common in the distant past ($z > 1$) is consistent with observations. There is some indirect evidence for them in quasar absorption lines and in bent radio jets as well as in optical emission lines. Cooling flows may be important for the formation and evolution of the largest galaxies, as well as for the evolution of their active nuclei.

References

Arnaud, K.A., 1985. *PhD thesis*, University of Cambridge.
Arnaud, K.A., 1988. These Proceedings.

323

Bregman, J.N., 1988. These Proceedings.
Canizares, C.R., Clark,G.W., Markert, T.H., Berg, C., Smedira, M., Bardas, D., Schnopper, H. & Kalata, K., 1979. *Astrophys. J.*, **234**, L33.
Canizares, C.R., Clark, G.W., Jernigan, J,G. & Markert, T.H., 1982. *Astrophys. J.*, **262**, L33.
Canizares, C.R., Stewart, G.C. & Fabian A.C., 1983. *Astrophys. J.*, **272**, 449.
Cowie, L.L. & Binney, J., 1977. *Astrophys. J.*, **217**,723.
Cowie, L.L., Hu, E.M., Jenkins, E.B. & York, D.G., 1983. *Astrophys. J.*, **272**, 29.
Crawford, C.S., Crehan, D.A., Fabian, A.C. & Johnstone, R.M., 1987. *Mon. Not. R. astr. Soc.*, **224**, 1007.
Crane, P.C., Van der Hulst, J.M. & Haschick, A.D., 1982. In: *Proc. IAU Sympn. 97*, eds. Heeschen, D.S. & Wade, C.M.,307.
Crawford, C.S., 1988. These Proceedings.
Fabian, A.C., Nulsen, P.E.J., Atherton, P.D. & Taylor, K., 1982. *Astrophys. J.*, **201**, 17P.
Fabian, A.C., Nulsen, P.E.J. & Canizares, C.R., 1982. *Mon. Not. R. astr. Soc.*, **201**, 933.
Fabian, A.C., Nulsen, P.E.J. & Canizares, C.R., 1984. *Nature*, **310**, 733.
Fabian, A.C., Nulsen, P.E.J. & Arnaud, K.A., 1984. *Mon. Not. R. astr. Soc.*, **208**, 179.
Fabian, A.C., Arnaud, K.A., Nulsen, P.E.J. & Mushotzky, R.F., 1986a. *Astrophys. J.*, **305**, 9.
Fabian, A.C., Arnaud, K.A. & Thomas, P.A., 1986b. In: *Proc. IAU Sympn. 117*, eds. Knapp, G. & Kormendy, J.
Fabian, A.C., Crawford, C.S., Johnstone, R.M. & Thomas, P.A., 1987. *Mon. Not. R. astr. Soc.*, **228**, 963.
Geller, M.J., 1984. *Comments on Astrophys. Space. Sci.*, **10**,47.
Gioia, I.M. *et al.* 1984. *Astrophys. J.*, **283**,495.
Heckman, T.M., 1981. *Astrophys. J.*, **250**, L59.
Henry, J.P. & Henriksen, M.J., 1986. *Astrophys. J.*, **301**, 689.
Hintzen, P. & Romanishin, W., 1986. *Astrophys. J.*,**311**, L11.
Hu, E.M., Cowie, L.L. & Wang, Z., 1985. *Astrophys. J. Suppl.*, **59**, 447.
Johnstone, R.M., Fabian, A.C. & Nulsen, A.C., 1987. *Mon. Not. R. astr. Soc.*, **224**, 75.
Kent, S.M. & Sargent, W.L.W., 1979. *Astrophys. J.*, **230**, 667.
Lea, S.M., Mushotzky, R.F. & Holt, S.S., 1982. *Astrophys. J.*, **262**, 24.
McCarthy, P.J., Spinrad, H., Van Breugel, W., Djorgovski, S., Strauss, M.A. & Dickinson, M., 1988. These Proceedings.
McKee, C.F. & Cowie, L.L., 1977. *Astrophys. J.*, **215**, 213.
McGlynn, T.J. & Fabian, A.C., 1984. *Mon. Not. R. astr. Soc.*, **208**, 709.
Miley, G., 1987. In: *Observational Cosmology*
Mushotzky, R.F., Holt, S.S, Smith, B.W., Boldt, E.A. & Serlemitsos, P.J., 1981. *Astrophys. J.*, **244**, L47.
Mushotzky, R.F. 1984. *Phys. Scripta*, **T7**,157.
Mushotzky, R.F., 1988. These Proceedings.
Norgaard-Nielsen, H-U., Jorgensen, H. & Hansen 1984. *Astr. Astrophys.*, **135**, L3.
Nulsen, P.E.J., Stewart, G.C. & Fabian, A.C., 1984. *Mon. Not. R. astr. Soc.*, **208**, 185.
Nulsen, P.E.J., 1986. *Mon. Not. R. astr. Soc.*, **221**, 377.
O'Connell, R.W., 1987. in *Proc. I.A.U. Symp. No. 127*, ed De Zeeuw, T., 167, Reidel.
Romanishin, W., 1986. *Astrophys. J.*, **301**,675.
Sarazin, C.L. & O'Connell, R.W., 1983. *Astrophys. J.*, **258**, 552.
Schwarz.J., McCray, R. & Stein, R.F., 1972. *Astrophys. J.*, **175**,673.
Schwartz, D.A., Schwarz, J. & Tucker, W.H., 1980. *Astrophys. J.*, **238**, L59.
Spinrad, H. & Djorgovski, G., 1984a. *Astrophys. J.*, **280**, L9.
Spinrad, H. & Djorgovski, G., 1984a. *Astrophys. J.*, **285**, L49.

Stewart, G.C., Fabian, A.C., Jones, C. & Fabian, A.C., 1984. *Astrophys. J.*, **285**, 1.
Thomas, P.A., Fabian, A.C., Arnaud, K.A., Forman, W. & Jones, C., 1986. *Mon. Not. R. astr. Soc.*, **222**, 655.
Thomas, P.A., Fabian, A.C. & Nulsen, P.E.J., 1987. *Mon. Not. R. astr. Soc.*, **228**, 973.
Ulmer, M., Cruddace, R.G., Fennimore, E.E., Fritz, G.G. & Snyder, W.A.,1987. *Astrophys. J.*, **319**, 118.
Valentijn, E.A., 1987. In:*Proc. IAU Sympn. 127*, ed De Zeeuw, T., 433.
Valentijn, E.A., 198 . These Proceedings.
Van Breugel, W. McCarthy, P.J.& Van Gorkom, J., 1988. These Proceedings.
Yee, H.K.C. & Green, R.F., 1984. *Astrophys. J.*, **280**, 79.
White, R.E., 1988. These Proceedings.

Extended Emission-Line Gas in Distant 3CR Radio Galaxies*

Patrick J. McCarthy, Hyron Spinrad and Wil van Breugel
Astronomy Department, University of California
S. Djorgovski
Astronomy Department, California Institute of Technology
Michael A. Strauss and Mark Dickinson
Astronomy Department, University of California

Abstract

We report on the results of long-slit spectroscopy and interference filter imaging of powerful 3CR radio galaxies. Narrow–band [OII]λ3727 imaging often reveals extensive regions of extranuclear emission associated with these objects. The extended [OII] emission is distributed on scales of several tens of kpc, it is often more luminous than the nuclear emission, and shows large velocity gradients. We discuss the physical conditions in these nebulæ and their relation to the radio source structure. In particular, we examine the ionization mechanism for the gas, and find that photoionization models reproduce the observed emission line strengths reasonably well. The source of the ionizing photons, however, is not clear; the large luminosities of the extended gas suggesting that some form of *in situ* ionization may be occurring. Compared to the low-z cooling flow filaments, the 3CR emission regions are larger, more luminous, and generally show higher excitation. Futhermore, the distant radio galaxies show evidence for much stronger interactions with their radio sources than is seen in nearby cooling flow filaments.

The first spatially resolved spectroscopic observations of distant 3CR radio galaxies at $z \sim 1$ showed that they often have associated regions of extended emission, mainly [OII]λ3727 (Spinrad and Djorgovski 1984a,b). Spinrad and Djorgovski (1984c) and Lilly and Longair (1984) noted that the distant 3CR galaxies tended to be multi-modal in both the stellar continuum and in [OII]. Furthermore, the extended [OII] emission lines tend to show large velocity gradients. These properties have generally been attributed to violent dissipative mergers (e.g., Djorgovski *et al.* 1987). We discuss the results of further spectroscopy and interference filter imaging, aimed at studying these objects in some detail. In what follows, we adopt the following cosmological parameters: $H_0 = 50$, $q_0 = 0$, and $\Lambda = 0$.

We have imaged 25 3CR radio galaxies with redshifts between 0.4 and 1.5 in [OII]λ3727 using interference filters with FWHM $\sim$ 70Å with the Lick Observatory's Shane 3-m telescope. Of these, 21 show extended emission on a scale larger than 2″, corresponding to a linear size of $\sim$ 8 kpc for a redshift of 0.6. This rate of incidence of extended emission ($\sim$ 80%) is higher than that of low-redshift radio galaxies (van Breugel 1986), QSOs (Stockton and MacKenty 1987), and cooling-flow clusters (Hu *et al.*1986). We will illustrate some of the general properties of these objects with an example.

In Figure 1 we show redshifted [OII]λ3727 and continuum images of the radio galaxy 3C 265 (z=0.81). The positions of the radio core and hotspots are indicated. This object exhibits the type of diffuse extended [OII] emission characteristic of the high-z 3CR galaxies. The [OII] emission extends over roughly 40″, corresponding to 220 kpc in the above cosmology. The total luminosity of

---

*Based in part on observations performed at Lick Observatory, University of California.

A. C. Fabian (ed.), Cooling Flows in Clusters and Galaxies, 325–329.
© *1988 by Kluwer Academic Publishers.*

the extended [OII] emission is 2.7×10^{43} ergs s^{-1}. The nuclear emission spectrum is rich in high excitation lines (Smith *et al.* 1979). This object is particularly interesting because it exhibits extended [NeV]λ3426 and [NeIII]λ3869 in addition to [OII]. The spatial variations in the [NeV]/[OII] will be used to investigate the ionization mechanism below. This example, 3C 265, illustrates many of the general properties of the extended narrow line regions (ENLRs) associated with distant radio galaxies. They are typically $50 - 200$ kpc in extent, have [OII] luminosities of $\sim 10^{43}$ erg s^{-1} and often have weak underlying continua. The gas typically has velocity gradient amplitudes of ~ 800 km s^{-1}, and usually bears some morphological relation to the large scale radio source structure. For comparison the radio source luminosities are typically $10^{45} - 10^{46}$ erg s^{-1} and their median size is ~ 300 kpc (in the redshift range 0.5 to 1.8).

A key question in the physical study of the ENLRs in the ionization mechanism for the gas. Diagnostic line ratios can be used to discriminate between various possible mechanism, particularly when one has line ratios for extended emission. In Figure 2 we display a line ratio diagram in the manner of Baldwin, Phillips and Terlevich (1981). The particular line ratios that we use are [NeV]/[OII] and [NeIII]/[OII]. The choice of the two ratios is motivated by the fact that they extend over a small range in wavelength, and hence suffer little from reddening, they span a large range in ionization energy, and they are observable over a very large range in redshift. Seven 3CR galaxies with z > 0.5 are shown in the figure, the open symbols denote the nuclei, while the filled symbols denote the extended emission for the same object. A typical Seyfert narrow line region is also plotted along with the results of a power law ($\alpha = -1$) photoionization calculation by Halpern and Filippenko (1984, unpublished). The 3CR galaxies show a large range in the [NeV]/[OII] ratio and follow the photoionization model calculations over a decade

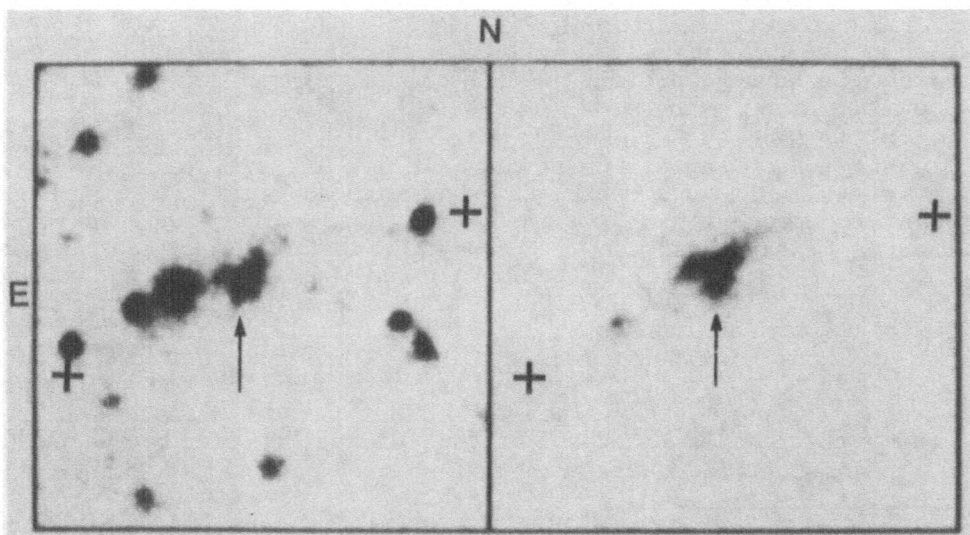

Figure 1. Broad band r_S (left) and [OII]λ3727 (right) images of 3C 265. The size of each panel is $73''$ on a side. The radio lobes (crosses) and the host galaxy (arrows) positions are marked.

in the ionization parameter, U. More importantly, the regions of extranuclear emission in 3C 265, 3C 280, and 3C 330 also fall along the photoionization trend, and in each case give lower values of the ionization parameter than their corresponding nuclei. This is what one naturally expects if photoionization from the nucleus is the dominant ionization mechanism. Figure 2 then suggests that the ENLRs are similar to the NLRs of Seyfert galaxies, but they tend to have lower ionizing photon densities compared to the gas density. The similarity between the ENLRs in low-z galaxies and Seyfert NLRs have been pointed out by a number of authors (e.g., van Breugel et al. 1985). The source of the ionizing photons is, however, unclear in many low-z cases. In a number of objects (e.g., 4C 29.30, van Breugel et al. 1986; PKS 2152–069; Tadhunter et al. 1987) a local source of ionizing photons (or beamed UV radiation from the nucleus) is required to reproduce the line ratios observed at large distances from the nucleus. The large ratio of extended [OII] luminosity to nuclear luminosity in the high-z cases, and the observed underlying continua also argue for an extended source of ionizing photons, perhaps associated with the radio source itself. Robinson et al. (1987) and Robinson (this volume) show that photoionization models reproduce the observed line-ratios for low-z radio galaxies and cooling-flow filaments quite well. The high-z cooling flow galaxy, 3C 295 (Henry and Henriksen 1986) , is also shown in Figure 2, falling to the extreme low ionization end of the diagram. The excitation mechanism in cooling flow filaments is unclear, as Heckman et al. (this volume) show that shock-models fit the cooling flow filaments better that photoionization models, particularly when the spatial variations in the line ratios and the radial surface brightness variations are taken as constraints.

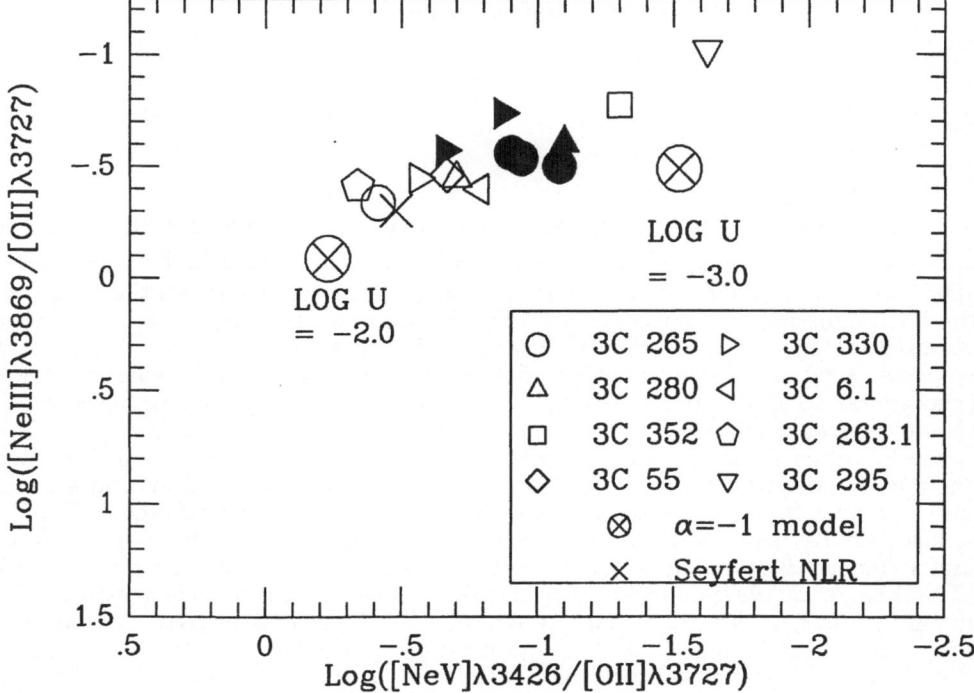

Figure 2. Line ratio plot for 3CR radio galaxies. The open symbols refer to the nuclei while the filled symbols refer to the extended gas for the same source.

In Table 1 we list some of the properties of a representative sub-sample of distant radio galaxies, along with a number of cooling flow galaxies. Table 1 shows that in most respects the high-z objects tend to be more extreme in their properties when compared to their local counterparts: their luminosities are a factors of 10 larger, the velocity fields have amplitudes that are typically 5 times larger, and their radio sources are typically a factor of 10 to hundreds of times more luminous (3C 295 being an exception). A particularly spectacular example of an extended high-z system, 3C 368, has recently been described in detail by Djorgovski *et al.* (1987).

Table 1.

3CR	z	L([OII]) x 10^{42} Nuc.	Ext.	Size(kpc)	Delta V(km/s)	Comment
330	0.55	3.6	2.5	40	--	Cluster
337	0.64	1.5	2.0	70	200	Cluster
277.2	0.77	4.0	40.0	200	1000	Group
352	0.81	--	30.0	40	800	
265	0.81	12.0	15.0	220	600	This paper
280	0.99	13.0	10.0	70	800	McCarthy et al.
356	1.08	2.0	4.0	60	1500	
368	1.13	--	60.0	85	1000	Djorgovski
267	1.14	--	10.0	40	1500	
324	1.21	--	7.0	100	1500	
Cooling Flow Galaxies						
274	0.004	--	0.01	6	200	M87
84	0.017	--	1.0	20	250	Perseus A
P0745	0.10	5.0	--	12	250	Fabian et al.
295	0.46	1.3	--	15	200	

Fabian *et al.* (1985, 1986) have suggested that the extended [OII] regions associated with distant 3CR galaxies are indications of distant high $\dot{M}$ ($\sim$ 500 – 1500 $M_\odot$ yr^{-1}) cooling flows. The observed luminosities and spatial scales of the [OII] emission are consistent with this hypothesis. The high-z radio galaxies, however, differ from the locally observed cooling flows in a number of important respects. First, the distant 3CR galaxies show evidence for *massive* star formation (Djorgovski 1988 and references therein), unlike the nearby cooling flows which require almost exclusively *low mass* star formation (see, however, Silk *et al.* 1986). Secondly, as is shown in Table 1, the [OII] emission in the distant 3CRs often show large velocity gradients, in contrast to the lack of ordered motion seen in low-z cooling flow filaments (Heckman *et al.* 1988, and this volume). Finally, it has recently been shown that distant radio galaxies have their extended [OII] and *stellar continuum* aligned with their radio source axes (McCarthy *et al.* 1987; Chambers *et al.* 1987; and van Breugel, this volume). These facts suggest that the distant radio galaxies are strongly influenced by their radio sources. The overall properties of the high-z ENLRs are quite similar to the low redshift radio galaxies described by Danziger (this volume), Fosbury *et al.* (this volume) and van Breugel (1986), the high-z objects being more extreme in their properties. Entrainment and compression of ambient gas by the radio source may be responsible for the observed velocity gradients and the large scale star formation along the radio source axis. Such strong interactions may well serve to inhibit cooling flows of the type seen locally. None of the arguments given above, however, rule out the possibility that we are seeing extreme examples of distant cooling flows, but in which the radio source has imposed conditions that are quite far from those seen (and modeled) at low redshifts.

We would like to thank Prof. A. Fabian and the Institute of Astronomy for their hospitality during the workshop. We also thank the staffs of the Lick Observatory, and the Kitt Peak National Observatory. The Lick Observatory is operated in part under the NSF grant AST-86-14510. H. S., P. McC. and M. D. acknowledge support from NSF grant AST85-13416, W.v.B. acknowledges support from NSF grant AST 84-16177, S. D. acknowledges partial support from Harvard University and M. A. S. acknowledges an NSF graduate fellowship.

References:

Baldwin, J. A., Phillips, M., and Terlevich R. 1981, *P. A. S. P.* **83**, 5.

Chambers, K. C, Miley, G. K, and van Breugel, W. J. M. 1987,*Nature* in press.

Djorgovski, S., Spinrad, H., Pedelty, J., Rudnick, L., and Stockton, A. 1987a, *A. J.* **91**, 1267.

Djorgovski, S. 1988, in *Starbursts and Galaxy Evolution*, Th. Montmerle (ed.), Paris: Editions Frontieres, in press.

Fabian, A. C. *et al.* 1985, *M. N. R. A. S.* **216**, 923.

Fabian, A. C., Arnaud, K. A., Nulsen, P. E. J., and Mushotzky, R. F. 1986, *Ap. J.* **305**, 9.

Heckman, T. M., Baum, S., van Breugel, W. J. M., and McCarthy, P. J., 1988 *Ap. J.*, submitted.

Henry, J. P., and Henriksen M. J. 1986, *Ap. J.* **301**, 689.

Hu, E., Cowie, L., and Wang, Z. 1985 *Ap. J. Supp.* **59**, 447.

Lilly, S. and Longair, M. 1984, *M. N. R. A. S.* **151**, 421.

McCarthy, P., van Breugel, W., Spinrad, H., and Djorgovski, S. 1987, *Ap. J. Lett.* in press.

Robinson, A., Binette, L., Fosbury, R. A. E., and Tadhunter C. N. 1987, *M. N. R. A. S.* **227**, 97.

Silk, J., Djorgovski, S., Wyse, R., and Bruzual, G. 1986, *Ap. J.* **307**, 415.

Smith, H. E., Junkkarinen, V., Spinrad, H., Grueff, G., and Vigotti, M. 1979, *Ap. J.* **231**, 307.

Spinrad, H., and Djorgovski, S. 1984a, *Ap. J. Lett.* **280**, L9.

Spinrad, H., and Djorgovski, S. 1984b, *Ap. J. Lett.* **285**, L49.

Spinrad, H., and Djorgovski, S. 1984c, *P. A. S. P.* **96**, 795.

Stockton, A., and MacKenty, J. 1987, *Ap. J.* **316**, 584.

Tadhunter, C. N., Fosbury, R. A. E., Binette, L., Danzinger, I. J., and Robinson, A. 1987, *Nature* **325**, 504.

van Breugel, W. J. M., Miley, G. K., Heckman, T. M.,Butcher, H., and Bridle, A. 1985,*Ap. J.* **290** 496.

van Breugel, W. J. M., Heckman, T. M., Miley, G. K., and Filippenko, A. V. 1986, *Ap. J.* **311** 58.

van Breugel, W. J. M. 1986, in *Jets from Stars and Galaxies*, eds. R. N. Henriksen and T. W. Jones; *Can. J. Phys.* **64**, 392.

COOLING FLOWS AROUND QUASARS

C. S. Crawford
Institute of Astronomy
Madingley Road
Cambridge CB3 0HA
U.K.

Abstract. 3C 48 is shown to possess a cooling flow of $\sim 100\,M_\odot\,yr^{-1}$ from observations of the extended oxygen line emission 3" from the nucleus. Preliminary results are included for four other quasars that indicate a similar behaviour.

Introduction

A large fraction of systems at the present epoch contain cooling flows, so it would be surprising if cooling flows did not exist out to at least intermediate redshift. The smooth X-ray profiles are consistent with flows of lifetimes $\geq 5.10^9$ years. Examples of cooling flows have been observed out to $z = 0.5$: 3C 295, with $L_x \leq 10^{45}\,erg\,s^{-1}$, possesses a flow of $\dot{M} \leq 200\,M_\odot\,yr^{-1}$, and is associated with a cluster of richness 1 (Henry & Henriksen 1986); 3C 275.1 has been shown to be centred on a large ($\sim 100\,kpc$) low-ionization nebulosity displaying solid body rotation out to 40 kpc, in a rich cluster of galaxies (Hintzen & Stocke 86; Hintzen & Romanishin 1986). It is not possible to use the present X-ray waveband instruments to search for further cooling flows at higher redshift, so we must develop alternative techniques. Extensive optical line filaments have been observed round some radio galaxies and quasars (McCarthy et al. 1987; Heckman et al. 1986; Stockton & McKenty 1987), and this emission gas is often interpreted as having its origin in a disruptive interaction between galaxies. The appearance of this emission, however also resembles the filamentary structure seen in the emission line gas around cooling flow cluster central galaxies, e.g. NGC 1275 (Lynds 1970).

We have been continuing the search for evidence of cooling flows around moderate redshift quasars using spectroscopic data taken in 1986 with the 2.5m INT on La Palma. Our sample of targets was drawn from the previously published work of Hutchings, Malkan, Spinrad, and collaborators (see references) , concentrating mainly on the radio-loud or X-ray luminous quasars that show signs of clustering or of an underlying nebulosity, as well as observing some comparison radio-quiet QSO. The sample actually observed was constrained by conditions of atmospheric dust.

3C 48

3C 48 is a steep spectrum radio source at a redshift of 0.37, and was the first QSO to

A. C. Fabian (ed.), Cooling Flows in Clusters and Galaxies, 331–335.

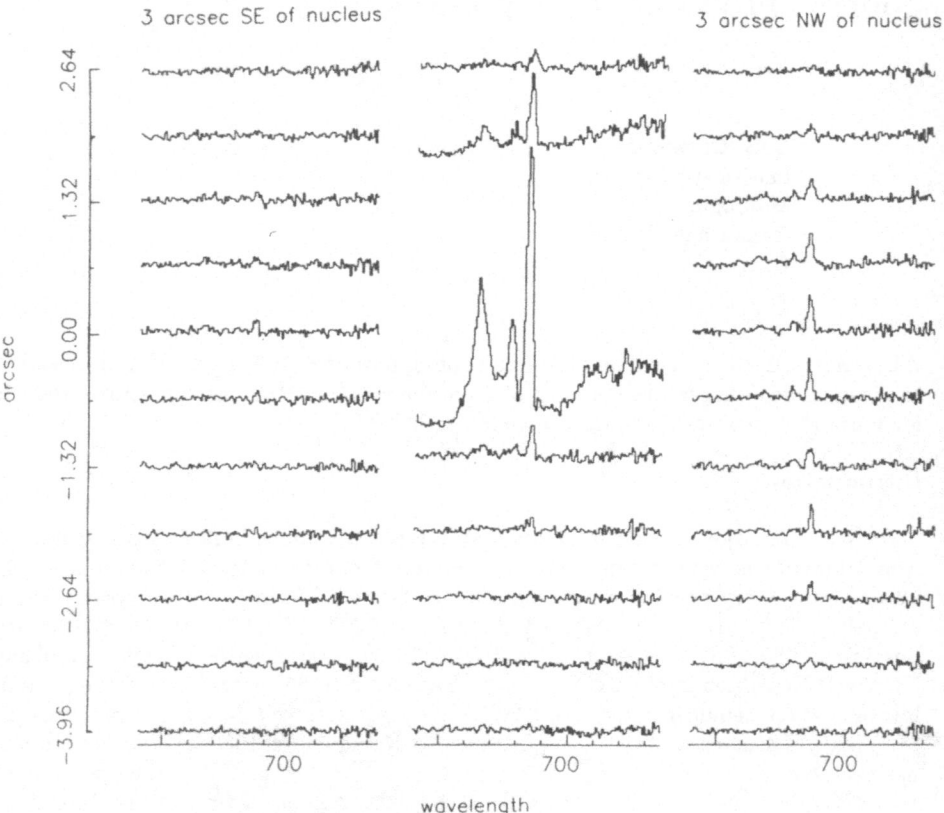

Figure 1. The region corresponding to 4670-5360Å(in the rest frame of the quasar) for three 2000 sec exposures of the quasar 3C 48: i) 3" south-east of the nucleus ii) directly across the nucleus and iii) 3" north-west of the nucleus. The spectra have not been corrected for scattered nuclear light, and the nuclear spectrum has been rescaled and offset.

show convincing evidence of a surrounding nebulosity (Sandage & Miller 1966; Wampler *et al.* 1975; Boroson & Oke 1984; Malkan 1982; Stockton & McKenty 1987), and has an X-ray luminosity $L_x \sim 3.10^{45} \, \mathrm{erg\,s^{-1}}$. We observed it with the CCD detector in subarcsecond seeing in three long slits; one across the nucleus, one at 3" NW, one at 3" SE , all at p.a. 54° (see *Fig. 1*, also Fabian *et al.* 1987).

The [OIII] emission lines are obviously extended in the NW slit over 8 cross-sections (> 5", $\sim$ 35 kpc for $H_o = 50 \, \mathrm{km\,s^{-1}\,Mpc^{-1}}$). We removed any contamination of the fuzz by spill-over of nuclear light by subtracting a nuclear spectrum from each cross-section, the amount of which was determined by scaling to the broad FeII blends on either side of the Hβ and [OIII] complex. This was only necessary for the central few cross-sections in the NW slit. In the resulting fuzz spectra, the λ5007 lines display a slight velocity trend of $\sim 660 \, \mathrm{km\,s^{-1}}$ across 35 kpc along the slit (and at a slightly higher velocity to the nucleus). We were only able to place upper limits to the amount of continuum present at 0.01 mJy in the NW slit, and 0.06 mJy in the SE slit, consistent with the values of Boroson & Oke (1982). We do not have a reliable observation of the density-dependent [SII] lines, so we were forced to use an indirect method similar to that of Bergeron (1976; also Robinson *et al.* 1987) to estimate these quantities.

We are able to measure the intensity of the $\lambda5007$ and $\lambda3727$ oxygen lines in the NW nebulosity, giving an intensity ratio [OIII]$\lambda5007$/[OII]$\lambda3727$ of ~ 1.5 at a distance of 30 kpc from the nucleus, close to the outer radial distance in our slit. We used Ferland's photo-ionization computation programme CLOUDY to match the observed line ratios in the fuzz to model predictions for various ionization parameters L_q/nr^2. Since the quasar has a luminosity $L_q = 10^{46}$ erg s^{-1} and a UV to X-ray spectrum similar to 3C 273 (Wilkes & Elvis 1987) - *i.e.* unabsorbed beyond 912Å, we assume that the gas at 30 kpc radius is ionized by radiation originating at the QSO nucleus. For a power-law of slope $\alpha_{ox} = -1.3$ (Worrall *et al.* 86) of $L_x = 10^{46}$ erg s^{-1} the ratios and intensities of lines observed show the gas density to be $n \geq 30$ cm^{-3}, with a radial depth of stable ionized gas of 10pc. If we add to the power-law a blackbody bump of excess emission of equal energy, the density could be as high as 80 cm^{-3} in a depth of 3pc.

Since the emission in the NW slit is extended over ≥ 35 kpc, from the total $\lambda5007$ luminosity in the slit and the source model, we can deduce a minimum amount of cooled gas present to be 3.10^8 M$_\odot$. This is very much a lower limit as our slit will cover only a fraction of the emission-line gas. It is not easy to explain this mass of gas by an interaction between normal galaxies without a problem with the lifetime of the system. The high density emission-line gas at 10^4 K is not confined and expands rapidly at its local sound speed, and so depth of ionized gas observed dissipates on a sound crossing time $t_{diss} \leq 10$ pc/10^4 m s^{-1} $\sim 10^6$ yr. If the interaction occurs at a relative velocity of 100 km s^{-1}, the timescale to establish the 35 kpc length of emission observed is $t_{dyn} \sim 35$ kpc/100 km s^{-1} $= 3.5 \times 10^8$ yr. There must be more cold gas than we observe to compensate for that continually evaporated away by the photo-ionization, by a factor of t_{dyn}/t_{diss}, giving a total mass of $\sim 1.1 \times 10^{11}$ M$_\odot$. If the calculations are repeated using the density and depth given with an extra ionizing blackbody, this total mass value is boosted considerably. This amount of cold dense gas appears to rule out an interaction of two normal galaxies.

Other than the extended X-ray emission, the best observational discriminant for the presence of a cooling flow is a high gas pressure $nT \geq 10^5$ cm^{-3} K (Fabian *et al.* 1986). The high gas density inferred in 3C 48 implies a cooling time in any confining gas of less than a Hubble time for $T < 10^9$ K. A cooling flow would provide a ready source of cooled gas, and the outer hotter gas supplies a confining pressure so that there need not be an excessive amount of gas present. If the profile of the gas density at 10^4 K in pressure equilibrium with the hotter gas is considered next to known cooling flows of varying strength (see *Fig. 2*), we find 3C 48 comparable to a cooling flow of $\dot{M} \sim 100$ M$_\odot$ yr^{-1}, similar to the poor group MKW3s (Canizares *et al.* 1983).

Similar techniques to those described here for 3C 48 were applied to our single-slit observations of 4c 37.43 and 4C 11.72, giving gas densities of > 20 cm^{-3} for 4c 37.43 and ~ 100 for 4C 11.72. We can apply our technique also to 3C 249.1, using the observations of Boroson & Oke (1984) to give a density of 15 cm^{-3}; also to those of 3C 275.1 by Hintzen & Stocke (1986) to give > 10 cm^{-3}. For all these quasars we take

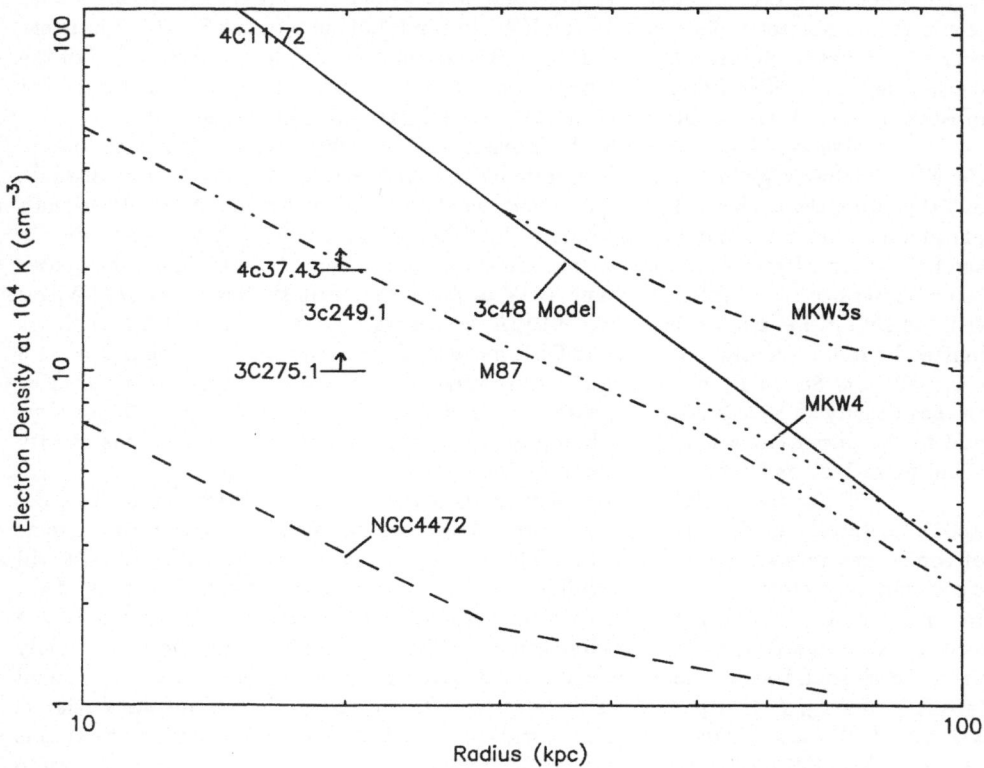

Figure 2. Electron density of confined gas at 10^4 K versus radius. Plotted are nearby cooling flows (NGC 4472 has $\sim 1\,\mathrm{M_\odot\,yr^{-1}}$; M87 $\sim 10\,\mathrm{M_\odot\,yr^{-1}}$; MKW4,$\sim 30\,\mathrm{M_\odot\,yr^{-1}}$; MKW3s, $\sim 100\,\mathrm{M_\odot\,yr^{-1}}$), and the quasars. The power-law photo-ionization model for 3C 48 requires that the observed gas lies on the r^{-2} line shown.

observations at 20kpc from the nucleus, and where a value for the index α_{ox} is not given in Worrall *et al.* (1986), we scale to 3C 48 using the quasars' visual magnitude and X-ray luminosity. The results are also shown in *Fig. 2*. Note that the value for 3C 275.1 is a lower limit, as its faint optical magnitude suggests that much of its nuclear light may be absorbed. 4C 37.43 is also shown as a lower limit, as the observations of Stockton (1976) and Boroson & Oke (1984) imply a higher density of $20\,\mathrm{cm^{-3}}$.

Conclusion

A cooling flow of $\dot{M} \sim 100\,\mathrm{M_\odot\,yr^{-1}}$ seems a preferred explanation for our observations of 3C 48. Results consistent with this hypothesis are also found for 3C 249.1, 3C 275.1, 4C 37.43 and 3C47 - which is not surprising if radio-loud quasars are inferred to lie in poor groups of galaxies (Yee & Green 1984, 1987). It is intended to publish a more comprehensive report on our sample of quasars soon.

Acknowledgements

Andy Fabian and Roderick Johnstone are thanked for many helpful discussions. SERC is thanked for financial support.

References

Bergeron, J., 1976. *Astrophys. J.*, **210**, 287.
Boroson, T.A. & Oke, J.B., 1982. *Nature*, **296**, 397.
Boroson, T.A. & Oke, J.B., 1984. *Astrophys. J.*, **281**, 535.
Canizares, C.R., Stewart, G.C. & Fabian A.C., 1983. *Astrophys. J.*, **272**, 449.
Fabian, A.C., Arnaud, K.A., Nulsen, P.E.J. & Mushotzky, R.F., 1986. *Astrophys. J.*, **305**, 9.
Fabian, A.C., Crawford, C.S., Johnstone, R.M. & Thomas, P.A., 1987. *Mon. Not. R. astr. Soc.* in press.
Heckman, T.M., *et al.*, 1986. *Astrophys. J.*, **311**, 526.
Henry, J.P. & Henriksen, M.J., 1986. *Astrophys. J.*, **301**, 689.
Hintzen, P. & Romanishin, W., 1986. *Astrophys. J.*, **311**, L11.
Hintzen, P. & Stocke, J.S., 1986. *Astrophys. J.*, **308**, 540.
Hutchings, J.B., Crampton, D., Campbell, B. & Pritchet C., 1981. *Astrophys. J.*, **247**, 743.
Hutchings, J.B., Crampton, D. & Campbell, B., 1984. *Astrophys. J.*, **280**, 41.
Lynds, C.R., 1970. *Astrophys. J.*, **159**, L151.
Malkan, M.A., 1984. *Astrophys. J.*, **287**, 555.
Malkan, M.A., Margon, B. & Chanan, G.A., 1984. *Astrophys. J.*, **280**, 66.
Malkan, M.A. & Sargent, W.L.W., 1982. *Astrophys. J.*, **254**, 22.
McCarthy, P. *et al.*, 1987. Preprint
Robinson, A., Binette, L., Fosbury, R.A.E. & Tadhunter, C., 1987. *Mon. Not. R. astr. Soc.* in press.
Sandage, A. & Miller, W.C., 1966. *Astrophys. J.*, **144**,1238.
Spinrad, H., 1982. *Publ. astr. Soc. Pacif.*, **94**, 397.
Stockton, A., 1976. *Astrophys. J.*, **205**, L113.
Stockton, A. & MacKenty, J.W., 1987. *Astrophys. J.*, **316**, 584.
Wampler, E.J., Robinson, L.B., Burbidge, E.M. & Baldwin, J.A., 1975. *Astrophys. J.*, **198**, L49.
Wilkes, B.A. & Elvis M., 1987. *Astrophys. J.* in press.
Worrall, D.M., Giommi, P., Tananbaum, H. & Zamorani, G., 1987. *Astrophys. J.*, **313**, 596.
Yee, H.K.C. & Green, R.F., 1984. *Astrophys. J.*, **280**, 79.
Yee, H.K.C. & Green, R.F., 1987. *Astrophys. J.*, **319**, 28.

A SEMI-ANALYTIC APPROACH TO COOLING FLOW EVOLUTION

Edmund Bertschinger
Department of Physics
MIT, Room 6-207
Cambridge, MA 02139
U.S.A.

ABSTRACT. Piecewise similarity solutions are used to model the evolution of a cooling flow in the absence of star formation, heating, or heat conduction. In these solutions the cooling radius R_c increases as a power of time and the character of the X-ray emission changes qualitatively across the cooling radius. Application to M87 shows that time-dependence does not change the conclusion obtained by others that gas drops out of the flow; it is shown that inside $\approx 0.2R_c$ local thermal instabilities can plausibly lead to star formation. The similarity solution fit to M87 implies that the cooling flow was more vigorous in the past.

1. INTRODUCTION

Most of the theoretical work on cooling flows to date has assumed that the flows are in a steady state. This assumption is likely to be valid well inside of the cooling radius R_c, defined here as the radius at which the cooling time derived from the initial density and temperature profiles equals the age of the cooling flow. However, near the cooling radius the flow cannot be steady because, in effect, gas elements are discovering that they must cool and begin a slow fall into the center. This qualitative change in the structure of a cooling flow across R_c does not occur because the gas is out of hydrostatic equilibrium, since the gas motions are extremely subsonic outside of the center of the flow. Rather, the gas is out of thermal equilibrium and the decrease in entropy of a given gas element causes the element to move inward to maintain hydrostatic equilibrium.

Because the density decreases outward from the center of a cooling flow, the cooling time increases with radius. Conversely, the cooling radius must increase with time. Thus, time-dependence has two complementary effects: At a fixed time, the structure of a cooling flow changes across R_c; as a function of time, the cooling flow expands to encompass more and more gas. I will refer to this expanding region of cooling as a "cooling wave." The goal of the present work is to derive the structure and evolution of a cooling wave using a simplified model. From comparison with observations of the cooling flow onto M87 I will demonstrate the importance of time-dependence on the structure and will estimate the global evolution of $\dot{M}$ and the X-ray luminosity.

2. SIMILARITY SOLUTIONS

To solve the time-dependent cooling flow problem semi-analytically, a number of simplifying assumptions are required. They are spherical symmetry, a one-phase medium, no star formation, no mass or energy injection, no heat conduction, and power-law forms for the cooling function $\Lambda(T)$,

A. C. Fabian (ed.), Cooling Flows in Clusters and Galaxies, 337–341.
© 1988 by Kluwer Academic Publishers.

binding mass profile $M(r)$, and initial entropy distribution $p \propto \rho^\Gamma$. Undoubtedly this results in an enormous over-simplification of a real cooling flow, but the problem now becomes tractable and the effect of time-dependence is made clear by its isolation from other complications.

Given the above assumptions, the procedure is to solve the fluid equations by obtaining piecewise similarity solutions as will be described below. The basic idea is that as the cooling wave expands, the profiles of the fluid variables retain a constant form with the normalization changing in a simple time-dependent fashion. When the time-dependence of the scalings is removed the cooling wave appears steady. The fluid equations are then as easy to solve numerically as they are for an ordinary steady flow.

The fluid equations to be solved are

$$\frac{\partial \rho}{\partial t} + \frac{1}{r^2}\frac{\partial}{\partial r}\left(r^2 \rho v\right) = 0 \, , \tag{1}$$

$$\frac{\partial v}{\partial t} + v\frac{\partial v}{\partial r} + \frac{1}{\rho}\frac{\partial p}{\partial r} = -\frac{GM(r)}{r^2} \, , \tag{2}$$

$$\left(\frac{\partial}{\partial t} + v\frac{\partial}{\partial r}\right)\ln\left(p\rho^{-\gamma}\right) = \frac{(\gamma-1)}{p}\rho^2 \Lambda(T) \, , \tag{3}$$

where $\gamma = 5/3$ is the adiabatic exponent. From the power-law assumptions the initial density and pressure profiles are $\rho_0 \propto r^{-\alpha}$ and $p_0(r) \propto r^{-\beta}$ and the cooling function is $\Lambda(T) \propto T^\lambda$. From the initial gas distribution one may define the cooling radius such that $t_c(R_c) = t$, yielding

$$R_c \propto t^\eta \, , \quad \eta = [\alpha + (\alpha - \beta)(1 - \lambda)]^{-1} \, . \tag{4}$$

An independent length scale is given by the distance a sound wave travels in the age of the flow,

$$R_s(t) = \left(\frac{\beta - \alpha + 2}{2}\right)\left[\frac{\gamma p_0(R_s)}{\rho_0(R_s)}\right]^{1/2} t \, . \tag{5}$$

If there were only one relevant length scale or if the others occurred as constant multiples of one time-varying scale, then the evolution of the flow would be self-similar. In this way Chevalier (1987) has solved the problem of fully self-similar cooling flows in which $R_c(t)/R_s(t) = $ constant. Unfortunately, this condition is not met in cluster cooling flows. For example, if the initial gas distribution is isothermal, the density profile must be $\rho_0 \propto r^{-1}$, which is too shallow. When there are multiple length scales the problem can still be solved approximately, provided that the flow can be separated into regions where only one physical process is of dominant importance. Similarity solutions are obtained separately in each region and are then joined, yielding piecewise similarity solutions. In the present case this is possible since

$$R_c(t) \ll R_s(t) \quad \text{and} \quad R_c/R_s \to 0 \quad \text{as} \quad t \to \infty \, . \tag{6}$$

For cluster cooling flows, typically $R_c \sim 0.1\,\mathrm{Mpc}$ and $R_s \sim 10\,\mathrm{Mpc}$. There is a third characteristic length scale, the sonic radius R_* (where the flow is transsonic), which is not independent of R_c and R_s. Typically $R_* \lesssim 10^{-3}\,\mathrm{Mpc}$ so that $R_* \ll R_c \ll R_s$ and piecewise similarity solutions should be an excellent approximation. I note that in the limit where the above inequalities are strongly satisfied but the ratios R_*/R_c and R_c/R_s are constant, the solutions I obtain agree exactly with those of Chevalier (1987).

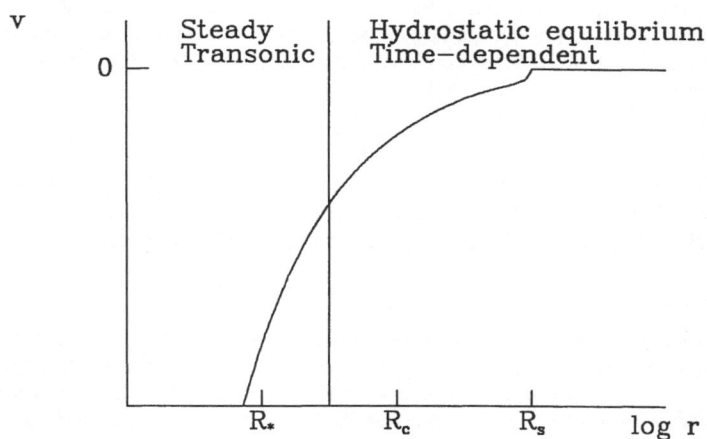

Figure 1. Velocity profile of a cooling flow (schematic). Characteristic scales R_* (sonic radius), R_c (cooling radius), and R_s (rarefaction wave radius) are indicated. The flow is divided into two regions for a piecewise similarity solution.

Figure 1 shows schematically the velocity profile of the ideal cooling flow considered here. There is a kink at R_s because a rarefaction wave is traveling outward at the local sound speed signaling the gas to begin falling in. This behavior is similar to the rarefaction wave in the isothermal sphere collapse problem solved by Shu (1977). Obviously the flow in this region is time-dependent but, because it is extremely subsonic for $R_s \gg R_c$, hydrostatic equilibrium is an excellent approximation (i.e., the velocity terms in eq. [2] are negligible). Since I am not concerned with the small departures from hydrostatic equilibrium at large radius (they have an insignificant effect on the cooling flow) I assume hydrostatic equilibrium applies until the flow nears R_*. Thus, in the outer parts a similarity solution is valid with R_c providing the unique scale length. The fluid equations are reduced to ordinary differential equations by nondimensionalizing the fluid variables as follows:

$$r = xR_c(t) , \quad \rho = \rho_0(R_c)D(x) , \quad p = p_0(R_c)P(x) , \quad v = -\frac{dR_c}{dt}xW(x) . \tag{7}$$

Inside the cooling radius the Mach number increases and the assumption of hydrostatic equilibrium breaks down. However, for $R_* \ll R_c$ the flow becomes approximately steady ($t_{\text{flow}} \equiv r/|v| \ll t$) far outside the sonic point, so that the flow may be solved neglecting time derivatives. The flow does change slowly with time, but this enters only into the outer boundary conditions for the region of steady flow. The velocity terms are now included in equation (2) so that a sonic point occurs. There is a sufficiently large region over which the flow is both quasi-steady and subsonic so that the solutions obtained in the two regions can be accurately matched. The mathematical details may be found in Bertschinger (1988).

3. APPLICATION TO M87

The similarity solutions have been applied to the cooling flow onto M87 to try to determine the effects of time-dependence. The electron density profile is assumed to be

$$n_e(r) = 1.46 \times 10^{-2} \frac{(r/30\,\text{kpc})^{-0.8}}{1 + (r/30\,\text{kpc})} \text{cm}^{-3} , \tag{8}$$

which provides a good fit (to within ≈5% for 2 kpc < r < 200 kpc) to the profile obtained from observations by Stewart *et al.* (1984). At large r, $n_e \propto r^{-1.8}$, which is assumed here to reflect the initial density profile. The binding mass profile is less certain, but a reasonable fit (to within ≈15% for 5 kpc < r < 100 kpc) to the model C profile of Stewart *et al.* is

$$M(r) = 2.4 \times 10^{10} \, (r/1\,\text{kpc})^{1.3} \, M_\odot \, . \tag{9}$$

The cooling rate is taken to be $1.2 \times 10^{-23} \, n_e n_p \, \text{erg cm}^{-3} \, \text{s}^{-1}$. From equation (4), the cooling radius is expanding as $R_c = 82 \, (t/10^{10}\,\text{yr})^{0.48}$ kpc. The luminosity within R_c is therefore slowly declining, $L \propto t^{-0.29}$. The similarity solution fitting equation (8) at large r has central accretion rate

$$\dot{M}(r = 0) = 15 \, (t/10^{10}\,\text{yr})^{-0.43} \, M_\odot \, \text{yr}^{-1} \, . \tag{10}$$

Because gas is not allowed to drop out of the flow, all of the mass is deposited in the center. Equation (10) may be a good approximation to the maximum accretion rate, but it does not give $\dot{M}$ correctly well inside of R_c because of the neglect of star formation in the cooling flow. Regardless of star formation, however, equation (10) implies that the M87 cooling flow is slowly declining with time.

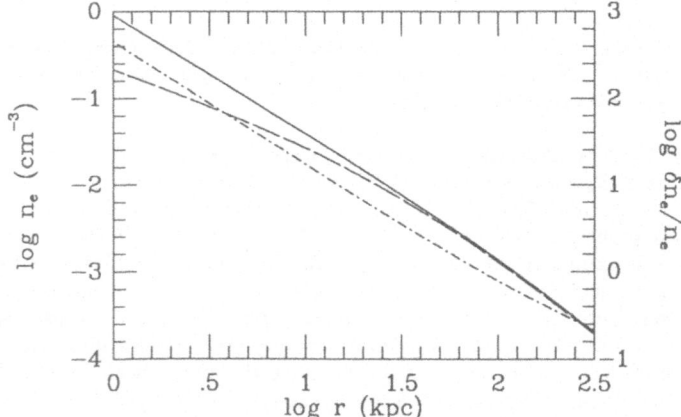

Figure 2. Electron density profiles (solid curve: similarity solution, dashed curve: observations) for the cooling flow onto M87. The discrepancy at $r \lesssim 15$ kpc may be due to the neglect of star formation in the similarity solution. The dashed-dotted curve gives the linear growth factor for comoving isobaric density perturbations.

Figure 2 shows the density profile obtained from the similarity solution compared with equation (8). The two are matched at $r \gg R_c$. The observed $n_e(r)$ bends over around the cooling radius in a way suggestive of the cooling wave, but is significantly less than that of the similarity solution inside ~ 10 kpc. This reflects the well-known fact that without heating or star formation the X-ray emission would be excessive in the centers of cooling flows (Stewart *et al.* 1984). Bertschinger and Meiksin (1986), following several other groups, suggested that heat conduction might decrease $\dot{M}$ enough to solve this problem, but the Fe XVII line emission reported by Canizares (1988) suggests that heat conduction transports little heat globally. The most likely explanation for the decreased central density compared with the similarity solution is star formation (Stewart *et al.*).

Star formation is a plausible solution in the context of the cooling wave solutions found here. The cooling waves are unstable to small wavelength radial thermal instabilities, which grow fastest if they are comoving and isobaric (White and Sarazin 1987). For such a density perturbation of amplitude $\delta \equiv \delta\rho/\rho$, the perturbation at r is given to within an arbitrary normalization factor by (Mathews and Bregman 1978)

$$\ln\delta = \frac{2}{5}\int_r^\infty \left(2 - \frac{d\ln\Lambda}{d\ln T}\right)\frac{\rho^2\Lambda(T)}{p}\frac{dr}{|v|}. \tag{11}$$

Although equation (11) suggests exponential growth, δ grows only as a power of radius (at both small and large r, but with different slopes) because the gas flows in about as quickly as it cools.

The dashed-dotted curve in Figure 2 shows the thermal instability growth factor vs. radius for the self-similar cooling wave. The normalization has been chosen so that $\delta = 1$ at $r = R_c$. Note that the growth is slow for $r \gg R_c$, so that pre-existing large density perturbations must be present if gas is to drop out near R_c. A more conservative view is that $\sim 10\%$ density perturbations are present at R_c due to entropy variations produced during the collapse and shock heating of an initially irregular cluster or due to entropy variations in the gas ejected from galaxies. These perturbations would become nonlinear within $r \approx 15\,\mathrm{kpc}$, dropping out of the X-ray emitting gas and potentially solving the problem of excessive hot gas in the similarity solution. Demonstrating this in detail will require numerical hydrodynamic simulations. One could use a star-formation law like that of White and Sarazin (1987), $\dot{\rho} = -q\rho/t_c$, but this would not be justified if only small ($\delta \lesssim 0.1$) perturbations are present initially since gas would not leave the flow until $t_c(r) \ll t$. Simulations of cluster collapse and virialization with galaxies and gas dynamics are needed to determine the magnitude of the initial perturbations.

4. CONCLUSIONS

Although the self-similar cooling wave model does not provide a good fit to the X-ray emission around M87 at small radius, presumably because it neglects star formation, it has nevertheless been instructive. From the similarity solutions one may safely conclude that time-dependent effects are small well inside the cooling radius, but that they are important near R_c and they cause a bend in the X-ray luminosity profile even if the initial density and pressure profiles are pure power laws of radius. The time-dependence does not change the need for star formation suggested by the X-ray luminosity profile. However, thermal instabilities can plausibly grow enough to make the cooling flow model consistent with X-ray observations. It may not be necessary for gas to drop out at $r \gtrsim 0.2R_c$. Indeed, large amplitude perturbations must be present if the gas is to drop out early. Finally, the similarity solutions imply that, at least in the case of M87, even neglecting cluster dynamical evolution, cooling flows should have been more vigorous in the past.

REFERENCES

Bertschinger, E. 1988, submitted to *Ap.J.*
Bertschinger, E., and Meiksin, A. 1986, *Ap.J. (Letters)*, **306**, L1.
Canizares, C. R. 1988, these proceedings.
Chevalier, R. A. 1987, *Ap.J.*, **318**, 66.
Mathews, W. G., and Bregman, J. N. 1978, *Ap.J.*, **224**, 308.
Shu, F. H. 1977, *Ap.J.*, **214**, 488.
Stewart, G. C., Canizares, C. R., Fabian, A. C., and Nulsen, P. E. J. 1984, *Ap.J.*, **278**, 536.
White, R. E., and Sarazin, C. L. 1987, *Ap.J.*, **318**, 612.

THE CURRENT EVOLUTION OF X-RAY CLUSTER COOLING FLOWS

Raymond E. White III
Institute of Astronomy
Madingley Road
Cambridge CB3 0HA
England

ABSTRACT.   The shape of X-ray surface brightness profiles of galaxy clusters with cooling flows may provide enough information to estimate how cooling flow accretion rates are evolving now. I find that most cooling flows have accretion rates which are increasing with time, so they are unlikely to have formed the bulk of the (mostly dark) mass in their associated accreting galaxies. Masses comparable to those in the luminous components of these accreting galaxies could have been generated, however.

1. INTRODUCTION

If observed X-ray cluster cooling flows are to be significant sites of ongoing galaxy formation, they must be long-lived, with past accretion rates at least as large as their current values. The history of cooling flow accretion rates may be coarsely estimated by continuity with estimates of current behavior. For example, if all cooling flows have accretion rates which are currently increasing rapidly with time, they are probably young or sporadic and unlikely to have formed the galaxies they are centered upon. Alternatively, if all cooling flows have accretion rates which are decreasing slowly with time, they are likely to be long-lived and may have formed their central galaxies (since their accretion rates were even higher in the past). I show that currently available X-ray surface brightness data from the *Einstein Observatory* may provide enough information to estimate how cooling flow accretion rates are evolving now. I find that most cooling flows have accretion rates which are increasing with time, so they are unlikely to have formed the bulk of the (mostly dark) mass in their associated accreting galaxies. Masses comparable to those in the luminous components of these accreting galaxies could have been generated, however.

2. ESTIMATING ACCRETION RATE EVOLUTION

As a cooling flow evolves, its outer edge will be that radius where the cooling time of the gas equals the system age. Steady-state flow obtains within this cooling radius (r_c) and a quasi-hydrostatic atmosphere is maintained exterior to r_c. The cooling radius increases (in a Langrangian sense) over time as gas progressively farther out satisfies this cooling criterion. As long as the cooling flow consumes only a small fraction of the intracluster gas mass (as is the case in observed cooling flows), there will be little readjustment in the temperature

343

A. C. Fabian (ed.), Cooling Flows in Clusters and Galaxies, 343–348.

and density distributions of the gas beyond r_c. In this case, the Langrangian advance of r_c will be closely approximated (for short times) by an Eulerian advance through a fixed atmosphere. Then, as r_c marches out, the accretion rate $\dot{M}_c$ at r_c depends on the density and temperature structure beyond r_c, which are products of the details of cluster collapse, energy-injection, etc.

In a steady-state cooling flow, the accretion rate at a radius r is given by $\dot{M} = -4\pi r^2 \rho v$, where ρ is the gas density and v is its (inherently negative) velocity. The velocity can be related to the density and temperature structure of the gas by using the steady-state energy equation:

$$\frac{3}{2}\rho v \frac{d\theta}{dr} - \theta v \frac{d\rho}{dr} = -\rho^2 \Lambda. \tag{1}$$

Here $\theta \equiv kT/\mu m$ is the temperature parameter, μm is the mean mass per particle and $\Lambda(T)$ is the collisional cooling function. Thermal conduction, sources and sinks of gas, and energy sources such as supernovae, cosmic rays, etc., are ignored. Evidently, the velocity is given by

$$-v = \frac{\rho \Lambda r}{\theta}\left(\frac{3}{2}\Delta_r \theta - \Delta_r \rho\right)^{-1} \approx \frac{r}{t_c}, \tag{2}$$

where $\Delta_r \equiv d\ln/d\ln r$ is a logarithmic differential operator and $t_c \simeq \theta/\rho\Lambda$ is the local cooling time. Logarithmic derivatives are of order unity, so equation (2) essentially states that the flow time $t_{flow} \equiv -r/v$ is comparable to the cooling time in steady state flow.

If the flow variables are power-laws in radius, the logarithmic derivatives in equation (2) are constants, so $\dot{M} \propto r^3 \rho^2 \Lambda/T$. This relation holds (in a piecewise sense) even for non-power-law flow variables, since the logarithmic derivatives in equation (2) vary slowly in observed clusters. The accretion rate variation is then connected to the density and temperature structure via:

$$\Delta_r \dot{M} = 3 + 2\Delta_r \rho + (\Delta_T \Lambda - 1)\Delta_r T. \tag{3}$$

Since the temperature profiles of intracluster atmospheres are uncertain, I assess the possible extremes for the temperature profiles by assuming the gas is polytropic. In a polytrope, $P \propto \rho^\gamma$, so $\Delta_r T = (\gamma - 1)\Delta_r \rho$. The possible extremes for the temperature profile correspond to $\gamma = 1$ (isothermal) and $\gamma = \frac{5}{3}$ (adiabatic). Positive values of $\Delta_r T$ ($\gamma < 1$) occur only within cooling flows or in intracluster gas with unlikely initial cluster conditions; temperature gradients steeper than adiabatic ($\gamma > \frac{5}{3}$) are convectively unstable. With the polytropic $\Delta_r T$, equation (3) gives

$$\Delta_r \dot{M} = 3 + [2 + (\gamma - 1)(\Delta_T \Lambda - 1)]\Delta_r \rho. \tag{4}$$

The density distribution can be derived from the X-ray surface brightness distribution $S_x(r)$. If S_x is a power-law in radius, then

$$S_x(r) \propto \rho^2(r)\Lambda_x(T(r))r \propto r^{-\sigma}, \tag{5}$$

where Λ_x is the instrument-convolved spectral cooling function, $\rho^2\Lambda_x$ is the volume X-ray emissivity and $-\sigma$ is the surface brightness slope. Thus, for a polytropic gas, the density structure is related to the surface brightness slope (and γ) through

$$\Delta_r \rho = -\frac{1+\sigma}{2 + (\gamma - 1)\Delta_T \Lambda_x}. \tag{6}$$

The variation of the accretion rate with radius is then given by

$$\Delta_r \dot{M} = 3 - (1 + \sigma) \left[\frac{2 + (\gamma - 1)(\Delta_T \Lambda - 1)}{2 + (\gamma - 1)\Delta_T \Lambda_x} \right]. \tag{7}$$

To estimate qualitatively the time-evolution of cooling flow accretion rates, apply equation (7) to the cooling radius over time. For the two limiting extremes of the temperature behavior, isothermal and adiabatic, the accretion rate variation at r_c is given by

$$\Delta_r \dot{M}_c = \begin{cases} 2 - \sigma, & \text{isothermal } (\gamma = 1); \\ 3 - (1 + \sigma) \left[\frac{2 + \Delta_T \Lambda}{3 + \Delta_T \Lambda_x} \right], & \text{adiabatic } (\gamma = \frac{5}{3}). \end{cases} \tag{8}$$

From equation (8) we see that the accretion rate evolution of *isothermal* gas depends only on the slope of the X-ray surface brightness profile. As the cooling radius moves out, the accretion rate increases (decreases) if the surface brightness slope is shallower (steeper) than -2. The accretion rate is independent of r_c if the surface brightness slope is equal to this transitional slope.

The accretion rate evolution in *adiabatic* gas evidently depends upon the gas temperature, through the dependence of $\Delta_r \dot{M}_c$ on $\Delta_T \Lambda$ and $\Delta_T \Lambda_x$ in equation (8). There are two temperature regimes of interest. For $T \gtrsim 3 \times 10^7$ K, cooling is dominated by bremsstrahlung (for the half-solar abundances characteristic of intracluster gas), so $\Delta_T \Lambda = 0.5$. In this temperature range, the instrument-convolved cooling functions for the IPC and HRI of the *Einstein Observatory* are nearly temperature-independent (Fabricant et al. 1980), so $\Delta_T \Lambda_x \simeq 0$. This temperature range is appropriate for gas in rich clusters of galaxies. Thus, for adiabatic atmospheres in rich clusters

$$\Delta_r \dot{M}_c = (13 - 5\sigma)/6, \tag{9}$$

so the transitional surface brightness slope is -2.6.

Intracluster gas in poor clusters has temperatures in the range $10^7 \lesssim T \lesssim 3 \times 10^7$ K, for which $\Delta_T \Lambda \simeq \Delta_T \Lambda_x \simeq -1$ (White and Sarazin 1987). Gas with temperatures in this range cannot be adiabatic because the cooling time t_c of the gas would then decrease with radius, leading to Rayleigh-Taylor instabilities at large radii. The maximum polytropic index γ is then given by the requirement that $\Delta_r t_c = \Delta_r(T/\rho\Lambda) > 0$. With $\rho \propto T^{1/(\gamma-1)}$ and $\Lambda \propto T^{-1}$, we see that $\gamma < \frac{3}{2}$ for gas with $10^7 \lesssim T \lesssim 3 \times 10^7$ K. Substituting $\Delta_T \Lambda = \Delta_T \Lambda_x = -1$ into equation (7), we find

$$\Delta_r \dot{M}_c = 3 - 2(1 + \sigma) \left[\frac{2 - \gamma}{3 - \gamma} \right] \tag{10}$$

in this case, where $1 < \gamma < \frac{3}{2}$. The transitional slope is then between -2 and -3.5.

To estimate how the accretion rate evolves with time, we need to find how the cooling radius evolves with time. That is, we ultimately want $\Delta_t \dot{M}_c$, where $\Delta_t \dot{M}_c = \Delta_r \dot{M}_c \Delta_t r_c$ and $\Delta_r \dot{M}_c$ is given by equation (7). To determine $\Delta_t r_c$, note that in our Eulerian approximation, the time evolution of the cooling radius r_c is simply related to the radial evolution of the cooling time t_c:

$$\frac{dr_c}{dt} = \left[\frac{dt_c}{dr} \right]^{-1}, \tag{11}$$

where the cooling time is given below equation (2). For polytropic gas with power-law flow variables and cooling function, $t_c \propto \rho^{(\gamma-1)(1-\Delta_T\Lambda)-1}$. With equations (6) and (11), this leads to

$$\Delta_t r_c = \frac{2 + (\gamma - 1)\Delta_T\Lambda_x}{(1 + \sigma)[1 - (\gamma - 1)(1 - \Delta_T\Lambda)]}. \qquad (12)$$

The time-evolution of the accretion rate is then given by the product of equations (7) and (12):

$$\Delta_t \dot{M}_c = \frac{4 - 2\sigma + (\gamma - 1)[3\Delta_T\Lambda_x + (1 + \sigma)(1 - \Delta_T\Lambda)]}{(1 + \sigma)[1 - (\gamma - 1)(1 - \Delta_T\Lambda)]}. \qquad (13)$$

Thus, for the various extremes of the temperature profile,

$$\Delta_t \dot{M}_c = \begin{cases} \frac{2(2-\sigma)}{1+\sigma}, & \text{isothermal;} \\ \frac{13-5\sigma}{2(1+\sigma)}, & \text{adiabatic, } T \gtrsim 3 \times 10^7 \text{ K;} \\ \frac{3-(2-\gamma)(2\sigma-1)}{(1+\sigma)(3-2\gamma)}, & \gamma < \frac{3}{2}; 10^7 \lesssim T \lesssim 3 \times 10^7 \text{ K.} \end{cases} \qquad (14)$$

3. APPLICATION TO OBSERVED COOLING FLOWS

For cooling flows with power-law X-ray surface brightness profiles, the behavior of their accretion rates is independent of r_c. However, many X-ray surface brightness profiles are not power-laws: they frequently become monotonically steeper at increasing radii. In such cases r_c must be known in order to assess the current behavior (which depends on the slope of the surface brightness at r_c). Unfortunately, cooling radii are uncertain by over an order of magnitude, due largely to ambiguity in cooling flow ages, the cooling criterion for the onset of cooling flows, and the value of H_0 (White and Sarazin 1988). This uncertainty may span regions where the accretion rate is increasing, constant or decreasing. Despite this uncertainty, qualitative estimates can still be made for a number of cooling flows. In Table 1 is a list of 20 cooling flow clusters for which I could find surface brightness structural information from the literature. Jones and Forman (1984) fit surface brightness profiles of the form $S_x \propto (1 + r^2/a^2)^{-3\beta+1/2}$ to many X-ray clusters, where a is a core radius and β is a slope-fitting parameter. Using the Jones and Forman (1984) estimates of a and β (when available), I list in Table 1 the range of radii r_t where the local surface brightness slope may equal -2 (the transitional slope for isothermal gas). The poor clusters at the bottom of Table 1 have asymptotic slopes which are shallower than -2, so no transitional radii are indicated. Also listed in Table 1 are estimates of accretion rates, cooling radii, the slopes at the cooling radii and the sign of $d\dot{M}_c/dt$.

From Table 1 it is clear that cooling radii in rich clusters are smaller than the lower limits to their transitional radii (with the sole exception of A2029). This suggests that most cooling flow accretion rates are increasing with time. The cooling radius estimates listed could conceivably be increased somewhat, but since they already tend to be optimistically large, a plausible increase would be less than a factor of ~ 2. Such an increase would place cooling radii in a region where the accretion rate is constant or decreasing in six more cases: A1060, A1795, A1991, A2063, A2199, and A2626 (flagged by question marks in the last column of Table 1). If intracluster gas were adiabatic, the transitional radii would be much larger (corresponding to a slope of -2.6) and all rich clusters would have $\dot{M}$ increasing.

The situation is less ambiguous for the *poor* clusters listed below the Abell clusters in Table 1. Since their surface brightness profiles tend to be power-laws with slopes shallower than -2, their accretion rates are increasing with time in all cases, regardless of cooling radius uncertainties.

Although most clusters have accretion rates increasing with time, the evolution is probably slow. From Table 1 and equation (14) we see that a typical surface brightness slope is $-\Delta_r S_x \approx -1.5$, which corresponds to $\dot{M} \propto t^{0.4}$ for isothermal gas and $\dot{M} \propto t^{1.1}$ for adiabatic gas (with $T \gtrsim 3 \times 10^7$ K).

Table 1. Cooling flow parameters

Object	$\dot{M}$[†]	r_c[♭]	r_t[#]	$\Delta_r S_x \vert_{r_c}$[‡]	$\frac{d\dot{M}_c}{dt}$[§]
A 85	101	140	340-390	0.7-0.9	+
A 426	99	123	600-610	0.3-0.5	+
A 576	6	72	> 570	0.4-0.7	+
A 592	8	75	260-300	0.2-0.5	+
A1060	10	81	130-160	1.2	+?
A1795	337	239	300-440	1.0-1.7	+?
A1983	7	71	> 200	0.4-1.3	+
A1991	78	143	> 160	1.9	+?
A2029	321	200	130-300	1.2-2.4	−
A2063	100	200	280-300	1.5-1.6	+?
A2142	55	111	> 400	1.0-1.5	+
A2199	129	170	190	1.8-1.9	+?
A2626	30	118	180-250	0.6-1.7	+?
A2657	54	165	> 420	1.0-1.5	+
AWM7	40	—	—	1.1	+
MKW4	9	74	—	1.4	+
Virgo	11	74	—	1.6	+
Cen	41	126	—	1.5	+
MKW3s	75	154	—	1.7	+
AWM4	25	17	—	1.6	+

[†]Accretion rate (in $M_\odot$ yr^{-1}). From Arnaud and Fabian (1987) except AWM7, AWM4 (Canizares *et al.* 1983) and Virgo (Stewart *et al.* 1984).

[♭]Cooling radii (in kpc, assuming $H_0 = 50$). From Arnaud and Fabian (1987), except Virgo (Stewart *et al.* 1984).

[#]Transitional radii (in kpc) for isothermal gas. Derived from Jones and Forman (1984).

[‡]Slope of S_x at r_c. S_x from: Jones and Forman (1984) [Abell clusters]; Kriss *et al.* (1983) [AWM and MKW clusters]; Fabricant and Gorenstein (1983) [Virgo]; Matilsky *et al.* (1985) [Cen].

[§]Sign of $d\dot{M}_c/dt$.

348

4. DISCUSSION

Most X-ray cluster cooling flows have accretion rates which are increasing with time, although a few rich clusters may have $\dot{M}$ decreasing or constant. The time evolution is probably slow in most cases, so the cooling flows may be long-lived. Despite this longevity, those cooling flows with increasing accretion rates (the majority) are unlikely to have formed the bulk of the (mostly dark) mass in their associated accreting galaxies: because their average accretion rates are less than their current rates, less than $\sim 10^{12}$ $M_\odot$ would be generated in a Hubble time, while the total masses of the accreting galaxies tend to exceed $\sim 10^{13}$ $M_\odot$. However, the accumulated masses could be comparable to those of the luminous components in the accreting galaxies (but the mass so generated need not be luminous). Only those few rich cluster cooling flows with $\dot{M}$ decreasing are likely to have had accretion rates large enough to secularly form the total mass of their accreting galaxies.

These qualitative conclusions are independent of the uncertainties in the temperature profiles of intracluster gas. They do depend, however, on there being little readjustment in the gas distribution beyond cooling radii as cooling flows evolve. This is a reasonable assumption if a cooling flow consumes only a small fraction of the intracluster gas mass (as is the case currently).

REFERENCES

Arnaud, K. A. and Fabian, A. C. 1987, in preparation.
Canizares, C. R., Stewart, G. C. and Fabian, A. C. 1983, *Astrophys.J.*, **272**, 449.
Fabricant, D. and Gorenstein, P. 1983, *Astrophys.J.*, **267**, 535.
Fabricant, D., Lecar, M. and Gorenstein, P. 1980, *Astrophys.J.*, **241**, 552.
Jones, C. and Forman, W. 1984, *Astrophys.J.*, **276**, 38.
Kriss, G. A., Cioffi, D. F. and Canizares, C. R. 1983, *Astrophys.J.*, **272**, 439.
Matilsky, T., Jones, C. and Forman, W. 1985, *Astrophys.J.*, **291**, 621.
Stewart, G. C., Fabian, A. C., Jones, C. and Forman, W. 1984, *Astrophys.J.*, **285**, 1.
White, R. E. III and Sarazin, C. L. 1987, *Astrophys.J.*, **318**, 621.
White, R. E. III and Sarazin, C. L. 1988, preprint.

NON-STEADY CONSIDERATIONS FOR COOLING FLOWS

Michael Loewenstein
Institute of Astronomy
Cambridge, England

1. Introduction

Most of the recent work on cluster cooling flows, both in theory and in analysis of observations, has assumed the cluster gas is in a steady-state. This is only an approximation, however, since the most important timescale - the cooling time - is longer than the age of the cluster for most of the gas. Since the sound crossing time is relatively short, the approach to equilibrium of gas at large radii will alter the nearly steady flow further in so that, at best, a cooling flow is in a quasi-steady state.

Of particular interest is the time-evolution of the mass accretion rate because of its implications for the formation of central-dominant galaxies in clusters. The observations seem to imply that, generally, $\dot{M}$ is approximately proportional to r (Arnaud, this volume) so that mass is being deposited with the appropriate distribution for the formation of the dark halos of these central galaxies. However, producing halos with enough mass requires values of $\dot{M}$, time-averaged over the cluster age, typically ten times what they are now (assuming that cooling flows are long-lived).

This contribution addresses two questions of time-dependent effects in cluster cooling flows. Firstly, how can the present time-variation of the accretion rate $(\frac{\partial \dot{M}}{\partial t})$ be inferred from observable physical quantities and what does this imply for the early evolution of cooling flows? And secondly, might the apparent radial dependence of the accretion rate be a non-steady effect rather than a result of mass-deposition as is generally supposed?

2. The Accretion Rate Evolution Equation

The analytical tool used to answer these questions is an exact expression for the time-variation of $\dot{M}$ - the "accretion rate evolution equation" (AREE) - derived from the usual (sourceless and sinkless) non-steady cooling flow equations:

$$\frac{\partial \dot{M}}{\partial t} = \frac{3}{2} \frac{Pr^2}{\varepsilon} \frac{\partial \phi}{\partial r} - \frac{4}{3} \frac{\dot{M}^2 \varepsilon}{Pr^3} + 3 \frac{\dot{M} P \Lambda(\varepsilon)}{(5/3 - \gamma)\varepsilon^2} - \kappa Pr \left(\gamma + \frac{2}{3} \frac{\dot{M}^2 \varepsilon}{P^2 r^4} \right).$$

The variables are defined as follows: t is the temporal and r the radial coordinate, P the gas pressure, ε the specific internal energy, Λ the cooling function, $\dot{M}$ a quantity proportional to the mass flux ($\dot{M} \equiv -\rho u r^2$), κ the (local) slope of the density gradient, and γ the (local) polytropic index.

Since it is of interest to consider states close to equilibrium, the following two "departure coefficients" are defined:

A. C. Fabian (ed.), Cooling Flows in Clusters and Galaxies, 349–352.

$$\alpha \equiv \frac{\varepsilon}{\varepsilon_{hse}}$$

and

$$\beta \equiv \frac{P}{P_{dte}}.$$

Here, ε_{hse} is the specific internal energy for gas in hydrostatic equilibrium in the cluster potential ϕ,

$$\varepsilon_{hse} = \frac{3}{2}\frac{\partial \phi}{\partial r}\frac{r}{\gamma \kappa},$$

and P_{dte} is the pressure when dynamic-thermal equilibrium obtains (compression and advection balancing cooling),

$$P_{dte} = \frac{2}{3}\left(\frac{\kappa \dot{M}(5/3 - \gamma)}{\Lambda(\varepsilon)}\right)^{\frac{1}{2}}\left(\frac{\varepsilon}{r}\right)^{\frac{3}{2}}.$$

In a steady state, $\beta = 1$ and

$$\frac{1-\alpha}{\alpha} = \frac{2}{3}\frac{\dot{M}^2 \varepsilon}{\gamma \kappa P^3 r^4}(2 - \kappa)$$

which is much less than one for subsonic flow.

The AREE can be expressed in terms of α and β and evaluated where non-steady effects are expected to be most important – at some "cooling radius" (r_{cool}), here defined as the radius where

$$P = \frac{2}{3}\frac{\varepsilon^2}{\Lambda(\varepsilon)t}$$

for a cluster age t. The resulting expression is

$$\frac{\partial \dot{M}}{\partial t}(r_{cool}) = \frac{2}{3}\gamma(5/3 - \gamma)^{\frac{1}{3}}\kappa^{\frac{2}{3}}\beta^{\frac{2}{3}}\left(\frac{1-\alpha}{\alpha}\right)\frac{\varepsilon^{\frac{5}{3}}}{\Lambda^{\frac{1}{3}}}\left(\frac{\dot{M}}{t}\right)^{\frac{1}{3}} + (5/3 - \gamma)^{-1}\left\{\frac{\kappa(2\beta^2 - 1) - 2}{\beta^2 \kappa}\right\}\frac{\dot{M}}{t}.$$

The feature of this equation to note is the extreme sensitivity of $\frac{\partial \dot{M}}{\partial t}(r_{cool})$ to α and β in the sense that as α varies from less than to greater than one, and as β varies from less than to greater than $((1 + 2/\kappa)/2)^{1/2}$, the signs of the two terms in the equation flip. In other words, at the cooling radius where cooling is just becoming important, slight deviations from equilibrium (resulting either from initial conditons or transient phenomena) rather than the smooth onset of cooling can determine the local time-variation of $\dot{M}$.

For reasonable present-day cluster parameters, the first term (proportional to $1 - \alpha$) dominates the second implying that, in principle, the sign of $\frac{\partial \dot{M}}{\partial t}(r_{cool})$ could be determined from a measurement of α. This requires accurate independent measurements of the gravitational potential and gas temperature.

Determination of $\frac{\partial \dot{M}}{\partial t}(r_{cool})$ does not supply direct information about cooling flow evolution, but instead indicates how the transition from the initial conditions to the steady state cooling flow proceeds. If $\dot{M}$ is increasing with time at r_{cool} ($\alpha < 1$) the transition to a quasi-steady cooling flow is from a more nearly hydrostatic configuration implying that the dark matter content of the central galaxy is unlikely to have been formed from the cooling flow, but is presumably a result of processes occuring on an initial dynamical timescale. On the other hand, if $\dot{M}$ is decreasing with time at r_{cool} ($\alpha > 1$) the transition is from a state of higher accretion rate which is suggestive of continuing infall and the larger past accretion rates that are necessary if the dark matter is to have formed from the cooling flow.

3. The Radial Dependence of $\dot{M}$ – Non-Steady Effect?

In a steady state cooling flow $\dot{M}$ is constant with radius. This is not consistent with the accretion rates inferred from the observations if a steady state energy balance is assumed. Instead, a more or less linear decrease of $\dot{M}$ inwards is inferred – implying either that mass is "dropping out" of the flow or that the cooling flow is in a non-steady state. The AREE can be used to examine the latter possibility.

The accretion rate evolution equation in the subsonic, quasi-steady limit has the following very simple form:

$$\frac{\partial \dot{M}}{\partial t} = \frac{\dot{M}}{t_{cool}} + \frac{\dot{M}^2}{\rho r^3}\left(2 - \kappa - \frac{\partial log \dot{M}}{\partial log r}\right)$$

where

$$t_{cool} = \frac{A}{\rho}, \quad A \equiv \frac{(5/3 - \gamma)\varepsilon}{\Lambda(\varepsilon)}.$$

To the extent to which A (or equivalently the temperature) remains constant, the above equation for $\frac{\partial \dot{M}}{\partial t}$ can be combined with the equation of continuity,

$$\frac{\partial \rho}{\partial t} = \frac{\partial log \dot{M}}{\partial log r}\frac{\dot{M}}{r^3},$$

to yield ρ and $\dot{M}$ as functions of time. This provides a sequence of time-varying configurations close to hydrostatic and dynamic-thermal equilibrium consistent with the time-dependent cooling flow equations in the subsonic, constant-temperature regime.

The solution for times close to some reference time, t_o, is

$$\dot{M}(r,t) = \dot{M}(r,t_o)[1 + 2q\rho_o(r,t_o)(t - t_o)],$$

where

$$q = \frac{2\frac{\partial log \dot{M}}{\partial log r}}{A\left(5\frac{\partial log \dot{M}}{\partial log r} - 1\right)}.$$

This means that the timescale for any non-constant $\dot{M}(r)$ to breakdown in functional form is, not surprisingly, on the order of the local cooling time (i.e. $q\rho_o(r)(t - t_o) \sim 1$). Therefore, the apparent $\dot{M} \propto r$ relationship that arises naturally from mass deposition if reasonable assumptions about the gravitational potential and the form of the mass sink due to thermal instability are made, cannot be a non-steady effect unless the age of the cluster is less than the cooling time for the bulk of the flow inside $r(t_{cool} = t_{Hubble})$ – roughly 10^9 years or less.

This leaves one alternative scenario to that of a long-lived, steady cooling flow with extended mass deposition. That is a cooling flow which is young, unsteady beyond $r(t_{cool} \sim 10^9$ yr), and has cooling below $\sim 10^7$ K and mass deposition only at small radii (e.g. inside some sonic radius). $\dot{M} \propto r$ might still be inferred if, for example, the gas is approximately isothermal with an r^{-1}

density profile. However, in this case, the value of $\dot{M}$ inferred from soft X-ray line emission would be related to $\dot{M}_{ss}$ – the accretion rate inferred from the bremsstrahlung continuum under the mistaken assumption of a steady state out to $r(t_{Hubble} = t_{cool})$ by

$$\dot{M}_{lines} = \dot{M}_{ss}(t/t_{Hubble}) < 0.1\dot{M}_{ss}.$$

It has been pointed out (*e.g.* Canizares, this volume) that this is not consistent with the agreement between these two inferred values of the accretion rate.

4. Conclusions

There are two primary conclusions of this contribution.

1) The absolute value of $\frac{\partial \dot{M}}{\partial t}(r_{cool})$ is not determinable, but its *sign* is in principle derivable from independent measurements of the gas temperature and the cluster gravitational potential at r_{cool}. The implications of this for cooling flow evolution at early times are uncertain; however, a decreasing $\frac{\partial \dot{M}}{\partial t}(r_{cool})$ would be suggestive of the higher accretion rates in the past required to explain the dark matter of the central galaxy as originating in the cooling flow.

2) The radial dependence of the accretion rate inferred from X-ray observations under the steady state assumption must be a result of mass deposition as opposed to a non-steady effect unless the cluster age is less than the cooling time for most of the gas. However, this is inconsistent with the agreement between the mass accretion rates inferred from observations of the X-ray continuum and those of the soft X-ray line emission.

COSMOLOGY AND COOLING FLOWS

B. J. Carr & K. M. Ashman
School of Mathematical Sciences
Queen Mary College
Mile End Road
London El 4NS

ABSTRACT. A proper understanding of cooling flows may require cosmology and a proper understanding of cosmology may require cooling flows. Cosmological nucleosynthesis constraints suggest that a large fraction of the baryons in the Universe are dark; we argue that at least the dark matter in galactic halos could be baryonic. This would happen naturally if a lot of gas was processed into jupiters via pregalactic or protogalactic cooling flows. These would be analogous to the cooling flows observed at the present epoch but on a smaller scale.

1. INTRODUCTION

In this talk we will argue that there is an important link between cosmology and cooling flows. In the first place, cosmology could shed light on cooling flows. This is because even present epoch cooling flows can only be understood if they are studied in the context of a cosmological model which describes how galaxies and clusters of galaxies form. Such a model must be utilized if one is to have a complete understanding of the relationship between galaxy and cluster cooling flows and of the likely evolution of cooling flows. Another point is that the details of fragmentation within cooling flows are very dependent on the substructure of the inflowing gas and this substructure may (at least in part) be a result of processes which occurred at a much higher redshift than those presently observable.

In the second place, cooling flows could shed light on cosmology. This is because pregalactic or protogalactic analogues of present day cluster flows would be *expected* in many cosmological scenarios and this could help to resolve various cosmological problems. In particular, the proposal that cooling flows make dark matter (Fabian et al. 1982, Sarazin & O'Connel 1983) raises the question of whether cooling flows could explain the prevalence of dark matter in the Universe. This is the prime issue on which we focus here. We will argue that cooling flows could at least generate the dark matter in galactic halos. Thus the cooling flows we see today may only represent the endpoint of a process that began at a much earlier phase in the history of the Universe.

A. C. Fabian (ed.), Cooling Flows in Clusters and Galaxies, 353–359.
© 1988 by Kluwer Academic Publishers.

2. BARYONIC DARK MATTER

It is well known that, while ordinary visible material has a density $\Omega_v \approx 0.01$ in units of the critical density, there is evidence for a much larger density of invisible material (Faber & Gallagher 1979). In fact, there are four contexts in which dark matter seems to arise: (i) there is *local* dark matter, associated with our galactic disk, with a mass comparable to that of the visible disk; (ii) there is dark matter associated with galactic *halos*, with a density parameter of at least $\Omega_h \approx 0.1$ and possibly more, depending on the (presently uncertain) radius to which the typical halo extends; (iii) there is dark matter in *clusters*, with a density parameter in the range $\Omega_c \approx 0.2-0.3$; (iv) finally, if one accepts the inflationary scenario, there may have to be unclustered *background* dark matter in order to make the total cosmological density parameter unity.

Some of these dark matter components may be the same. For example, if one believes that individual galaxies are stripped of their halos when they aggregate to form clusters (thereby forming a collective cluster halo), it would be fairly natural to identify (ii) and (iii) providing the original halos were large enough. Likewise (iii) and (iv) could be identified if one invoked some form of biassed galaxy formation in which galaxies form in only a small fraction of the volume of the Universe (Kaiser 1984). On the other hand, it is equally possible that all the dark matter components are different.

In assessing how much of this dark matter could be baryonic, a crucial constraint comes from cosmological nucleosynthesis arguments (Yang *et al.* 1984). These require that the baryon density parameter lie in the range $0.014h^{-2} \leqslant \Omega_b \leqslant 0.035h^{-2}$ (where h is the Hubble parameter in units of 100 km/s/Mpc). If $H_0=100$, then the upper limit on Ω_b suggests that only the local dark matter could be baryonic in origin. Thus one would have to invoke some non-baryonic explanation, presumably an elementary particle relic of the early Universe, to explain the rest. On the other hand, if $H_0=50$ (as seems most likely if one wants the Universe to be old enough to explain the ages of globular clusters), Ω_b could be large enough to explain at least (ii) and possibly (iii). Indeed, if $H_0=50$, the discrepancy between Ω_b and Ω_v would imply that a large fraction of baryons *must* have gone into some dark form, although the closure dark matter would still need to be non-baryonic.

The suggestion that the halo and possibly cluster dark matter could be baryonic goes against the current trend to assume that all forms of dark matter except (i) are non-baryonic. However, in our view, the arguments advanced in support of this trend (Hegyi & Olive 1986) are not very convincing but just reflect a prejudice that the number of forms of dark matter should be as small as possible. There is really no reason why dark matter should not take on as many different forms as visible matter, so it is no more implausible that baryons should turn into dark material with high efficiency than that they should turn into visible material with high efficiency. Thus the fact that the dark matter required for closure (if such exists) has to be non-baryonic does not exclude the halo dark matter being baryonic. Admittedly, it might seem strange that baryonic material and non-baryonic material should have comparable densities (Turner & Carr 1987) but this is a coincidence which pertains independent of whether or not the baryonic material remains in mainly visible or invisible form.

Although the halo dark matter may in principle be baryonic, it cannot be in the form of ordinary gas else it would generate too many X-rays. The gas must therefore have been converted into some dark form. There are only two ways of doing this: it must have turned into either *jupiters* or the *black hole* remnants of massive stars. Low mass stars seem to be excluded by source counts limits (Gilmore & Hewitt 1983) and other stellar remnants are excluded by nucleosynthesis and background light constraints (McDowell 1986). Carr et al. (1984) have argued for the black hole option because of its more dramatic cosmological consequences, but there is no observational evidence that the large stars required can form with the required efficiency (at least at the present epoch). On the other hand, we have seen that there is evidence that jupiters can be made efficiently via cooling flows. Thus one is lead naturally to consider the possibility that dark halos are made by cooling flows.

Of course, the cooling flows observed at the present epoch are mainly confined to the central galaxies in clusters and therefore could not in themselves be responsible for either the cluster dark matter (since this is distributed throughout the cluster) or the halo dark matter in galaxies outside clusters. However, we will argue that one could expect analagous high pressure flows to occur at earlier cosmological epochs and these could have been on much smaller scales than clusters. This conclusion pertains in at least three scenarios for the origin of cosmic structure: the hierarchical clustering scenario, the pancake scenario and the explosion scenario. The details of the different models are given elsewhere (Ashman & Carr 1987); here we summarize the main results.

3. COOLING FLOWS IN THE HIERARCHICAL CLUSTERING MODEL

In the "hierarchical clustering" scenario, the first objects to separate out from the Hubble flow and collapse have a scale of about $10^6 M_\odot$, with galaxies and large-scale structure forming through subsequent gravitational clustering. This scenario was originally proposed in the context of a baryon-dominated Universe (Peebles & Dicke 1968) but it is now usually studied in the context of Universes dominated by cold dark matter (Blumenthal et al. 1984). In this picture bound regions will lose their identity as they are subsumed within larger bound regions unless they can cool on a dynamical time; for only then will they be able to collapse fast enough to avoid being disrupted by collisions.

If one considers a cloud of mass M which binds at a redshift z, the dynamical time will just be of order the Hubble time at that redshift, whereas the cooling time will depend upon the density and virial temperature of the cloud (which are themselves determined by M and z). Thus one can specify a region in the (M,z) plane in which bound clouds will cool within a dynamical time. This region is indicated by the shaded line in Figure (1). One sees that the cooling condition is satisfied provided the clouds lie within a certain mass range. The lower mass limit is associated with molecular hydrogen cooling and is 10^4-$10^9 M_\odot$; the upper mass limit is associated with hydrogen-helium cooling and is around $10^{11} M_\odot$. [The amount of molecular hydrogen formation is somewhat model-dependent. If there were none at all, the appropriate lower limit would be given by the broken line; this corresponds to the Lyman-α temperature, below which H/He cooling turns off.] The cooling curve in Figure (1) also has a boundary at z=0, corresponding to the requirement that the clouds bind by the present epoch, and at z≈10, corresponding to the Compton cooling of the microwave background.

Now if one considers a cloud well inside the cooling curve in Figure (1), one expects it to fragment immediately, with very little global collapse. In this case, star formation may be efficient but one would anticipate a standard IMF since fragmentation should proceed *isothermally*, as in the Hoyle (1953) hierarchical fragmentation picture. On the other hand, a cloud well outside the cooling curve will not fragment at all. Neither of these situations would be conducive to dark matter production. However, if one considers a cloud which is close to the cooling curve, one can have a situation where the cooling time within the *inner* part of the cloud is less than the Hubble time but greater than the local dynamical time. (This is because, when a cloud virializes, it develops a density profile in which the density decreases with distance from the centre.) This is analagous to the situation with present epoch cooling flows: the fraction of the cloud which can cool in a Hubble time will flow inwards under the pressure of the outer (uncooled) regions and fragmentation will proceed *isobarically* since the sound-crossing time is less than the cooling time. The fraction of the cloud involved will be maximized with a value somewhat less than 1 just outside the cooling curve. This corresponds to what we term a "pervasive pregalactic cooling flow" (PPCF).

In any particular version of the hierarchical clustering scenario, one can specify the mass which is binding as a function of redshift. This corresponds to a line $M_{bind}(z)$ in Figure (1). It is interesting to consider what happens to bound clouds as one follows the $M_{bind}(z)$ trajectory. If the first ones to bind are sufficiently small (as indicated by the dotted part of the line), one expects to start off to the left of the H_2-cooling curve. In this case, the first clouds will be unable to cool and so they will just be obliterated at later stages of the hierarchy. As $M_{bind}(z)$ approaches the cooling curve, one enters the cooling flow regime, with a PPCF occurring as one crosses it. When $M_{bind}(z)$ has penetrated well inside the cooling region, fragmentation becomes efficient but one no longer expects to make dark fragments, since the stars form isothermally rather than isobarically. As $M_{bind}(z)$ crosses the H/He cooling curve, one can have another PPCF phase (at least if enough gas remains) but cooling will cease altogether when M gets sufficiently large. Note that cluster-scale clouds would still be undergoing cooling flows at the present epoch (as observed) but the fraction of mass involved would be small.

The crucial prediction of our model is that, once the form of $M_{bind}(z)$ is specified, there are only two possible epochs at which PPCFs can occur. The associated mass-scales are always of order $10^6 M_\odot$ or $10^{11} M_\odot$ but the redshifts depend on the particular scenario. In the "cold dark matter" picture, for example, the associated redshifts are 30 and 10, respectively. How do we determine which of these alternatives is more plausible? Providing $M_{bind}(z)$ starts off to the left of the H_2-cooling curve, and providing H_2 does in fact form, the smaller scale PPCF is inevitable. In this case, it seems likely that this is the scale at which most of the dark matter will be made since much of the gas will have been consumed by the time $M_{bind}(z)$ crosses the H/He cooling curve. On the other hand, if $M_{bind}(z)$ starts off inside the cooling region, it is possible that isothermal fragmentation will also deplete the gas too much for a PPCF to occur when it reaches the H/He cooling line. In this case, it would be difficult to make dark matter through cooling flows at any stage. Of course, this conclusion would be avoided if H_2-cooling were never important; in this case, the first PPCF phase would occur when $M_{bind}(z)$ crosses the "Lyα" line in Figure (1).

Even if most of the baryons in the Universe are processed through cooling flows, they will only be turned into jupiters if the pressure in the cooling region is large enough. It is not really clear *how* high the pressure has to be, since we cannot claim to have a proper understanding of all the factors which go into determining the fragment mass. Figure (1) shows the line corresponding to $P=10^5 cm^{-3}K$, the sort of pressure associated with cluster cooling flows. If $M_{bind}(z)$ intersects the cooling curve above this line, one at least has empirical reasons for supposing that the PPCFs make jupiters.

An interesting constraint on the scale of PPCFs comes from X-ray background observations. In the hierarchical clustering picture, one can show that the present (redshifted) energy and density of the radiation generated by pregalactic cooling flows depends only on the mass-scale

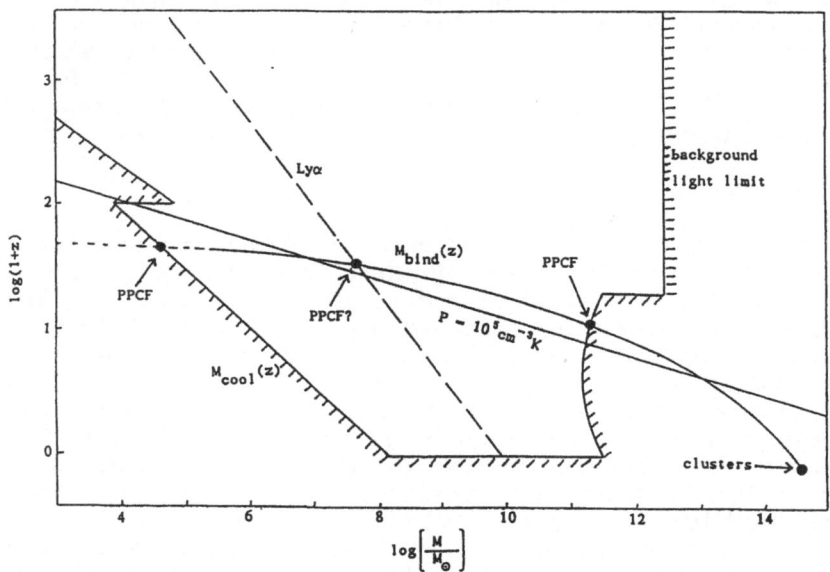

Figure (1). Showing the (M,z) domain in which clouds cool on a Hubble time at high pressure without producing too much background radiation.

and not the redshift. If the temperature exceeds 0.2 keV, the X-ray background constraints imply that only a small fraction of the halo density can have been processed through cooling flows. This applies for mass-scales larger than about $10^{12}M_\odot$, as indicated by the shaded line on the right of Figure (1), so only galactic or subgalactic scale flows could explain galactic halos.

4. COOLING FLOWS IN OTHER SCENARIOS

In the "pancake" scenario (Zeldovich 1970), the first objects to separate out from the Hubble flow are much larger than galaxies. A currently popular variant of this model arises in the context of a neutrino-dominated universe (Bond et al. 1984), when the pancakes have masses $\sim 10^{15}M_\odot$, characteristic of superclusters. In this case, some fraction of the gas in the pancake necessarily undergoes a cooling flow; the situation is analogous to the previous one in the sense that the cooling time is comparable to the Hubble time when the pancake forms and fragmentation occurs isobarically. However, there are differences from the previous situation: the pressure is now imposed by an external shock (rather than an outlying uncooled region) and the fraction of the pancake involved in the cooling flow is rather small (below 20%).

Another scheme to explain the large-scale structure in the Universe is the "explosion" scenario (Ostriker & Cowie 1981; Ikeuchi 1981), in which the shocks generated by explosive seeds sweep up vast shells of matter. In a recent version of this theory (Allen & Carr 1987), the shells eventually overlap, so that most of the gas in the Universe is compressed into slabs, very similar to the "pancakes" discussed above. In both the "pancake" and "explosion" picture, one expects the gas to fragment into clumps, with the clump mass depending on the dominant cooling process: if H/He cooling dominates, one may get galactic-scale clumps, but clumps as small as $10^6 M_\odot$ could arise if H_2 cooling dominates. In either case, one needs the clumps to subfragment into low mass objects at some stage if one wants to make dark matter.

5. DISCUSSION

In this talk, we have made no attempt to justify the assumption that high pressure cooling flows make jupiters, beyond reiterating the usual argument that the high pressure reduces the Jeans mass. This argument is not completely convincing [see, for example, Silk et al. 1986] and, in any case, our ignorance about star formation and the IMF in general should make us wary of any particular explanation for why the stellar mass is reduced in cooling flows. Nevertheless, there does seem to be considerable evidence that such a reduction occurs. The fact that cosmologists have been rather slow to acknowledge this seems to stem from a rather conservative tendency to assume that the solar neighbourhood IMF must apply in all circumstances.

An important feature in all the scenarios discussed above is the coherence scale of the cooling flows, because this should be the scale on which the dark matter aggregates. In some cases, this scale would be comparable to that of a galaxy, in which case cooling flows could make galactic halos directly. In other scenarios, the scale would be around $10^6 M_\odot$, in which case the first objects to form would be $10^6 M_\odot$ dark clusters. Galactic halos would then form as a result of the agglomeration of these objects. Possible observational evidence for the second suggestion has been presented by Carr & Lacey (1987), who argue that the $10^6 M_\odot$ objects postulated by Lacey & Ostriker (1985) to explain the observed disk heating are more likely to be dark clusters than supermassive black holes.

Once galactic halos have formed, it would be fairly natural to generate the dark matter in clusters by tidally stripping the halos from those galaxies within the cluster. Partial support for this view comes from the fact that the mass-to-light ratio of galaxies in cluster cores is less than that for galaxies in the cluster as a whole (Sarazin 1986). On the other hand, it is also possible that some of the cluster dark matter (perhaps most of it) is formed from the non-baryonic background material. Indeed this is obligatory if the cosmological nucleosynthesis upper limit on Ω_b is less than Ω_c. We note that if the background dark matter consists of a "cold" elementary particle (such as the photino or the axion), then it will behave dynamically just like the jupiters. Thus there would be no objection to the cluster dark matter being composite.

REFERENCES

Allen, A.J., and Carr, B.J., 1987. Preprint.
Ashman, K.M., and Carr, B.J., 1987. Preprint.
Bond, J.R., Centrella, J., Szalay, A.S., and Wilson, J.R., 1984. *Mon.Not.R. astr.Soc.*, 210, 515.
Blumenthal, G.R., Faber, S.M., Primack, J.R., and Rees, M.J., 1984. *Nature*, 311, 517.
Carr, B.J., and Lacey, C.G., 1987. *Astrophys.J.*, 316, 23.
Carr, B.J., Bond, J.R., and Arnett, W.A., 1984. *Astrophys.J.*, 277, 445.
Faber, S.M., and Gallagher, J.S., 1979. *Ann.Rev.Astron.Astrophys.*, 17, 135.
Fabian, A.C., Nulsen, P.E.J., and Canizares, C.R., 1982. *Mon.Not.R.astr.Soc.*, 201, 933.
Gilmore, G., and Hewitt, P., 1983. *Nature*, 306, 669.
Hegyi, D.J., and Olive, K.A., 1986. *Astrophys.J.*, 303, 56.
Hoyle, F., 1953. *Astrophys.J.*, 118, 513.
Ikeuchi, S., 1981. *Pub.Astron.Soc.Japan*, 33, 211.
Kaiser, N., 1984. *Astrophys.J.Lett.*, 284, L9.
Lacey, C.G., and Ostriker, J.P., 1985. *Astrophys.J.*, 299, 633.
McDowell, J., 1986. *Mon.Not.R.astr.Soc.*, 223, 763.
Ostriker, J.P., and Cowie, L.L., 1981. *Astrophys.J.Lett.*, 243, L127.
Peebles, P.J.E., and Dicke, R.H., 1968. *Astrophys.J.*, 154, 891.
Sarazin, C.L., 1986. *Rev.Mod.Phys.*, 58, 1.
Sarazin, C.L., and O'Connel, R.W., 1983. *Astrophys.J.*, 268, 552.
Silk, J., Djorgovski, S., Wyse, R.F.G., and Gustavo, A.B., 1986. Preprint.
Turner, M.S., and Carr, B.J., 1987. *Mod.Phys.Lett.A.*, 2, 1.
Yang, J., Turner, M.S., Steigmann, G., Schramm, D.N., and Olive, K.A., 1984. *Astrophys.J.*, 281, 493.
Zeldovich, Ya.B., 1970. *Astr.Astrophys.*, 5, 84.

MULTIPHASE COOLING FLOWS AND GALAXY FORMATION

P. A. Thomas,
Institute of Astronomy,
Cambridge, CB3 0HA,
U. K.

ABSTRACT. Multiphase cooling flow models are investigated. Deprojections of the data give results in agreement with a single-phase analysis. Theoretical models for the evolution of the density distribution suggest an explanation for the similarity of mass deposition profiles in different clusters. Cooling flows can play an important role in the formation of giant elliptical galaxies.

1. INTRODUCTION

Methods of analysis of the X-ray data from galaxies (Thomas *et al.* 1986; Thomas 1986; Thomas, this volume) and from clusters of galaxies (Arnaud & Fabian 1987; Arnaud, this volume) which incorporate a single density phase, give mass deposition rates which increase with radius. This implies that there must be a range of densities present in the gas and renders the analysis inconsistent. In this paper I briefly describe some results of a multiphase treatment of cooling flows which handles the mass deposition in a consistent manner. (see also the paper by Nulsen in this volume).

A homogeneous flow would result in a central singularity whereas in a non-uniform flow thermal instability amplifies density variations and causes the gas to cool at finite radii; very little matter need then reach the centre. I should emphasize that the density perturbations required to cause deposition of gas at large radii lie in the non-linear regime. There has been much discussion about the rate of growth (or lack of it) of small, linear perturbations (Cowie, Fabian & Nulsen 1980; Nulsen 1986, this volume; White & Sarazin 1987; Malagoli, Rosner & Bodo 1987; Balbus 1987, David, this volume); these would anyway be insufficient for the purpose.

The sound crossing time for density perturbations (or 'blobs') up to a kpc in size is so short that they will remain in pressure equilibrium until they cool to below 10^6 K. In general one might expect that the different buoyancies of the various blobs would cause them to have different dynamics; the physics of the interactions between them must then be considered in solving the flow equations. In fact, a detailed study of multiphase flows by Nulsen (1986) showed that this is not necessary. He considered the forces acting on density perturbations and concluded that relative motions will tend to disrupt gas blobs and reduce their length scales so that their terminal velocities are small. Magnetic fields also help to pin blobs to the mean flow. This justifies the simplifying assumption, which I make throughout the rest of this article, that the density phases comove as they flow inwards. I also assume that the state of the flow is determined by initial conditions; in steady-state flows this is equivalent to specifying the density distribution at the outer boundary. An alternative approach would be to continually introduce density variations throughout the flow, for example as a result of galaxy stirring.

A. C. Fabian (ed.), Cooling Flows in Clusters and Galaxies, 361–365.
© *1988 by Kluwer Academic Publishers.*

2. MULITPHASE DEPROJECTIONS

Thomas, Fabian & Nulsen (1987), with the help of Keith Arnaud, have developed a method of data analysis which treats the equations of motion in a consistent manner. The data are first deprojected as in Fabian *et al.* (1981) to yield the count emissivity in spherical shells. Within each of these the pressure is taken to be constant, so that the inflow is modelled as successive cycles of constant pressure cooling followed by adiabatic compression as the gas moves from one shell to the next. The inhomogeneous gas is treated as a finite number of homogeneous phases which assumed to be comoving and in pressure equilibrium. Mass injection is taken as negligible throughout the flow, so that the assumption of a steady state makes the mass flow rate in each of the phases independent of the radius. The radial variation of the total mass flow rate is then modelled by the removal of each phase from the flow as it cools to zero temperature.

The method starts at the centre where the counts are attributed to a single phase and adds one more phase to represent the gas cooling to zero at the inner boundary of each successive shell. The flow equations are integrated numerically to find the contribution of each phase to the count rate in terms of one free parameter, the flow time across the shell; hence the mass of the newly-added phase can be determined. Finally, the answer is iterated until the filling factor of the shell is unity and the equation of hydrostatic support is satisfied.

Because the counts are apportioned into density as well as spatial bins, this method can only be applied to the high-quality data from the centre of bright clusters. It gives results in close agreement with single-phase analyses. The mass deposition profiles are found to rise almost linearly with radius (as for the results described by Arnaud, this volume). For the best data-sets the density distribution does not agree with theoretical expectations unless a massive halo, similar to that found around M 87 (Fabricant & Gorenstein 1983; Stewart *et al.* 1984; Mould, Oke & Nemec 1987; Huchra & Brodie 1987), is added to the gravitational potential. Unfortunately the data are not good enough to investigate the detailed form of the density distribution.

3. A MULTIPHASE DYNAMICAL MODEL

Why are the observed mass deposition profiles of clusters all so similar in appearance? To answer this question we must consider the details of the density variation within the hot gas. Analytical progress has been made in this area by Paul Nulsen. He has derived expressions for the evolution of a range of density phases which cool in pressure equilibrium (Nulsen 1986; this volume). For a wide range of analytically tractable distributions there is a corresponding wide range in the form of the mass deposition profiles. I have developed a multiphase, dynamical model in order to verify and extend the analytic results.

The numerical scheme is adapted from a spherically-symmetric Lagrangian code. Each cell is split into several density phases which are taken to be in pressure equilibrium and comoving; without this latter assumption such an approach would not be possible. Phases are removed from the flow as they cool below 10^5 K and the mass deposition is thus returned without the need for any further assumptions.

To investigate the development of density distributions in the inflowing gas I supply a given distribution at the outer boundary and integrate until a steady state is attained. The results, of which an example is shown in Figure 1, are in close agreement with Paul Nulsen's theoretical models. Cooling tends to drive distributions towards a universal high-density form as they cool; the distribution consisting solely of this form

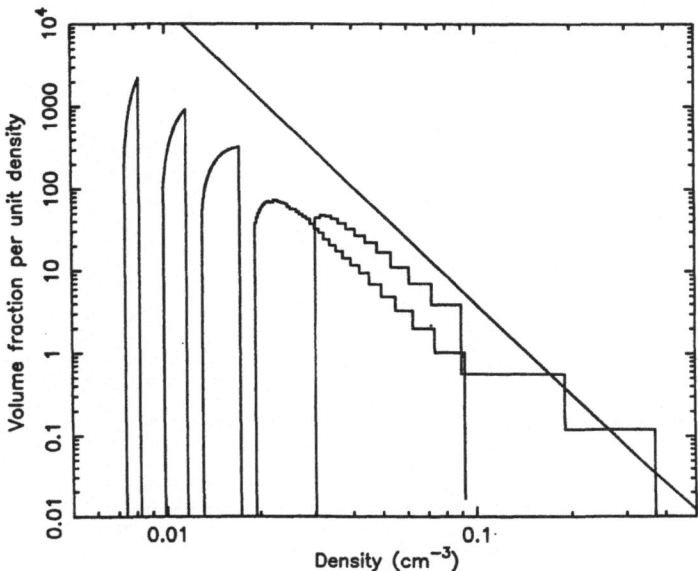

Figure 1. The volume fraction $f(\rho)$ of the density distribution defined such that $f(\rho)d\rho$ is the filling factor of density phases in the range $[\rho, \rho + d\rho]$. Five stages in the evolution are shown, with the initial distribution being on the left.

evolves self-similarly and is deposited in a manner which resembles that in a cluster. Sufficiently narrow initial distributions become dominated by such a high-density tail by the time matter begins to be deposited and so these too have mass deposition profiles of the correct form.

Not all distributions behave in this way, however; those which have a large fraction of low-density gas are deposited in a much more extended manner. It would seem, therefore, that the observations require density distributions with a sharp cut-off to low densities. One possible explanation for this is that the processes which act to tie density perturbations to the mean flow might be inefficient at preventing convection of low-density blobs; for example, overdense perturbations rapidly condense out of the flow once they become non-linear, whereas underdense perturbations survive for a much longer time.

These results suggest that the observed similarity between mass deposition profiles in different clusters can be explained if the well-observed central regions of cooling flows arise from a narrow density distribution; there may be high-density gas which is deposited at larger radii. In general, however, the current cooling times are too long for substantial evolution to have occurred. This may not be a problem if the flows we observe nearby are residuals of a much larger flows in the past.

4. GALAXY FORMATION

In this section, I consider the collapse of primordial gas clouds to form elliptical galaxies. Previously Silk (1977) and Rees & Ostriker (1977) have investigated the effect of the ratio of the cooling time, t_{cool}, to the gravitational free-fall time, t_{grav}, on the evolu-

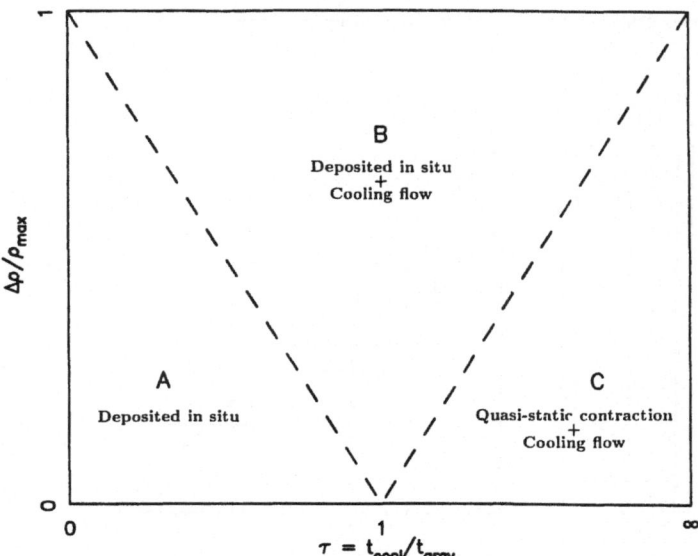

Figure 2. Behaviour of gas clouds with respect to their position in the $\tau - \Delta\rho/\rho_{max}$ plane, as described in the text.

tion of such clouds. I use the hydrodynamics code discussed above to extend this analysis to include local density variations within the gas. We expect such inhomogeneities to be present since the early Universe will contain density fluctuations on all scales which will be enhanced by, for example, the violent activity associated with metal enrichment of the gas. There is certainly evidence of large density variations throughout cooling flows in present day clusters.

The results are best interpreted with the aid of Figure 2. This diagram shows the range of possible values of $\tau = t_{cool}/t_{grav}$ (for the mean density) along the x-axis and the width of the density distribution, $\Delta\rho/\rho_{max}$, along the y-axis. The evolution of the gas clouds splits the diagram into three regions, although the boundaries between these are not sharp and the properties vary continuously from one to the next. Individual clouds as they evolve move up and to the left on the diagram. Given time they will eventually reach the upper boundary and mass will begin to be deposited. (This is true only for the cores; gas in the outer regions may be depleted by inflow into the core). The higher the value of τ, the more important will be the effects of inflow.

In region A, which corresponds to the smallest mass scales, the cooling time of all phases is less than t_{grav}, there is no time to initiate a cooling flow, and so most of the gas is deposited *in situ*. In fact the gas clouds will not virialize in the first place and the models discussed here do not apply.

On larger mass-scales for which $\tau \gtrsim 1$, region B, significant dissipative infall will occur. The single-phase models are uninteresting; they have a single stage of mass deposition which either deposits the gas uniformly (as in the above low-mass models) or drags it into the centre of the galaxy. For models with a wide range of densities, however, there is a balance between cooling and inflow which allows mass deposition over an extended period and in a manner not characterised by a single radius (*i.e.* with a power law profile). Firstly a small amount of high-density gas is deposited *in situ*, followed by a prolonged period of inflow with the central density increasing and the core radius decreasing at later times. The deposition of matter is thus extended at first but gradually becomes more centrally condensed. It is possible in this way to build an r^{-2} halo which is similar in magnitude to those around central cluster galaxies. However, the degree of central concentration of the dark matter depends upon the form of the density distribution; more work is necessary to examine the size of parameter space that gives rise to deposition profiles of the correct form and to investigate the expected form of the initial density distribution.

Finally, in region C where the cooling time of all the gas is initially long, the gas cloud at first undergoes a period of quasi-static contraction during which the mean density and temperature increase (roughly as $T \propto n^{1/3} \propto r^{-1}$), as does the range of densities at each radius. Eventually the highest density gas begins to be deposited and the ensuing cooling flow transports gas into the centre of the system as for region B. Because of the high value of τ, dynamical evolution may prevent large cooling flows from becoming established in cluster-sized systems (see White & Rees 1978). Observationally, dynamically young systems with no single central-dominant galaxy do not contain cooling flows.

The transition mass for which $\tau \sim 1$ corresponds to 2–30×10^{12} M$_\odot$, depending on the metallicity. If elliptical galaxies have mass-luminosity ratios as high as 70 $(M/L)_\odot$ or more, as suggested by the lower mass limit of Fabian et al. (1986; see also Thomas, this volume), then a typical galaxy with luminosity L^* ($\approx 5 \times 10^{10}$ L$_\odot$) may lie in this range.

Thus cooling flows play little part in the formation of low-mass galaxies but become increasingly important in giant ellipticals, and the largest central cluster galaxies may result almost entirely from cooling flow activity.

REFERENCES

Arnaud, K. A. & Fabian, A. C., 1987. Preprint.
Balbus, S. A., 1987. Preprint.
Cowie, L. L., Fabian, A. C. & Nulsen, P. E. J., 1980. *Mon. Not. R. astr. Soc.*, **191**, 399.
Fabian, A. C., Hu, E. M., Cowie, L. L. & Grindlay, J., 1981. *Astrophys. J.*, **248**, 47.
Fabian, A. C., Thomas, P. A., Fall, S. M. & White, R. E., 1986. *Mon. Not. R. astr. Soc.*, **221**, 1049.
Fabricant, D. & Gorenstein, P., 1983. *Astrophys. J.*, **267**, 535.
Huchra, J. & Brodie, J., 1987. *Astrophys. J.*, in press.
Malagoli, A., Rosner, R. & Bodo, G., 1987. *Astrophys. J.*, **319**, 632.
Mould, J. R., Oke, J. B. & Nemec, J. M., 1987. *Astr. J.*, **92**, 53.
Nulsen, P. E. J., 1986. *Mon. Not. R. astr. Soc.*, **221**, 377.
Rees, M. J. & Ostriker, J. P., 1977. *Astrophys. J.*, **179**, 541.
Silk, J., 1977. *Astrophys. J.*, **211**, 638.
Stewart, G. C., Canizares,. C. R., Fabian, A. C. & Nulsen, P. E. J., 1984. *Astrophys. J.*, **278**, 536.
Thomas, P. A., 1986. *Mon. Not. R. astr. Soc.*, **220**, 949.
Thomas, P. A., Fabian, A. C., Arnaud, K. A., Forman, W. & Jones, C., 1986. *Mon. Not. R. astr. Soc.*, **222**, 655.
Thomas, P. A., Fabian, A. C. & Nulsen, P. E. J., 1987. *Mon. Not. R. astr. Soc.*, in press.
White, S. D. M. & Rees, M. J., 1978. *Mon. Not. R. astr. Soc.*, **183**, 341.
White, R. E., III & Sarazin, C. L., 1987. *Astrophys. J.*, **318**, 612.

THE ROLE OF MAGNETIC FIELDS IN CLUSTER COOLING FLOWS

Noam Soker and Craig L. Sarazin.
Department of Astronomy
University of Virginia
P. O. Box 3818 University Station
Charlottesville, VA 22901 U. S. A.

ABSTRACT. We show that the magnetic field may become very important in the inner regions of cluster cooling flows. The magnetic pressure becomes comparable to the thermal pressure within a radius of typically $1 - 10$ kpc. Within this region, reconnection of the magnetic field becomes efficient and magnetic energy is released at a rate of $10^{41} - 10^{43}$ erg s^{-1}. Buoyancy is ineffective in transporting the magnetic field out of the cooling flow.

1. INTRODUCTION

The magnetic field can have an number of effects on the energetics and dynamics of cooling flows. Even a very weak magnetic field, if it is tangled on small scales, can effectively inhibit transport processes in the gas, such as thermal conduction, viscosity, and the settling of heavy ions (Sarazin 1986). Since thermal conduction can effectively compete with the inflow of enthalpy in the gas, cooling inflows may be greatly reduced unless conduction is supressed. Thus, the common occurence of cooling flows is probably due to the effects of weak, tangled magnetic field. Unfortunately, it is very difficult to calculate reliably the small scale properties of the magnetic field. Estimates and limits on the strength of the magnetic fields in clusters suggest that the field is *not* dynamically important there. However, when gas cools and flows into the center of cooling flow clusters, the magnetic pressure is greatly increases by the compression and shear in the flow. The thermal gas pressure does not increase as rapidly, since there is no large gradient in the gravitational potential. Then, at some point in the inflow the magnetic pressure and gas pressure can become comparable. A similar process may occur in miniature in thermally unstable blobs of cooling gas. In this paper, we consider the dynamical effects of the large scale magnetic field. We calculate the growth of the magnetic pressure under some simple assumptions. We consider several loss mechanisms for the field. We conclude that field line reconnection is probably the dominant loss mechanism, and that buoyancy effects are probably not important. We consider the effect of the magnetic field on thermal instabilities. We discuss briefly the effects of the release of magnetic energy through reconnection.

There is almost no direct observational evidence concerning the magnetic field

A. C. Fabian (ed.), Cooling Flows in Clusters and Galaxies, 367–371.

in clusters or cooling flows. As a result, the strength and geometry of the field are essentially unknown. Thus, our calculations are meant only as first–order estimates of role of the magnetic field.

2. LARGE SCALE BEHAVIOR OF THE MAGNETIC FIELD

Observational estimates of or limits on the intracluster magnetic field come from limits on the strength of diffuse Inverse Compton X-ray emission from clusters with diffuse radio emission, and from Faraday rotation measurement towards radio sources located within or behind clusters (Sarazin 1986). The Faraday rotation measurement have been variously interpreted as giving actual detections or only upper limits on the field. The Faraday rotation measurements give the integral of the component of the magnetic filed parallel to the line–of–sight, and are thus affected by tangling in the field. Faraday rotation observations in clusters include Jaffe (1980), Dennison (1980), Lawler and Dennison (1982), Vallee *et al.* (1986,1987), and Dreher, Carilli and Perley (1987). All of these detections or limits, as well as the limits from Inverse Compton emission, are all more–or–less consistent with a field strength and coherence length of roughly

$$B = 1\,B_\mu\;\mu\mathrm{G}, \qquad l_B = 20\,L_{20}\;\mathrm{kpc}. \tag{1}$$

We will assume these values apply at the cluster cooling radius r_c at which the cooling time reaches 10^{10} years, and we will assume that $r_c \approx 100$ kpc. For the gas temperature and proton density at the cooling radius, we take the typical values

$$T = 7 \times 10^7 T_7\,\mathrm{K}, \qquad n_p = 2 \times 10^{-2} n_2\,\mathrm{cm}^{-3}, \tag{2}$$

respectively. At $r = r_c$, we have (assuming solar abundances)

$$\frac{P_{Bo}}{P_{go}} = 8.8 \times 10^{-5} B_\mu n_2^{-1} T_7^{-1}. \tag{3}$$

Let us now assume that the inward cooling flow is homogeneous, in steady–state, and spherically symmetric, and that the magnetic field lines are frozen-in to the cooling gas. (By homogeneous, we mean that no mass is dropping out of the flow.) Let l and w be the dimensions of a typical magnetic loop in the radial and perpendicular directions, respectively. Under our assumption of a frozen-in field, w changes as $w = w_o(r/r_c)$, and thus, the magnetic pressure varies as

$$P_B = P_{Bo} \left(\frac{r}{r_c}\right)^{-4}. \tag{4}$$

Since w decreases with decreasing radius, and l increases, the magnetic field lines become increasingly radial as the gas flows inward. In the homogeneous inflow models of White and Sarazin (1987), the thermal pressure at $r \approx 10$ kpc is roughly twice the pressure at the cooling radius, so the magnetic pressure becomes equal to the thermal gas pressure at r_B, where

$$r_B \approx 8 B_\mu^{1/2} \left(n_2 T_7\right)^{-1/4}\,\mathrm{kpc}. \tag{5}$$

Let us check our assumption of a frozen-in magnetic field by examining the rate of field line reconnection of the magnetic field. All the reconnection mechanisms which act when the magnetic pressure is much less than the thermal pressure are too slow to reconnect the magnetic field on the flow time. When the magnetic pressure becomes comparable to the thermal pressure, there are a number of rapid reconnection mechanisms which can occur. Typically, these involve reconnection at a fraction of the Alfven speed in the gas, so that a typical rapid reconnection time is (Parker 1979, pp. 428-437) $t_{rec} = \epsilon^{-1} l_B / v_A$, where $v_A \equiv B / \sqrt{4\pi\rho}$ is the Alfven speed, and we will assume that $\epsilon \approx 0.1$. Thus, rapid reconnection will only become effective at $r \lesssim 10$ kpc, with a typical timescale of

$$t_{rec} \approx 7.6 \times 10^7 \, \mathrm{yr} \left(\frac{\epsilon}{0.1}\right)^{-1} \left(\frac{l_B}{1\,\mathrm{kpc}}\right) \left(\frac{B}{100\,\mu\mathrm{G}}\right)^{-1} \left(\frac{n_p}{0.2\,\mathrm{cm}^{-3}}\right)^{1/2}. \tag{6}$$

The flow time at $r \approx 10$ kpc is

$$t_{flow} \equiv \frac{r}{v} \approx 2 \times 10^8 \, \mathrm{yr} \left(\frac{r}{10\,\mathrm{kpc}}\right) \left(\frac{v}{50\,\mathrm{km\,s}^{-1}}\right)^{-1}. \tag{7}$$

As the reconnection time is evidently shorter than the flow time for $P_B \gtrsim P_g$, reconnection will be effective and will tend to reduce the magnetic field until the magnetic pressure is roughly equal to the thermal pressure. We will therefore assume that $P_B \approx P_g$ in the flow within the radius r_B.

To check the consistency of this assumption, we calculate the variation of t_{rec} and t_{flow} with r. If we assume that $l_B \propto r$, $n_p \propto r^{-4/3}$ and $v \propto r^{-2/3}$ (the last are reasonable fits to the homogeneous cooling flow solutions of White and Sarazin [1987]), we find $t_{rec} \propto r^{5/6}$ and $t_{cool} \propto r^{5/3}$. The reconnection time becomes longer than the flowing time at $r \lesssim 2$ kpc. So in the region $2\,\mathrm{kpc} \lesssim r \lesssim 8\,\mathrm{kpc}$, the magnetic pressure is comparable to the thermal energy. Due to the fast reconnection in the region $2\,\mathrm{kpc} \lesssim r \lesssim 8\,\mathrm{kpc}$, the magnetic field is likely to become more nearly isotropic.

In a simple inhomogeneous model in which gas drops out of the flow due to thermal instabilities at a rate given by $\dot{M} \propto r$ (see Fabian, this volume), and in which one assumes that the remaining gas remains homogeneous, the magnetic field varies as $B \propto r^{-1}$. Then, the magnetic pressure becomes comparable to the thermal pressure only for $r \lesssim 1$ kpc. However, since the magnetic pressure tends to inhibit thermal instabilities, the gas that does not cool rapidly and drop out from the flow may preferentially consist of regions with higher magnetic field.

Let us consider now the energy released by the reconnection of the magnetic field. Initially, the flux of magnetic energy is

$$\dot{E}_{Bo} = \frac{B_o^2 \dot{M}}{8\pi\rho} = 5.3 \times 10^{39} \, \mathrm{ergs\,s}^{-1} B_\mu^2 n_2^{-1} \left(\frac{\dot{M}}{100\,M_\odot}\right), \tag{8}$$

which is much less than the X-ray luminosity of such a cooling flow and consequently of the enthalpy flux in the hot gas. For homogeneous inflow, the magnetic energy density varies in proportion to r^{-4}, and if we assume that the gas density varies as r^{-1}, the magnetic energy flux is found to vary as $\dot{E}_B \approx \dot{E}_{Bo}(r/r_c)^{-3}$. At

$r \approx 10$ kpc, this gives a magnetic energy flux of $\approx 5 \times 10^{42}$ ergs s^{-1}. In the simple inhomogeneous model we discussed previously, the magnetic energy flux is given by $\dot{E}_B \approx \dot{E}_{Bo}(r/r_c)^{-1}$, which implies a magnetic energy flux of $\approx 10^{41}$ ergs s^{-1} at $r \approx 10$ kpc with the parameters used in equation (8).

3. THERMAL INSTABILITIES

The magnetic field may influences the development of thermal instability by suppressing heat conduction and by the contributions of magnetic pressure and tension forces. Heat conduction can suppress very efficiently the growth of thermal instabilities. As the heat conduction across magnetic field lines is very small, the magnetic field lines in cooling blobs are probably disconnected from the surounding gas. Throughout the discussion here we will adopt this assumption and neglect heat transfer between the cooling blobs and the surounding.

The role of the magnetic field in the late evolution of thermal instabilities depends on the position of the cooling blobs and on the initial conditions in the cooling flow in the cluster. We will assume throughout our discussion that the magnetic field in cooling blobs is isotropic. First, let us look on blobs cooling in the outer regions of the cooling flow, that is, at $r \sim r_c \approx 100$ kpc. We will assume isobaric cooling down to $\sim 10^6$ K and isochoric cooling from $\sim 10^6$ K down to $\sim 10^4$ K. At $\sim 10^6$ K the density in the blob will be $n_p \approx 1.4 T_7 n_2$ cm^{-3}. Assuming magnetic flux conservation and an isotropic field, the magnetic pressure when the blob has cooled to $\sim 10^6$ K is $P_B \approx 300 P_{Bo} T_7^{4/3} P_{Bo}$, which implies that

$$\frac{P_B}{P_g} \approx 2.5 \times 10^{-2} B_\mu^2 n_2^{-1} T_7^{1/3}. \tag{9}$$

The magnetic pressure will become equal to the thermal pressure when the blob has undergone further isochoric cooling to a temperature of

$$T \approx 2.6 \times 10^4 B_\mu^2 n_2^{-1} T_7^{1/3} \text{ K}. \tag{10}$$

If the blob cools and drops out of the flow at smaller radii where P_B is much larger, the magnetic field may be even more important. For example, in a homogeneous cooling flow, the magnetic pressure is comparable to the gas pressure for $r \lesssim 10$ kpc even before the instability even starts to develop.

We see that in the outer regions of the cooling flow, the magnetic pressure probably plays a minor role in the development of thermal instabilities. In the inner regions (the radius of which depends on the nature of the cooling flow), the magnetic pressure can be very important. It can influences the cooling blobs in at least two ways. First, as the magnetic pressure becomes important during the isochoric phase, it can weaken or completely suppress the development of any repressurized shock. This has also been demonstrated in a numerical hydrodynamic calculations by David and Bregman (1987). Second, as the magnetic field undergoes reconnection, the magnetic field energy is released. This occurs on a time scale of

$$t_{rec}(blob) \approx 2.4 \times 10^6 \text{ yr} \left(\frac{\epsilon}{0.1}\right)^{-1} \left(\frac{l_B(blob)}{10\,\text{pc}}\right) \left(\frac{B}{100\,\mu\text{G}}\right)^{-1} \left(\frac{n_p}{2\,\text{cm}^{-3}}\right)^{1/2}. \tag{11}$$

This expression follows from equation (6). We assume that at least one of the blob dimensions is small ($\sim$ 10 pc), perhaps because the blob cools into a sheet configuration. Even with this assumption, the reconnection time is long compared to the cooling time in the blobs until the gas cools down to $T \approx 10^4$ K. Thus, if the magnetic energy mainly goes into heating of the gas, most of the energy is being deposited in gas at $T \approx 10^4$ K, and the magnetic energy may keep the blobs heated to this temperature for much longer than their cooling time.

4. BUOYANCY EFFECTS

We start by giving a crude estimate of the buoyancy velocity. For cylindrical magnetic flux tube the buoyancy velocity perpendicular to the cylinder axis, is given by (Parker 1979, p. 142)

$$v_b \approx v_A \left(\frac{\pi a}{C_D l_h} \right)^{1/2} = 0.75 B_\mu n_2^{-1/2} \left(\frac{a/l_h}{0.1} \right)^{1/2} \left(\frac{C_D}{30} \right)^{-1/2} \text{ km s}^{-1} \qquad (12)$$

where a is the radius of the cylinder, l_h is the pressure scale height in the gas, and C_D is the drag coefficient which depends on the Reynolds number (Parker 1979, p. 142). Here we assumed that the magnetic field in the flux tube is B and is zero in the ambient medium. As the inward velocity of the cooling flow is greater than 10 km s^{-1} for $r \approx 100 \text{ kpc}$ (White and Sarazin 1987), we conclude that buoyancy is not important in the outer regions of the cooling flow.

5. CONCLUSIONS

Our main conclusions are:
    (a) The magnetic field in cluster cooling flows must becomes dynamically important ($P_B \approx P_g$), at $r \lesssim 10$ kpc. In this region, field line reconnection becomes very efficient.
    (b) Buoyancy is ineffective in the outer regions of the cooling flow, and so it cannot transports the magnetic field out of cooling flows.

REFERENCES

David, L. P., and Bregman, J. N. 1987, in preparation
Dennison, B. 1980, *Ap. J.*, **236**, 761.
Dreher, J. W., Carilli, C. L., and Perley, R. A. 1987, *Ap. J.*, **316**, 611.
Jaffe, W. J. 1980, *Ap. J.*, **241**, 924.
Lawler, J. M., and Dennison, B. 1982, *Ap. J.*, **252**, 81.
Parker, E. N. 1979, *Cosmical Magnetic Fields*, Clarendon Press, Oxford.
Sarazin, C. L. 1986, *Rev. Mod. Phys.*, **58**, 1.
Vallee, J. P., MacLeod, J. M., and Broten, N. W. 1986, *Astr. Ap.*, **156**, 386.
Vallee, J. P., MacLeod, J. M., and Broten, N. W. 1987, *Ap. Lett. Comm.*, **25**, 181.
White, R.E. III, and Sarazin, C.L. 1987, *Ap. J.*, **318**, 629.

HI IN COOLING FLOW ELLIPTICALS

Joel N. Bregman
NRAO
Edgemont Road
Charlottesville, VA
USA

M.M. Roberts
NRAO
Edgemont Road
Charlottesville, VA
USA

R. Giovanelli
Arecibo Obs.
Box 995
Arecibo, PR
USA

ABSTRACT. About 2×10^8 $M_\odot$ of HI was detected in N 4406 with the properties expected for gas that condensed from a cooling flow. Upper limits to the HI mass in two other giant ellipticals are 10-30 times lower, suggesting that either HI is opaque or cooling flows are variable in time.

1. INTRODUCTION

If the standard scenario for cooling flows is correct, the cooled gas forms stars (e.g. Fabian, Nulsen, and Canizares 1984; Sarazin and White 1987). During the formation of protostellar clouds and their subsequent collapse, the gas is likely to be neutral and will emit 21 cm line radiation. The minimum amount of neutral gas in a system will be the product of the cooling rate and the collapse time of a protostellar cloud. We estimate that this leads to a mass of about 10^7 $M_\odot$, which can be detected with current radio telescopes. If star formation is not perfectly efficient, if the time needed to accumulate a Jeans unstable mass is not negligible, or if magnetic fields and stellar heating processes slow the collapse of the protostellar cloud, more HI should be detectable. We have used the Arecibo Observatory to make sensitive observations of HI in three giant elliptical galaxies with cooling flows (see Bregman, Roberts, and Giovanelli 1987 for details).

2. RESULTS AND INTERPRETATION

Neutral hydrogen was detected at the center and along both the major and minor axis in N 4406. The beam size has a radius (half power) of 12 kpc (distance to the Virgo cluster of 25.5 kpc); the X-ray emission and starlight from the galaxy extend to several beam diameters. The HI spectrum of the central region has a narrow component of width 130 km/s and a broader component of width 400 km/s. The central velocities of spectra along the major and minor axis are nearly the same as for the nuclear spectrum, indicating that the gas is not rotationally supported. Also, the amount of HI appears to decrease with distance from the center. These are the properties expected for

A. C. Fabian (ed.), Cooling Flows in Clusters and Galaxies, 373–374.

HI condensing out of a cooling flow. The gas cools and forms into stars before it can fall inward and become rotationally supported. Because the hot gas density rises toward the center, the cooling rate per unit volume and the rate at which cold gas is produced is greatest there. In contrast, for the few elliptical galaxies that are abundant in gas (it is not known whether these have cooling flows), the gas is distributed in a rotationally supported torus (Raimond et al. 1981; van Gorkom et al. 1986; van Driel 1987), and was most likely captured.

The other two giant elliptical galaxies that we observed, N 4472 and N 5846 have X-ray and optical luminosities similar to N 4406 (Forman, Jones, and Tucker 1985; Canizares, Fabbiano, and Trinchieri 1987). It is therefore a surprise that no HI was detected, with 3σ upper limits of 6×10^6 $M_\odot$ (N 4472) and 1.7×10^7 $M_\odot$ (N 5846). The upper limit for HI in N 4472 is below the minimum expected theoretical mass. This suggests that either the HI is opaque or that cooling flows have broad variation in the cooling and star formation rates with time. Neither possibility can be ruled out.

3. REFERENCES

Bregman, J.N., Roberts, M.M., and Giovanelli, R. 1987, in preparation.
Canizares, C.R., Fabbiano, G., and Trinchieri, G. 1987, Ap.J., 312, 503.
van Driel, V., 1987, Ph.D. Thesis, Groningen Univ.
Forman, W., Jones, C., and Tucker, W. 1985, Ap.J., 293, 102.
van Gorkom, J.H., Knapp, G.R., Raimond, E., Faber, S.M., and Gallagher, J.S. 1986, A.J., 91, 791.
Raimond, E., Faber, S.M., Gallagher, J.S. III, Knapp, G.R. 1981, Ap.J., 246, 708.

DISCUSSION PAPERS

E. Bertschinger

The Fe XVII line emission reported by Canizares for the Perseus cluster is very important because it provides an independent estimate of $\dot{M}$ which does not rely on detailed theoretical modelling. Comparison with the conduction-dominated models I made with Avery Meiksin suggests that heat conduction must be suppressed by a factor of 10 or more from the classical Spitzer value. However, these models assumed a temperature profile similar to that obtained by Fabian et al. (1981) using deconvolution, which evidently has too little cool ($kT < 1\,\mathrm{keV}$) gas in the center of the Perseus cluster. More careful modelling should be done to obtain a better estimate of the conduction suppression factor. While conduction may not transport enough heat to significantly reduce cooling flows, it is likely still to be important in small-scale thermal instability. The theoretical models of cluster cooling flows undertaken to-date have been over-simplified in treating clusters as spherical systems with a smooth, static gravitational potential. From both observations (e.g. Coma and the discovery of two clumps in A2029 reported here by Bower) and theoretical modelling of cosmological structure formation it is clear that clusters are dynamically complicated and that substructure cannot be ignored. Even presently relaxed clusters (if there are any) must have suffered a chaotic collapse, with large entropy fluctuations (10 per cent) imprinted on the hot gas. These fluctuations could be important as seeds for thermal instabilities.

James Binney

It has become fashionable to study cooling flows under the assumption that the thermal conductivity $\kappa = 0$. While I agree that magnetic fields probably suppress κ to a value so small that conductivity does not significantly modify the large-scale structure of cooling flows, I am confident that conductivity sets the scale at which matter condenses out of the flow. If this conviction is correct, it is pointless to study mass deposition with κ set equal to zero. In an unpublished Oxford D.Phil thesis completed January of this year, Ian Pallister studied thermal instability in Perseus-type cooling flows with $\kappa \neq 0$. Pallister first constructed cooling flow models of Perseus by assuming that the cluster's potential is that of the $r^{1/4}$ model, and that the flow is in a steady state at $r < r_{cool}$ and isothermal at $r > r_{cool}$. He found that the observations could be well fitted for $\kappa < 0.01\kappa_{Spitzer}$. Then he performed a WKB analysis of the growth within the flow of perturbations with initial spectrum $\delta\rho \propto l^{-(n+3)/2}$. He concluded that the perturbations that dominate at late times are highly elongated in the radial direction - typically $a/b \approx 10$. These perturbations have similar scales to the filaments observed around NGC 1275 if $\kappa \approx 10^{-5}\kappa_{Spitzer}$.

J. N. Bregman

I'm impressed with the agreement on a variety of issues by the participants of the meeting. In particular, the existence of cooling flows in many clusters of galaxies seems well established (although perhaps not to people outside of this room). The major unsolved issue in my mind is the evidence for star formation. It is my hope that intensive efforts in this direction improve the evidence for star formation in clusters (there is some evidence already) and in elliptical galaxies.

    Finally, I would like to stress the difficulty with stopping cooling in these systems by simply having a heat source balance the radiative losses. Provided that there is some small variation in the entropy of the gas, which could occur by the motion of galaxies through the gas or variation in the magnetic field strength, low entropy regions cool most rapidly through thermal instabilities. If heating equals cooling, then higher entropy regions become hotter and thermal conduction does not stop the instability, but only allows large wavelength modes to grow. To conclude, you can't stop the gas from cooling just by supplying heat.

C.R. Canizares

X-ray Deficiency of Coma Galaxies: All the elliptical galaxies studied by Forman, Jones and Tucker and by Trinchieri, Fabbiano and myself (see Trinchieri, this volume, for references) show X-ray emission or non- restrictive upper limits. These galaxies are in the "field" or in small groups or loose clusters like Virgo. Paula Blizzard and I decided to look at galaxies in a rich cluster. We chose Coma despite the problem of increased distance and the difficulty of detecting galaxies against the bright intracluster emission. We were unable to get useful results for the inner 15' (1 core radius), but we did obtain reliable 3 sigma upper limits (no detections) for two dozen early type galaxies in the range $L_x = 3 \times 10^{40}$ to 3×10^{41} erg s^{-1}. These galaxies have $L_B = 3 \times 10^{10}$ to $1 \times 10^{11} L_B$, so the limits fall well within the range of measured L_x for the field galaxies. The joint upper limit, treating all galaxies equally, is 2×10^{40} erg s^{-1}, which is just above the luminosity one would expect from point sources alone. The probability of drawing the upper limits from a sample like the field galaxies is 10^{-3}. So with 99.9 per cent confidence we conclude that the Coma galaxies are underluminous compared to the field, but they are still consistent with the point source estimate. Most likely the Coma galaxies have been stripped of much of their hot intergalactic medium by the ram pressure of the intracluster medium, as has long been discussed in the literature. We can reject the alternative possibility that galaxies in rich clusters have increased L_x resulting from pressure confinement by the intracluster gas.

W. Jaffe

From my point of view the most interesting problems are:
The source of turbulence in HII and HI regions (several hundred km/s stellar dispersion in galaxies)
Is formation of stars so fast that neutral components CO, H, HI correspond with observations?
Shouldn't compression of B give observable consequences?
Can radio/optical observations of spirals provide useful tracer of X-ray gas at intermediate z (0.3 - 0.5)?

P. McCarthy

I'd like to raise a few points concerning the ionization excitation mechanisms in the emission line filaments. The relative line strengths of the filaments are well fit by shock models, but they have serious energetic problems. The alternative proposed by Robinson *et al.* namely photo-ionization, does rather well in explaining the observed optical line ratios. Two problems with this model are that the filaments do not show a systematic radial excitation gradient, as one might expect for photo- ionization by a central source, and it fails to fit the [SIII] 9096, 9525 line strengths (see Heckman paper in this volume). A similar situation arises when considering the emission line filaments associated with starburst galaxies. In these systems (e.g. M82) the filaments have spectra characteristic of slow velocity shocks, presumably driven by the hot outflowing gas seen in the X-rays (e.g. Watson, Stanger and Griffiths 1986). The problem is that one needs to convert a large fraction of the mechanical energy of the supernovae driven winds (50 - 100 per cent) into the emission line filaments (McCarthy, Heckman, & van Breugel 1987). Thus we have an energetic problem with the filaments in starburst galaxies as well as in cooling flows. The difference is that we cannot appeal to a power law source of ionizing photons in the starburst case. Perhaps this is telling us that there is a problem with the energetics of multiple shocks, but that shocks should not be ruled out simply on the basis of the H_{rec} problem. Thus, I am not prepared to abandon shock models entirely yet.

B. McNamara

Ionized gas in cD galaxies: O'Connell and McNamara have obtained flux-calibrated spectra of cD galaxies with a range of mass accretion rates in both rich and poor clusters. A number of these galaxies show evidence of excess blue light at $\lambda = 3600$Å with respect to the average blue color of the non-accreting galaxies MKW1 and MKW2. The most extreme case in our sample is the cD galaxy in A1795 which has a blue excess with respect to the template non-accretors; $\delta(3600/4500) \sim 0.52$ mag. We find weak evidence for a correlation between the equivalent width of the [OII] 3727 emission line and the strength of $\delta(3600/4500)$. The three cases which show the strongest blue excesses and the largest E-W [OII] are A1795, M87, and A2052 (30 <EW(Å)> 70). A1795 and A2052 show evidence for hot stars (O-B) via the spectrum synthesis technique of O'Connell (1973). This suggests that the blue excesses observed for these galaxies are related to hot stars which implies that the hot stars are contributing to the ionization of the gas. The claim that the blue excesses are primarily due to a hot or warm population, and not a non-thermal nuclear source is strengthened by the blue ($\delta(3600/4500) \sim 0.4$) off nuclear (17 kpc) spectrum for A1795, and other clusters. The presence of [OII] (2 <EW(Å)> 20) in the galaxies A496, MKW3s and A262 with no evidence for a significant blue excess suggests a non-thermal or shock origin for the ionizing photons. We suggest that both power law shock and black body ionization mechanisms may be necessary to account for these observations. We appreciate that the EW's measured are sensitive to many physical parameters and that we cannot distinguish between these parameters on the basis of one point for each galaxy. This result is, however, suggestive.

A. Meiksin

A mechanism for creating optical filaments along lines of magnetic field: A new thermal instability mode is presented which may be responsible for the formation of optical filaments in cooling flow clusters. The mode has a growth rate boosted above the cooling instability rate by an amount dependent on the divergence of the heat conduction flux of the unperturbed flow. The mode is isobaric, and its wavevector is transverse to the direction of conductive heat flux in the unperturbed system. The paradoxical enhancement of the instability growth rate by heat conduction arises from an imbalance between the cooling, heat conduction, and adiabatic compression rates: an overdense region has a temperature dip for an isobaric perturbation, with the effect of both enhanced cooling and less efficient heat conductivity. The mode would be suppressed by heat-conduction along the wavevector for sufficiently strong conductivity. In the presence of a magnetic field, however, the instability would not be suppressed, because a magnetic field shuts off heat conduction perpendicular to, but not along, itself. Gas could then cool and condense along magnetic field lines. In the context of a cooling flow, filaments would condense along radial magnetic fields with the growth rate

$$\nu = \frac{1}{2}\frac{\gamma - 1}{\gamma}\left[\frac{2 - \frac{d\log\Lambda_0}{d\log T_0}}{(\gamma - 1)t_{cool}} + \frac{1}{p_0}\left(1 + \frac{\partial\log\kappa_{r0}}{\partial\log T_0}\right)\nabla\cdot(\kappa_{r0}\nabla T_0)\right] - \frac{\partial v_0}{\partial r} \pm i\omega_{BV}$$

where p_o = the unperturbed gas pressure, v_0 = the flow velocity, T_0 = the gas temperature, Λ_0 = the cooling function, κ_{r0} = the radial thermal conductivity coefficient, and t_{cool} = the local cooling time. The mode would be overstable in this context, oscillating at the Brunt-Vasaila frequency ω_{BV} for buoyant oscillations. For both Virgo and Perseus, it is found that the heat-conduction term in ν exceeds the sum of the cooling and flow terms, and that a linear perturbation would rapidly grow to a non-linear stage in the regions in which optical filaments are observed.

L. Miller
The case for a more complex picture of cluster gas dynamics
1. Where we can measure velocities, from HI and emission lines, gas motions are complex and show bulk motions with speeds comparable to the sound speed of the hot gas. HI data show both inflow and outflow of gas. If these motions are gravitationally induced we should expect that the X-ray gas is also being stirred around with bulk energy comparable to its thermal energy.
2. Most 'cooling flows' have radio sources. These must produce pressure changes on relatively short timescales, of order 10^8 years. And the disturbed region can be large and in the most important location - the cluster centre. The assumption of hydrostatic equilibrium is probably not correct, and observed values of $\dot{M}$ can change greatly with time.
3. There are plenty of viable heating mechanisms (conduction, cosmic rays, galaxy motions, radio sources) which could greatly change the thermal history of the gas, even if cooling still proceeds via thermal instabilities.

Paul Nulsen
It has frequently been suggested (e.g. Hu, these proceedings) that the lifetime of cluster cooling flows is much less than 10^{10} yr. The high fraction of clusters containing cooling flows places some constraint on this, but cannot rule out intermittent cooling or recent onset of cooling in many clusters. I argue here that the agreement between X-ray morphological and spectroscopic measures of the mass flow rate requires cooling flows to have continued uninterrupted for close to 10^{10} yr.

The region of steady flow extends to the radius, r_{cool}, where the cooling time $t_{cool} = \tau$, the age of the flow (Fabian, Nulsen & Canizares 1984). Since the estimated value for the mass flow rate varies with radius as $\dot{M} \sim r^\eta$, with η about 1 (Thomas, Fabian & Nulsen 1987), the morphological value for the total mass flow rate depends on r_{cool} and hence the assumed age. Throughout the steady flow the flow time $= t_{cool}$ (Fabian, Nulsen & Canizares 1984) so that

$$\dot{M}(r_{cool}) \sim \tau^{\frac{2\eta}{3-\eta}} T^{-\frac{\eta(1-\alpha)}{3-\eta}},$$

where T is the gas temperature at r_{cool} and the cooling function $\Lambda(T) \sim T^\alpha$.

X-ray spectroscopic determinations of the total cooling rate (Canizares these proceedings; Mushotzky these proceedings) are not sensitive to the assumed age of the flow or even the assumption of a steady flow (Canizares these proceedings; Nulsen these proceedings). The agreement between the two types of estimate therefore requires that the assumed age of about 10^{10} yr must be roughly correct. Uncertainty in the location of r_{cool} and possible non-steady mass deposition outside r_{cool} limit the precision of this statement, but it is difficult to imagine that the age of the flows observed can be nearly as short as 10^9 yr, as has been suggested. Ultimately we will be able to use this constraint to determine the ages of cooling flows.

A. Pedlar

I don't think the effect of radio sources on the cooling flows can be neglected simply because the energy in relativistic particles ($\sim 10^{59}$ erg in NGC 1275) is small compared with the thermal energy of the X-ray gas ($\sim 10^{61}$ erg). The overall efficiency for producing relativistic particles (via the nucleus, jets, shocks, entrainment, adiabatic expansion, etc.) surely cannot be more than a few per cent - with most of the energy input finishing up as hot shocked gas at a variety of temperatures. It seems to me that you can't go around dumping this sort of energy into the centres of clusters and not severely affect the cooling gas. Perhaps all the bright X-ray emission in the vicinity of radio sources is due to this rather than a cooling flow?

R. W. O'Connell

This conference will be found to have changed the views of most of our more sceptical colleagues with respect to cooling flows. Using remarks voiced here, I think their opinion will evolve from "patently absurd" to OK, OK. Cooling flows exist. But what are they good for?"
This is progress.

A. Robinson

It's clear that, because each proton would have to undergo about 100 recombinations in order to reconcile the observed $L_{H\beta}$ with the inferred $\dot{M}$'s, the optical emission lines in cooling flows cannot be understood simply as resulting from cooling condensations undergoing repressurizing shocks. Furthermore, if star formation really does occur in all cooling flows then I find it difficult to see how any combination of photoionization by hot stars with photoionization by thermal X-rays (or, indeed, shocks) can explain the excitation of the optical filaments since these are not always observed.

In my view the best bet remains that the condensations, having cooled to $\sim 10^4$ K, are photoionized by the radiation field of an active galactic nucleus (that these galaxies are active is demonstrated by the presence of relatively powerful radio sources, even though starlight dominates the nuclear optical continuum in most cases). Alternatively, it has been emphasized throughout the week that the extended radio source is an important part of the cooling flow factor and the presence of observable optical line emission may well be linked to the structure and/or power of this source. Direct heating by relativistic particles or photoionization by distributed or localized UV sources associated with the latter could provide plausible alternatives to central source photoionization.

C. L. Sarazin

1. I would like to emphasize the importance of observations of the coronal optical lines ([FeVIII], [FeX], [FeXIV]) in cooling flows. The lower ionization optical lines (H, [OII], ...) can be influenced by photo- ionization or cosmic ray heating, and may not reflect directly the rate of cooling of the gas. On the other hand, the coronal lines (particularly [FeXIV]) are very difficult to excite by photo-ionization and will therefore be directly indicative of moderately hot gas ($T = 3 \times 10^5 - 2 \times 10^6$ K). Since this is just below the temperatures of the observed X-ray emitting gas, detection of this emission would directly demonstrate the cooling of the gas. Mapping this emission would show where the hot gas cools. The X-ray observations require that the gas cool below X-ray emitting temperatures at large radii; coronal line observations could directly confirm this. Finally, at present we have no way to detect hot gas in X-rays, so optical techniques should be exploited fully.

2. Ray White and I have calculated an extensive grid of steady-state cooling flow models for normal elliptical galaxies. We find that only the models with 'heavy halo' galaxy potentials fit the X-ray surface brightness profiles (at large radii) or the observed X-ray spectra. Homogeneous models (no loss of gas by thermal instabilities) give X-ray luminosities which are somewhat (~ 3 times) too large, and X-ray surface brightness profiles are much too centrally peaked. We suggest that thermal instabilities cause the gas to cool below X-ray emitting temperatures at large radii. This makes the profile less peaked, and reduces the gravity heating and luminosity.

N. Soker

Cooling Flows and the Stability of Radio Jets: If the cooling flow is in a steady state and is homogeneous, it passes through a sonic radius, given by

$$r_s \approx 0.4\,\text{kpc} \left(\frac{\dot{M}_s}{1\,\text{M}_\odot\,\text{yr}^{-1}} \right) \left(\frac{T_s}{10^6\,\text{K}} \right)^{-2.6}$$

where $\dot{M}_s$ is the accretion rate and T_s is the temperature at r_s. At the sonic radius the cooling flow pressure drops rapidly inward. The condition that a radio jet passing through the sonic radius will not be disrupted is

$$\left(\frac{L_r}{10^{40}\,\text{erg\,s}^{-1}} \right) > 0.5 \left(\frac{v_j}{c} \right) \left(\frac{\eta}{0.1} \right) \left(\frac{\beta}{0.1} \right)^2 \left(\frac{\dot{M}_s}{1\,\text{M}_\odot\,\text{yr}^{-1}} \right)^{1.19} \left(\frac{r_s}{\text{kpc}} \right)^{-0.19}$$

where L_r is the radio luminosity in the jet, v_j the jet velocity, $\eta = L_r/L_{jet}$, L_{jet} the total energy flux in the jet, $\beta = R_j/r_s$, R_j the jet radius. As many radio jets in cD galaxies expand up to $r > 20\,\mathrm{kpc}$, we conclude that in most of those cases the radio jets are not disrupted at the sonic radius. However, there are cases in which it seems that there is disruption (D. Sumi, this meeting).

D. Sumi

Before full-heartedly endorsing the high mass accretion rates claimed at this conference ($1 - 1000\,\mathrm{M_\odot}\,\mathrm{yr}^{-1}$), one should investigate more fully the possible biases of the two X-ray methods used to infer these rates. Although there is no reason to assume that the local IMF is universal, it is, in fact, observed. In the context of a universal IMF, the optical observations accept only 1 per cent to 15 per cent of the X-ray method's accretion rates. The question is whether it is possible to lower the inferred mass accretion rate of these X-ray methods to the optical method rates.

Both the X-ray methods, the deconvolution method (Fabian et al. 1981) and the spectroscopic method (Canizares, Mushotzky) have uncertainties due to evolutionary effects. The deconvolution method is highly sensitive to the maximum radius of steady state cooling, r_{cool}. The spectroscopic method is sensitive to the total time gas emitting a certain emission line remains at the temperature where this gas emits this line. Energy inputs into the cooling flows will cause both methods to overestimate the mass accretion rate.

The apparent agreement of the two methods are compelling only if uncorrelated in some way. As stated above, energy inputs can play this correlating role. Mechanisms for delaying cooling (conduction, etc.) may play some role in heating the cooling gas, but do not produce the cusps in the X-ray distribution (i.e. the XD cluster morphology). In this regard, large dynamically important gravitational sources (massive halos or cluster cusps) may be important. These gravitational sources could heat the cooling gas via PdV work and provide large cusps in the X-ray distribution.

Cooling flows are more than likely to be inevitable, since they represent a transition of the system into its lowest energy state. The roles of various heating sources, however, are not negligible and not well understood. Consequently, systems like NGC1275 and A1795 are undoubtedly cooling but what is in doubt is the rate at which the gas is cooling.

P.A. Thomas

The X-ray picture is straightfoward, at least for clusters. Several hundred solar masses of gas per year is seen to be cooling from a few times $10^7\,\mathrm{K}$ down through $5 \times 10^6\,\mathrm{K}$ (Arnaud, Canizares, Mushotzky, this volume). Then it disappears. To a non X-ray observer cooling flows are very complex. They consist of bright, high-velocity line-emitting filaments associated with radio jets /halos, active nuclei, massive star formation ... It should be remembered that although such processes dominate our observations they represent perhaps of the radiated energy and involve perhaps of the cooled gas. They may have little to do with the bulk of the cooling flow. The theorists' cooling flow is a highly-ordered, non-interacting system. Added complexities are not required until observations are able to discriminate between them. One such is the large range of density variations which must be present in order to allow the mass-deposition profile to rise with radius as is observed. Do not forget the simplicity and power of the X-ray observations.

E.A. Valentijn

Bertschinger's paper:

The calculations you presented are obviously very interesting in understanding how a cooling hydrostate evolves as gas leaks out. I would like to emphasize that your calculations are entirely complementary to the considerations I presented. The difference is that I consider the cooling flows in an evolving universe, where the gas that is leaking out in cooling flows is continuously replenished by accretion from both the cluster and supercluster pervading gas, the latter in turn should expand with the Hubble flow. In itself, without considering the details of the hydrostate evolution, this resulted in a strong $\dot{M} \sim (1+z)^5$ dependence at the centres of cooling flows. I guess we would agree that a merging of these two approaches would eventually be desirable.

Thomas paper:

Some people might become confused about the relation between L_{opt} and L_x. The pure observation is that there is a continuous regression over 5 decades in L_x (E - cD) albeit with a large scatter. On interpreting this result one has to decide whether one wishes to assume that cDs by themselves are that massive, in some cases over 10^{14} $M_\odot$, that they alone can do the binding job. If one does not like that interpretation and decides to split the sample in bright and low L_x then the intrinsic spread of the L_x for a certain L_{opt} allows different regression for the elliptical galaxies sample. A L_x versus T plot of the total sample does not indicate whether this separation between E's and cD's is justified, especially not when the intermediate types (poor group cD's) are not included.

Star formation: Peletier, Jameson, Davis and myself are working on a sample of 20 gE's and have acquired radial B, V, R, J, M and K colour profiles, next to long slit spectra. All these objects do show different types of gradients especially when visual-infrared and metallicity indices are computed. A possible link between radial gradients and cooling flow driven star formation can apparently only be made after we have understood the very complex nature of stellar populations in gE galaxies.

W. van Breugel

Surprisingly, to me at least, it seems that the excitation mechanism of the (liner-type) emission line gas in cooling flow clusters is not well understood. After several presentations by various ionization-code 'pundits' it appeared that the original shock-ionization models might not work, primarily of the observed large emission-line luminosities. Subsequently, as the workshop progressed, it got the impression that there were secret hopes that, perhaps in some cases, some of the ionization might be caused by newly formed stars which supposedly condensed out of the cooling flow. Or, as is normally assumed, by a moderately active nucleus in the central galaxy.

I believe however that the SIII observations of M87, and the line ratio plots of NII/Hα vs. OI/Hα as presented by Heckman *et al.* are remarkably well consistent with shock models. I have, however, heard no comments from emission-line specialists on this! Why? What would be required to get the shock models in cooling flows to work? It seems that some additional source is required. Because of the good, overall spatial correlation of emission-line filaments and the radio sources in cooling flows (also shown by Heckman *et al.*) it would seem to me that this might be the source of ionization.

I think further ionization modelling including high energy particles, as done, for example, by Ferland and Mushotzky might be very fruitful.

Ray White

I want to emphasize five points:

1) The X-ray cluster spectroscopy of Canizares and Mushotzky shows that substantial mass fluxes are cooling, but this does not show that mass is dropping out of the flow (since these data have no spatial resolution). The X-ray surface brightness profile must be used, as well, if mass deposition is to be demonstrated. Having mass drop out prevents the surface brightness from being too sharply peaked, but this may also be achieved with a shallower gravitational potential. This degeneracy should be constrained more carefully.

2) We need to know whether the IR luminosity from dust in ellipticals pervades the galaxies or is merely associated with the (randomly oriented) HI disks. Otherwise we cannot tell whether the ongoing star formation inferred from the dust is associated with the galaxy as a whole or with a recently gobbled gas-rich companion.

3) Although most ellipticals have X-ray luminosities which are too low to be consistent with much SN energy injection, the brightest ellipticals' luminosities are consistent with them having heavy halos and SNe occurring at at least the usual Tammann rate.

4) We need numerical simulations of the evolution of cluster cooling flows. Bertschinger's self-similar solutions are gravity-dominated throughout. However, rich cluster cooling flows (and possibly those in poor clusters) are largely pressure-dominated (otherwise central dominant galaxies could initiate cooling flows).

5) We wrote in stone that "cooling flow" denotes a quasi-hydrostatic process, where the cooling time is much longer than the dynamical (free fall) time. If the cooling time is comparable to the dynamical time, it is galaxy formation in the usual sense.

OBJECT INDEX

Named Clusters

Centaurus / NGC 4696 19, 34, 63, 68, 70, 138, 268, 271, 347

Coma cluster (A1656) 2, 3, 5, 6, 7, 8, 71, 1165-167, 197, 240, 318, 322, 375, 376

Fornax cluster 168, 169, 268, 269

Ophiuchus cluster 169

Perseus (A426) / NGC 1275 / 3C84 5, 9, 26, 27, 47, 54, 56, 58, 59, 60, 63, 68, 69, 70, 71, 81, 82-85, 87, 90, 91, 96, 107, 110, 112, 121-124, 145-147, 149-154, 155-157, 159-163, 165, 167, 175, 203, 225, 247, 249, 258, 294 -297, 301 316, 319, 320, 327, 330, 347, 375, 378, 381

Virgo / M87 / NGC4486 / 3C274 5, 19, 47, 54, 56, 58, 59, 60, 61, 63, 70, 71, 87, 89, 91, 93, 95, 109, 128, 130, 147, 168, 192, 213, 235, 238, 240, 247, 249, 257, 294, 332, 335, 337ff, 347, 362, 373, 346, 377, 378, 382

Abell Clusters

A85 19, 20, 27, 56, 59, 81, 347

A194 / PKS 0123-016A / NGC 541/ Minko Object 251ff, 297

A262 / NGC 703 19, 20, 22, 23, 27, 63, 70, 110, 247, 377

A370 11, 12, 311

A401 56

A426 (see Perseus cluster)

A478 34

A496 34, 56, 59, 60, 63, 70, 81, 110, 192, 377

A576 27, 347

A592 27, 346

A1060 / NGC 3311 19, 27, 63, 70, 171, 344, 347

A1185 221, 222

A1367 5, 19, 20, 170, 235

A 1656 (see Coma cluster)

A1795 / 4C 26.42 27, 34, 56, 59, 78, 81, 107, 110, 122, 123, 203, 247, 258, 260, 301, 316, 320, 346, 347

A1890 27

A1927 221,222

A1983 27, 347

A1991 27, 81, 344, 347

A2029 / IC 1101 11, 27, 56, 59, 79, 109, 115ff, 257ff, 346, 347, 375

A2052 81, 110, 247, 377

A2063 347

A2142 56, 347

A2199 / NGC 6166 22, 27, 35, 39, 56, 59, 60, 81, 110, 130, 131, 346, 347

A2255 22, 23

A2256 19, 20, 22, 23

Α2410 27

A2415 27

A2593 27

A2597 77, 81, 247

A2626 27, 344, 347

A2657 27, 347

Poor clusters

AWM4 347

AWM7 109, 347

MKW1 377

MKW2 377

MKW3s 109, 110, 331, 332, 318, 347, 377

MKW4 332, 347

Other clusters

0335+096 / NGC 3311 59, 60, 63, 70, 122, 123, 170, 172, 318

CA 0340-53 318

C2242-02 11

NGC galaxies

NGC 315 94

NGC 541 (see A194)

NGC 545/547 252

NGC 703 (see A262)

NGC 1052 94, 98, 99, 266

IC galaxies

Messier galaxies

3C Radio sources

4C radio sources

Parkes radio sources